Computer Analysis of Electrophysiological Signals

BIOLOGICAL TECHNIQUES

A Series of Practical Guides to New Methods in Modern Biology

Series Editor
DAVID B SATTELLE

CLASSIC TITLES IN THE SERIES

Microelectrode Methods for Intracellular Recording and Ionophoresis
RD Purves
Immunochemical Methods in Cell and Molecular Biology
RJ Mayer and JM Walker

BIOLOGICAL TECHNIQUES

Computer Analysis of Electrophysiological Signals

JOHN DEMPSTER
Department of Physiology and Pharmacology
University of Strathclyde
Glasgow

ACADEMIC PRESS
Harcourt Brace Jovanovich, Publishers
London · San Diego · New York
Boston · Sydney · Tokyo · Toronto

ACADEMIC PRESS LIMITED
24–28 Oval Road
London NW1 7DX

United States Edition published by
ACADEMIC PRESS INC.
San Diego, CA 92101

Copyright © 1993 by
ACADEMIC PRESS LIMITED

All rights reserved
No part of this book may be reproduced in any form, photostat, microfilm, or by any other means, without written permission from the publishers

This book is printed on acid-free paper

A catalogue record for this book is available from the British Library

ISBN 0-12-208940-5

Typeset by J&L Composition Ltd, Filey, North Yorkshire
Printed and bound in Great Britain at The Bath Press, Avon

Series preface

The rate at which a particular aspect of modern biology is advancing can be gauged, to a large extent, by the range of techniques that can be applied successfully to its central questions. When a novel technique first emerges, it is only accessible to those involved in its development. As the new method starts to become more widely appreciated, and therefore adopted by scientists with a diversity of backgrounds, there is a demand for a clear, concise, authoritative volume to disseminate the essential practical details.

Biological Techniques is a series of volumes aimed at introducing to a wide audience the latest advances in methodology. The pitfalls and problems of new techniques are given due consideration, as are those small but vital details that are not always explicit in the methods sections of journal papers. The books will be of value to advanced researchers and graduate students seeking to learn and apply new techniques, and will be useful to teachers of advanced undergraduate courses, especially those involving practical and/or project work.

When the series first began under the editorship of Dr John E Treherne and Dr Philip H Rubery, many of the titles were in fields such as physiological monitoring, immunology, biochemistry and ecology. In recent years, most biological laboratories have been invaded by computers and a wealth of new DNA technology. This is reflected in the titles that will appear as the series is relaunched, with volumes covering topics such as computer analysis of electrophysiological signals, planar lipid bilayers, optical probes in cell and molecular biology, gene expression, and *in situ* hybridization. Titles will nevertheless continue to appear in more established fields as technical developments are made.

As leading authorities in their chosen field, authors are often surprised on being approached to write about topics that to them are second nature. It is fortunate for the rest of us that they have been persuaded to do so. I am pleased to have this opportunity to thank all authors in the series for their contributions and their excellent co-operation.

DAVID B SATTELLE SCD

Preface

This is a book about the principles and practice of the analysis of electrophysiological signals using the digital computer. It developed out of a research programme to produce computer software for the needs of a group of researchers with related interests within the Scottish universities.

When the author first started to work in this field, more than a decade ago, the impact of the microcomputer on our working practices had yet to become fully apparent. Since then we have seen a period of rapid development which has radically changed the role of the computer, from a situation where a computer was a rather special and expensive item, to one where it is ubiquitous, with at least one possessed by each researcher.

It is now necessary to take stock to some extent of the progress made and to consolidate the many methods that have been developed. Strangely, the techniques of laboratory data acquisition have not been widely documented over the past 30 years. Perhaps this is not surprising since it bridges quite different disciplines, such as the neurosciences, computing and electronics. A primary aim of this text is therefore to draw together what has been produced, from a number of disparate sources, and to present it in a context meaningful to the practising electrophysiologist.

Much of the focus of the book stems from experiences obtained through the development and support of the Strathclyde Electrophysiology Software package for the IBM PC, developed by the author over the past seven years. This package has been distributed free of charge by the author for a number of years, and is in use in over 150 laboratories throughout the world. In the course of this activity, it became clear that simply providing manuals describing how to use each program was insufficient. There was also a need to provide an insight into why the programs operated in particular ways, into the pitfalls awaiting the unwary and into the general philosophy behind signal analysis techniques. Hopefully, this volume will contribute something towards that goal.

The final aim is to present some examples of computer programming techniques used for analog data acquisition and analysis software, again from the perspective of the electrophysiologist. Although many researchers now purchase software from commercial sources, there are still many situations where it may be necessary to develop software within the laboratory. In any case, it is striking that currently successful commercial packages, such as Axon Instrument's pCLAMP, often have originated within an academic research laboratory. If we wish to continue to have good electrophysiology software, we must maintain such skills within our laboratories. (The program source code presented in this book, and the Strathclyde Electrophysiology Software package are available from the author at a nominal charge for disc media and postage.)

Acknowledgements

I would particularly like to thank the researchers who contributed the experimental recordings which illustrate parts of this book: Gabriel Boachie-Ansah, Alison Elliot, Oliver Holmes, Jerry Lambert, John Peters, Ian Marshall, Chris Prior, Tim Searl and Lijun Tian. I am also indebted to Francis Burton, Alisdair Gibb and Chris Prior for the unenviable task of reading the manuscript at various stages, and for suggesting changes which were invariably improvements. As always, the good ideas are other people's and the mistakes are mine. Finally, I would like to thank Ian G. Marshall for the staunch support and encouragement he has given to me over the years I have worked with him. In particular, the patience with which he has borne my endless demands for ever more expensive pieces of computer equipment.

This work has been generously supported by the Wellcome Trust and the Organon Scientific Development Fund.

JOHN DEMPSTER

Contents

Series Preface — v
Preface — vii

CHAPTER ONE
Introduction — 1

1.1 A short history of electrophysiology — 1
1.2 The digital computer in the laboratory — 2
1.3 Minicomputers – the LINC — 3
1.4 The PDP11 minicomputer — 4
1.5 The microcomputer — 5
1.6 The IBM Personal Computer — 6
1.7 The Apple Macintosh Family — 7
1.8 Performance benchmarks for laboratory computers — 7
1.9 The current state of the art — 8
1.10 Software and analysis methods — 9
1.11 The scope of this volume — 10

CHAPTER TWO
Digital Recording of Analog Signals — 13

2.1 Digitization of analog signals — 14
2.2 The analog-to-digital converter — 14
2.3 The laboratory interface — 15
2.4 Programming the laboratory interface — 16
2.5 Laboratory interface to host computer data transfer — 17
2.6 Programmed data transfer — 17
2.7 Limitations of the program data transfer methods — 18
2.8 Interrupt-driven data transfer — 19
2.9 Programming for interrupt-driven data transfer — 20
2.10 Limitations of the interrupt-driven method — 21
2.11 Direct memory access data transfer — 21
2.12 Programming for DMA data transfer — 23
2.13 Advantages of DMA data transfer — 24

2.14	Shared memory	25
2.15	Continuous sampling-to-disc	25
2.16	Detection and recording of spontaneous signals	27
2.17	Choosing a laboratory interface	27
2.18	Scientific Solutions Labmaster	29
2.19	Cambridge Electronic Design (CED) 1401	29
2.20	Data Translation interfaces	30
2.21	National Instruments LAB-PC	30
2.22	Laboratory interfaces for the Macintosh	31
2.23	Other laboratory interfaces	31
2.24	Summary	32

CHAPTER THREE
Analog signal conditioning — 42

3.1	Signal amplification	42
3.2	DC offset removal	43
3.3	Signal filtering	44
3.4	Aliasing	44
3.5	Signal smoothing and noise	45
3.6	Filter types and characteristics	46
3.7	The RC filter	47
3.8	Active filters	47
3.9	Bessel and Butterworth filters	49
3.10	Event detection and triggering	49
3.11	Sources of electrical interference	50
3.12	Stimulus isolation	52
3.13	Modular signal conditioning systems	52
3.14	Computer-controlled signal conditioning	53
3.15	Summary	54

CHAPTER FOUR
Signal analysis: measurement of waveform characteristics — 55

4.1	Primary and secondary signal analysis	55
4.2	Inspection of signal records	56
4.3	Computer graphic displays	57
4.4	Displaying digitized signal records	57
4.5	Hard copies of signal records	59
4.6	Measurement of signal characteristics	60
4.7	Cardiac action potential	60
4.8	Voltage-activated currents	60
4.9	Endplate currents	61
4.10	Semi-manual measurement using screen cursors	61
4.11	Calibration of measurements	62
4.12	Defining the zero reference level	62
4.13	Automated parameter measurement	63
4.14	Acceptance/rejection procedures	63
4.15	Software for automated measurement	64
4.16	The measurements list	65

4.17	Exporting data and results	66
4.18	Signal averaging	67
4.19	Averaging of detected spontaneous signal	68
4.20	Digital filters	69
4.21	The gaussian filter	70
4.22	Digital signal processors	71
4.23	Signal analysis software packages	71
4.24	Software development environments	72
4.25	Summary	72

CHAPTER FIVE
Statistical analysis and presentation of results — 77

5.1	Spreadsheets	78
5.2	Descriptive statistics	79
5.3	The normal distribution	80
5.4	Student's t-distribution	81
5.5	Statistical significance	83
5.6	Student's t-test	83
5.7	The paired t-test	85
5.8	Non-parametric tests: the Wilcoxon rank sum-test	85
5.9	A comparison of parametric and non-parametric tests	87
5.10	Limitations of simple pair-wise tests	88
5.11	Analysis of variance	88
5.12	Multiple comparison tests	89
5.13	Analysis of trends in measurements	92
5.14	Analysis of random distributions	92
5.15	Quantitative comparison of frequency distributions	94
5.16	Statistical software packages	95
5.17	Scientific graph-plotting programs	96
5.18	Summary	97

CHAPTER SIX
Mathematical modelling and curve fitting — 104

6.1	Mathematical models of ion currents	104
6.2	Curve fitting	105
6.3	Quantifying goodness of fit	105
6.4	Finding the least squares	106
6.5	Linearizing transforms	107
6.6	Iterative non-linear curve fitting	108
6.7	Direct search function minimization: the simplex method	109
6.8	Fitting an exponential curve with the simplex method	110
6.9	Gradient methods for function minimization	112
6.10	Method of steepest descent	112
6.11	Newton's method	113
6.12	Levenberg–Marquardt method	114
6.13	Quasi-Newton methods	114
6.14	A practical example using the NAG library	114
6.15	Parameter scaling and initial estimates	115

6.16	Estimation of the parameter standard error	116
6.17	Comparison of the simplex and gradient methods	116
6.18	Exponentials are 'ill posed' least-squares problems	117
6.19	Error profile for the exponential function	119
6.20	Error profile for the double exponential	120
6.21	Choosing the best model to fit the data	121
6.22	Chi-square and variance ratio tests	123
6.23	Distribution of residuals	124
6.24	Standard error of best-fit parameters	125
6.25	Summary	125

CHAPTER SEVEN
Analysis of voltage-activated currents 133

7.1	The voltage clamp	133
7.2	Membrane equivalent circuit model	134
7.3	Digital recording of voltage clamp signals	136
7.4	Simultaneous voltage generation and recording	136
7.5	Command voltage protocols – single step	138
7.6	Current–voltage curves	139
7.7	Voltage ramp protocols	140
7.8	Analysis of current time course	140
7.9	The Hodgkin–Huxley equations	141
7.10	Two-step command voltage protocols	142
7.11	Digital subtraction of leak currents	143
7.12	Leak subtraction protocols	145
7.13	Voltage clamp software packages	146
7.14	Axon Instruments pCLAMP	147
7.15	Cambridge Electronic Design	148
7.16	Strathclyde Software: VCAN	148
7.17	Other software packages	149
7.18	Summary	150

CHAPTER EIGHT
Analysis of single-channel currents 157

8.1	Recording single-channel currents	159
8.2	Analog signal storage and conditioning	159
8.3	Digitization of single channel currents	160
8.4	Voltage-activated single-channel currents	161
8.5	Analysis of channel current amplitudes	161
8.6	Current amplitude histograms	161
8.7	Open channel probability	162
8.8	Fitting gaussian curves to amplitude histograms	162
8.9	Average amplitude histogram	163
8.10	Patlak's running average amplitude	164
8.11	Analysis of channel kinetics	165
8.12	Transition detection	165
8.13	Threshold crossing methods	167
8.14	Two threshold methods	167

8.15	Compensating for zero current baseline drifts	168
8.16	Time course fitting	169
8.17	Analysis of channel dwell times	171
8.18	Linear dwell time histograms	171
8.19	Logarithmic dwell time histograms	172
8.20	Modelling of dwell time distributions	174
8.21	Maximum likelihood method	175
8.22	Practical aspects of likelihood maximization	176
8.23	Determining the number of components	177
8.24	Kinetic models of ion channel gating	178
8.25	The missing events problem	179
8.26	Single-channel analysis software	179
8.27	Axon Instruments software	180
8.28	Strathclyde software: PAT	180
8.29	Cambridge Electronic Design software	181
8.30	Other software	181
8.31	Summary	181

CHAPTER NINE
Analysis of ionic current fluctuations: noise analysis   191

9.1	Recording current fluctuations	192
9.2	Analysis of current variance	193
9.3	Computing the signal variance	194
9.4	Variance vs mean current plots	195
9.5	Spectral analysis of current variance	196
9.6	Estimating channel kinetics from the power spectrum	196
9.7	Computing the power spectrum	197
9.8	Non-stationary noise analysis	199
9.9	Non-stationary kinetic analysis	200
9.10	Non-stationary noise recording procedures	200
9.11	Software for noise analysis	201
9.12	Summary	201

APPENDIX
An introduction to computers   204

A1	The binary number system	204
A2	Computer systems architecture	205
A3	Input and output devices	206
A4	Magnetic disc storage media	206
A5	Computer interface bus	207
A6	Operating systems	207
A7	Programming languages	208
A8	FORTRAN	209
A9	Pascal	211
A10	The C language	211
A11	BASIC	211

A12 Choosing a programming language	212
A13 Graphics and GUI programming	213
References	214
Suppliers	222
Index	223

Listings

2.1	A/D conversion subroutine using programmed I/O	33
2.2	Interrupt-driven A/D conversion routine	35
2.3	DMA A/D conversion subroutine	38
2.4	Continuous A/D recording-to-disc routine	40
2.5	Spontaneous signal detection and recording routine	41
4.1	Automatic waveform measurement routine	73
4.2	Signal averaging routine	74
4.3	Signal average with alignment of rising phases	75
4.4	Gaussian digital filter	76
5.1	Unpaired Student's *t*-test (Borland Quattro spreadsheet formulae)	98
5.2	Paired Student's *t*-test (Borland Quattro spreadsheet formulae)	99
5.3	Wilcoxon rank-sum test for unpaired differences	100
5.4	Wilcoxon paired rank-sum test	101
5.5	One-way analysis of variance with multiple comparisons	102
6.1	Exponential function fitting routine	127
6.2	Simplex iterative curve-fitting routine	128
6.3	Quasi-Newton iterative curve-fitting using NAG library	132
7.1(a)	Simultaneous A/D recording and D/A waveform generation program	151
7.1(b)	Assembler interrupt routine for DAC sweep	152
7.1(c)	Routine to install DAC interrupt service routine into IRQ0 vector and to initiate the D/A output sweep	153
7.2	Leak current subtraction routine	155
8.1	Single-channel current amplitude histograms	182
8.2	Two threshold channel transition detection routine	185
8.3	Estimation of channel open/close time using time course fitting	187
8.4	Single-channel dwell time histograms	189
9.1	Routine to calculate average power spectrum of current noise	202

CHAPTER 1

Introduction

In recent years the digital computer has come to play an increasingly important role in the analysis of electrophysiological signals. This volume is intended to provide an introduction to the procedures involved in a wide range of signal acquisition and analysis methods that have been developed to date. The term electrophysiology has a certain imprecision in its definition – the study of the electrical activity associated with the living cells or tissue – and has been used to describe a wide range of studies from transmembrane ion currents, extracellular recording of nerve firing rate, to clinical electrocardiographic and electroencephalographic studies. In current usage, electrophysiology is generally restricted to referring to studies of the electrical currents or potentials associated with the movement of ions across cell membranes. This discipline encompasses the following topics: the action potentials observed in excitable tissue; the associated ionic current flow; pre- and post-synaptic potentials; currents associated with neurotransmission and neurosecretion; the study of current fluctuations through individual ion channels in cell membranes.

From the point of view of analysis procedures, all of these experimental methods involve the measurement of the amplitude and shape of the electrical waveforms associated with the various membrane phenomena. This use of the waveform to infer properties at a cellular level, distinguishes electrophysiology from neurophysiological studies of the rate of firing of neurones, where the waveshape is not of primary interest, or from clinical EEG or ECG signals used as diagnostic tools.

1.1 A SHORT HISTORY OF ELECTROPHYSIOLOGY

Electrophysiology can be considered to have begun in the late 18th century when the Italian Luigi Galvani noted that the application of an electrical potential to a frog's crural nerves evoked a twitch in the leg muscles. In the mid-19th century it became possible to observe, using the mercury galvanometer, an electrical signal associated with the activity of nerve or muscle. By the end of the 19th century an understanding of the physico-chemical properties of ions in solution was developed with the various works of Arrhenius, Nernst and Planck. In 1902, Bernstein laid the foundations of modern electrophysiology with his hypothesis of a semi-permeable cell membrane selective for K^+ ions, to explain the nerve impulse. These concepts were further advanced with the successful measurement of the resting and action potentials in the squid giant axon (Cole & Curtis, 1939; Hodgkin & Huxley, 1939). Cole's subsequent development of the voltage clamp (a feedback circuit allowing the experimental control of cell membrane potential) permitted, for the first time, the direct measurement of ionic current flow across the cell membrane (Cole, 1949). Hodgkin & Huxley then used the technique to deduce the key roles played by Na^+ and K^+ ions in the action potential (Hodgkin & Huxley, 1952a, b, c). At about the same time the development of the glass micropipette electrode allowed similar intracellular potential measurements to be extended to skeletal muscle. Since then, intracellular recording and the voltage clamp technique have been applied to almost all cell and tissue types. With the development of the patch clamp (Neher & Sakmann, 1976) it has been possible to observe the detailed function of the ion channel – the basic mechanism by which ions cross the cell membrane.

The primary recording technique used by Hodgkin & Huxley in their seminal work on the squid axon was photographic film. Current and voltage signals were displayed on an oscilloscope screen and recorded on 35 mm film using a motorized camera with its shutter synchronized to the oscilloscope sweep. Films were subsequently processed and measurements made manually from prints or enlarged images. The analysis of such experiments often took substantially longer than their performance, one day of experiments producing weeks of analysis work. This manner of working was the norm for many years until the mid-1960s, and common until the late 1970s. Since then, however, the collection and analysis of signals has been increasingly done using the digital computer. This enthusiastic use of the computer is in contrast with other physiological recording techniques which have changed little over the past 40 years. The recording of muscle tension, for instance, by a modern tension transducer and pen recorder differs little in principle from the kymograph.

A striking characteristic of electrophysiologists (in the wide sense here) among life scientists has been a preparedness to apply developments in computer systems to their laboratory work. There are a number of reasons for this. Many electrophysiological signals, being brief electrical events of small amplitude, often superimposed on a noisy background, have always been difficult to acquire using the more conventional recording devices. Since its inception the technique has always pushed the boundaries of available technology, particularly in the areas of high input impedance low noise amplifiers and of recording media such as FM magnetic tape. This orientation of researchers in the field towards new technology, and the limitations of existing methods, created an early willingness to explore the possibilities of the computer, in spite of the cost and difficulties then involved. In addition, interpretation of the signals often demanded detailed quantitative analysis, sometimes involving mathematical modelling. The large numbers of measurements that this required could be particularly burdensome. Again, the potential of the computer to automate these measurements was one of its main attractions.

1.2 THE DIGITAL COMPUTER IN THE LABORATORY

The development of digital computers can be divided into a series of phases closely related to the available electronics technology. The 1940–1952 period can be considered to be the experimental phase of computer development. Eckert & Mauchly completed ENIAC (Electronic Numerical Integrator and Calculator) in 1946 at the University of Pennsylvania.

It used 18 000 thermionic valves and was 35 m long. In the following decade numerous individual machines were hand crafted in computer research laboratories. These machines were seen as the prototypes for future powerful electronic calculating machines. During this period, all that was available for use in the laboratories of other scientific disciplines were mechanical calculators.

By 1951, the first commercially produced general purpose digital computers were becoming available, notably the UNIVAC 1 which was used by the US Census. The first IBM (International Business Machines) computer, the IBM 701, was produced in 1952. These machines were the forerunners of the class of computers known as *mainframes*. Such machines were very large, room-sized devices generating large amounts of heat requiring complex cooling systems. The Whirlwind computer, for instance, required 150 kW of power and was reputed to dim the lights of Cambridge, Massachusetts when it was used. They were very expensive, comparable in price with the modern supercomputer. Universities and large research organizations began to acquire them as central resources in the mid-1950s. The high cost and complexity of the mainframe class computer precluded its use within an individual laboratory, its use for data acquisition being largely ruled out. However, use could be made of the shared central resourse for simulation easing, for instance, the calculation of the Hodgkin–Huxley equations (Cole & Antosiewicz, 1955; Noble, 1966)

The initial stimulus to develop computer systems which could acquire and process data in real time (a key requirement for laboratory data acquisition) came from the military. In particular, the Whirlwind project at the Lincoln Laboratory at the Massachusetts Institute of Technology (MIT), which formed the core of the United States SAGE air defence system, was crucial in the eventual development of the laboratory computer. These MIT scientists were also probably the first to use the computer as a vehicle for laboratory experiments, albeit in psychophysiological studies associated with the air defence project (Green *et al.*, 1959). In addition, Kenneth Olsen, the founder of the Digital Equipment Corporation (DEC) was an alumnus of this laboratory. An interesting history of this period can be found in Mayzner & Goodwin (1978).

Computers continued to improve throughout the 1950s. In particular, the replacement of valves by transistors (IBM 7090, 1959) greatly reduced power consumption, and increased reliability. Transistors themselves were soon replaced by integrated circuits (ICs), as in the IBM 360 series of computers in 1964. The development of the integrated circuit – a complete multi-transistor device etched on single pieces of silicon – permitted the computer to increase in speed, complexity and reliability, while at the same time cutting the cost of construction. This process continues at an ever increasing pace to this day.

By the early 1960s a number of workers, particularly in the United States, were exploring the potential of the computer in the analysis of laboratory data. By then efforts had also been made to offset the enormous costs of the mainframe computer by providing several users simultaneous access to the computer using a process known as time sharing. Digital magnetic tape recorders were developed which could record the analog signals during experiments. Later, such tapes could be taken to another site and replayed into the mainframe for processing.

1.3 MINICOMPUTERS – THE LINC

In the early 1960s, the cost of the digital computer began to fall, due to the development of medium-scale integrated circuits, and the *minicomputer* came into existence. This class of computer was significantly slower than any mainframe but was also much cheaper. Such machines were still too expensive for individual laboratories but could be justified for a department. Consequently systems were developed which allowed experimental signals to be transmitted along telephone lines from a number of laboratories to the departmental computer. This approach seems to have been most easily justified for clinically related work such as the recording of EEG, ECG and evoked potentials (Stark *et al.*, 1964). H. K. Hartline was one of earliest to apply the computer in physiological experimentation, using it to record the frequency of nerve firing of *Limulus* (horseshoe crab) eye to a variety of computer-generated light stimuli (see Schonfeld, 1964, for a review). As usual, it was also used to compute the Hodgkin–Huxley equations (Dodge, 1963). Descriptions of the laboratory computer systems of this time can be found in the New York Academy of Sciences symposium 'Computers In Medicine and Biology' (1964).

In that symposium there was a report of what can be regarded as the birth of the laboratory computer, the LINC (Laboratory Instrument Computer). Clark & Molnar (1964) from the Lincoln laboratory at MIT described a small computer system equipped with the analog–digital (A/D) and digital–analog (D/A) converters necessary for digitally recording experimental signals and producing stimuli. Results could be displayed graphically on an oscilloscope screen via D/A converters. At the same time, DEC (Maynard, MA, USA) produced the first of its low cost minicomputers which would establish its dominance in that field, the PDP8. They developed a variant of the PDP8, the LINC-8 based upon the LINC concept. This computer became the first standard system to go into widespread use by electrophysiologists. Initially such computers often lacked magnetic discs. Programs were stored on punched paper tapes and data on magnetic tape. They were primitive by modern standards having 4 kbyte of RAM memory and teletypewriters, rather than visual display screens. However, they did have the necessary performance to be of use in the electrophysiological laboratory and were extensively used in clinical EEG work (Ackmann, 1979), neurophysiology (Ktonas *et al.*, 1975; Cohen & Myklebust, 1978) and membrane electrophysiology (e.g. Gage & McBurney, 1975; D'Agrosa & Marlinghaus, 1975; Witkowski & Corr, 1978; Pennefather & Quastel, 1981).

A notable use of such computer systems was the analysis of voltage-activated Na^+ and K^+ currents in the squid axon. Armstrong & Bezanilla (1974) developed the digital leak current subtraction system for extracting accurate records of specific currents from combined recordings contaminated with currents from other sources. This method proved crucial in the extraction of the gating or capacity currents associated with the charge movements involved with the opening and closing of ionic current gates. They developed their own digital waveform recorder (Bezanilla & Armstrong, 1977a) which was linked to a PDP8. Another system in use at this time was the LM^2 computer developed by T.H. Kehl at the University of Washington (Kehl *et al.*, 1975; Kehl & Dunkel, 1976).

1.4 THE PDP11 MINICOMPUTER

In 1970, DEC introduced a 16 bit minicomputer the PDP11 which could directly address the then enormous memory of 64 kbyte, and had a much more powerful and elegant instruction set. It rapidly became accepted as the successor to the PDP8, and remained the mainstay of laboratory computing until the mid-1980s. Both hard and floppy magnetic disc drives were becoming common as storage devices. One of the reasons for the success of the PDP11 was its range of excellent operating systems, in particular the RT11 (Real Time 11) operating system which, combined with a range of computer language compilers, for the first time produced a standard and powerful platform for the development of laboratory applications software. Details of the PDP11 in laboratory can be found in Cooper (1977) or Bourne (1981).

DEC opted not to provide specialized laboratory versions of the PDP11 like the LINC8 or PDP12. Instead, a number of small specialist companies both in the United States and the UK began to produce PDP11-based systems equipped with high performance analog–digital conversion sub-systems. In the UK the CED 502 system (Cambridge Electronic Design, Cambridge, UK) found widespread acceptance with Indec Inc. playing a similar role in the United States. These powerful new systems arrived at a fortuitous time for the electrophysiologist. Analysis techniques were being developed in which the computer played an essential part, in particular, the calculation of the conductance of single ion channels by power spectrum analysis of ion current fluctuations (Anderson & Stevens, 1973). In 1976, the development of the patch clamp method allowed the recording of single ion channels (Neher & Sakmann, 1976). The stochastic nature of such signals required the accurate measurement of many thousands of ion channel current pulses, thereby necessitating the use of powerful digital computers (Sachs et al., 1982; Colquhoun & Sigworth, 1983).

In the USA a specific version of the BASIC language, BASIC-23, was developed by Brown & Hobbs for use in the electrophysiological laboratory. This language combined the simplicity of BASIC with an extended set of instructions designed to capture, manipulate and display analog signals. BASIC-23 was used by a number of workers and proved a sucessful vehicle for the dissemination of useful applications software between laboratories.

In the UK, more use was made of the FORTRAN IV language and MACRO 11 (PDP11 assembler language), with software being developed in a number of laboratories. Examples include single-channel analysis software developed by Colquhoun (Colquhoun & Sigworth, 1983) at University College London and electrophysiological software by a number of other groups (D'Agrosa & Marlinghaus, 1975; Black et al., 1976; Crunelli et al., 1983; Quint et al., 1983; Dempster, 1985; Park, 1985; Hof, 1986; Re & Di Sarra, 1988; Re et al., 1989). By the mid-1980s the utility of this kind of software was being widely recognized but it was becoming obvious that its production was difficult and time consuming. Because of this, software was often exchanged informally between laboratories which either had existing links with the software developer or had been attracted by demonstrations of the software at scientific meetings.

1.5 THE MICROCOMPUTER

Even though costs continued to fall and performance improved steadily, the minicomputer remained, throughout its lifetime as a laboratory computer, a high-cost item. Laboratories often only had one shared between several users. In 1984, a PDP11/23 cost at least as much as the complete electrophysiological recording setup. Its use was restricted to well-funded laboratories or those which could find a specific justification. Radical change began to occur with the development of the *microprocessor* in the early 1970s – a complete central processing unit (CPU) on a single integrated circuit. Minicomputer CPUs were typically constructed as circuit boards with multiple integrated circuits. Initially, microprocessors were simple 4 bit computers (Intel 4040) suitable mainly for industrial process control applications and calculators, but the introduction of the Intel 8080 in 1974 and the Motorola 6502 8 bit microprocessors made possible the production of ultra low cost but nevertheless useful small computers – the first microcomputers. Notable among this first generation were the Commodore Pet and the Apple microcomputer. Their performance was not dissimilar to the old PDP8 minicomputer but their low cost (£1000–2000), resulted in their rapid and widespread introduction into the laboratory. For the first time, the concept of the personal computer, i.e. a computer used almost exclusively by one individual, was realized.

These new computers, by producing a mass market for the first time, stimulated innovation in software development; in particular, the spreadsheet program VisiCalc for tabulating and performing calculations on tables of numbers, and the word processor Wordstar. Needless to say, the Hodgkin–Huxley equations were immediately recalculated on these machines. The Apple II proved to be exceptionally suited to laboratory work due to its ability to accept a variety of specialized interface cards including A/D and D/A converters, disc drives, stepper motor controllers etc. Examples of the widespread laboratory uses for the 8 bit microcomputers can be found in Littler & Maher (1989), Kerkut (1985) and Mize (1985).

Paradoxically, the first generation of microcomputers had less impact on electrophysiological laboratories than might have been expected. This was due to the great disparity between the computing power of the microcomputer and the existing PDP11 minicomputers. The PDP11 was a 16-bit computer and was an order of magnitude faster, supporting

10 Mbyte hard discs and 12-bit A/D conversion at sampling rates of the order of 25 kHz. A substantial body of applications software had been developed which could not be ported easily to the smaller 8 bit machines. The first generation of 8 bit micros lacked satisfactory high level languages comparable in quality to FORTRAN on the PDP11. It was also clear that the 8 bit computers were insufficiently powered to tackle the power spectrum calculations and single-channel analysis tasks for which the minicomputers were often employed at the time.

Nevertheless, the low cost of the machines inspired many attempts to utilize the new technology, in spite of their limitations (e.g. Agarwal & Priemer, 1979; Enomoto & Maeno, 1981; Cottrell *et al.*, 1983, 1985; Fusi *et al.*, 1984; Ireland & Long, 1984; Bagust, 1985; Kerkut, 1985; Kits *et al.*, 1987). Attempts were made to split the signal acquisition and processing task between more than one 8 bit processor (Klein *et al.*, 1983). Cambridge Electronic Design took this approach and produced the CED 1401 – an intelligent laboratory interface unit with its own microprocessor (6502) and internal memory as well as the normal A/D, D/A converters and timers. Array processing tasks could be performed faster by the 1401 than by the host computer. Many uses were made of the 1401 in combination with Apple II or BBC Micro (e.g. Quayle *et al.*, 1986).

1.6 THE IBM PERSONAL COMPUTER

In the 1980s, a second generation of microcomputers began to appear based upon 16 bit microprocessors. Most significantly, in 1981 IBM, the dominant supplier in the computer industry, produced the IBM Personal Computer (IBM PC). This machine radically changed the whole nature of the industry. It was based upon the Intel 8088 microprocessor which had an instruction set every bit as powerful as the PDP11 and which could directly address 1 Mbyte of memory, as opposed to the 64 kbyte of the PDP11. It had a set of expansion slots like the Apple II, and was supplied with an effective operating system MS-DOS (called PC-DOS by IBM) developed by the then small independent software company Microsoft. Unusually, IBM revealed the detailed technical specifications for the PC allowing other companies to produce IBM PC-compatible computers capable of running the same software and accepting the same expansion cards. These factors combined with IBM's endorsement of the personal computer concept resulted in the IBM PC becoming the most successful computer ever produced.

The development of laboratory interface expansion cards for the PC allowed it to be used in the laboratory. A wide range of cards became available, notably the Labmaster from Scientific Solutions (Solon, OH, USA) and the DT2801A from Data Translation (Marlboro, MA, USA). Software specifically designed for the analysis of electrophysiological signals also began to be produced, notably the pClamp package developed by Kegel *et al.* (1985) at CalTech. Although the original IBM PC was clearly a useful laboratory computer it still did not exceed the PDP11 in performance. However, the IBM PC AT (Advance Technology) computer, based upon a 6 MHz Intel 80286 microprocessor, and introduced in 1984, was four times faster than the original PC and quite comparable with the PDP11/23. When good quality, yet inexpensive, computer language compilers became available for the IBM PC family of microcomputers, they rapidly began to replace the PDP11 in electrophysiological laboratories.

Since then there has been a constant improvement in performance and memory capacity of the IBM PC family. Currently, the fastest PC-compatible computers based upon 50 MHz 80486 microprocessors are almost 100 times faster than the original PC. Great strides have also been made in the development of applications software with highly sophisticated packages containing many features. A typical laboratory PC (*ca* 1991) has a 16 or 25 MHz 80386 microprocessor, 1–4 Mbytes of RAM, memory 40 Mbytes of magnetic disc storage and a display screen capable of display graphics to a resolution of 640 × 480 points in 16 colours.

1.7 THE APPLE MACINTOSH FAMILY

In this period, a second significant family of computers appeared: the Apple Macintosh PCs, based upon the Motorola 68000 series of microprocessors. The Macintosh was radically different in concept from the PC using a *graphical user interface* (GUI) where the user interacted with the computer by clicking on objects (icons) or selecting from menus using a mouse pointing device. The Macintosh family proved remarkably easy to use, and attracted many to the use of computer who initially had been inhibited by the complexity and diversity of the commands required to use a conventional, command-driven, operating system, such as MS-DOS.

Suprisingly, in spite of the role of expansion slots in the success of the earlier Apple II series of computer, the original Macintosh lacked such features. For this reason, it was difficult to add the A/D converter cards essential for use within the laboratory. Devices could be produced which used the Macintosh serial port (e.g. MacLab, World Precision Instruments, New Haven CT, USA), but these were limited by the slow speed of this communications channels. The Macintosh II series of computers remedied this, being fitted with a more conventional set of expansion slots, called the Nu-Bus. Since then an increasing number of interface cards have become available. However, although the Macintosh has established itself in the area of image analysis (where its GUI is probably a distinct advantage) it is much less used for the analysis of signals, in situations such as the electrophysiology laboratory.

1.8 PERFORMANCE BENCHMARKS FOR LABORATORY COMPUTERS

An illustration of the increase in power of the laboratory computer over the three decades since its inception can be seen in Figure 1.1, which shows a comparison of the computation speed and RAM memory capacity of some of the computers discussed so far. It is difficult to make comparisons between computers with widely different architectures, instruction sets, and word lengths. Performance also depends somewhat on the application, with certain machines often having unique features which give them comparative advantages in particular areas but not others. However, one useful (though approximate) benchmark for a laboratory computer is the time taken to add 16 bit integer numbers, since this operation (and many similar to it) is used extensively in the processing of digitized signals. Another important parameter is the amount of RAM memory that can be accessed easily. RAM memory is used to store programs while they are executing and also to hold the digitized signal while simultaneously recording experimental data. Figure 1.1 shows the time taken to perform a 16 bit addition, plotted against the maximum addressable RAM memory for the PDP8, PDP11, BBC Micro, IBM PC, IBM PC AT, and some modern high-speed IBM PC-compatibles. (Note that the scales are logarithmic.)

This simple plot illustrates the importance of the computer's word length in determining efficiency of calculations. The word length, described in terms of binary digits (bits), is the largest number that the CPU can operate upon in a single instruction (see Appendix for more details). For instance a computer with an 8 bit word can add numbers in the range 0–255. Addition of numbers larger than the word length must be handled by a series of separate instructions. Word length also has an effect on the amount of directly addressable memory since the memory address may also have to fit into a single word (or pair of words).

The first widely used laboratory computer, the PDP8 with a 12 bit word, could only address 4 kbyte (4096 locations) of memory and took over 10 μs to perform the addition. It can be seen that the PDP11 (16 bit) family provided more than an order of magnitude improvement in memory (64 kbyte) and a two-fold reduction in computation time. This was achieved because the PDP11 could perform the addition with a single add instruction whereas the PDP8 needed at least three instructions to do the same job. This is also why the BBC micro, like the other microcomputers based upon 8 bit CPUs, could not match the PDP11. However, the IBM PC

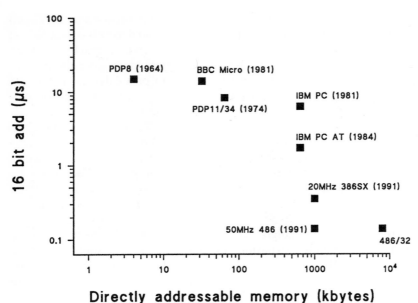

Figure 1.1 Comparison of the computational performance and addressable memory of some computers that have been commonly used in the laboratory. Computation speed measured as the time taken to add a 16 bit number.

family using a 16 bit CPU could not only perform the addition in a single instruction, matching the PDP11, but also address directly an order of magnitude more memory, by handling the address as a pair of words. While the PDP11 could be fitted with more than 64 kbyte of memory, this extra memory could not be used easily by a typical program. This fundamental design limitation quickly made the PDP11 obsolete. The original IBM PC, based on the Intel 8088, had a similar memory limitation (at 1 Mbyte) and this is currently a major problem for software running under the MS-DOS operating system. However, the latest IBM PC family computers, based upon the Intel 80386 and 80486 microprocessors, in addition to being faster, can operate using 32 bit words which allows many Mbytes of addressable memory.

It is likely that the next 5 years will see the development of new operating systems for the PC family which exploit these features. The beginnings of this process are already evident in the development of the Microsoft Windows (V3) programming environment. It is also worth noting that the Apple Macintosh family does not have these kind of memory limitations, being able to directly address whatever memory is installed within the system.

1.9 THE CURRENT STATE OF THE ART

At present, the IBM PC and Macintosh computer families constitute the majority of computers used in the laboratory. The wider availability of suitable expansion cards has resulted in a preponderance of IBM PCs dominating the field of electrophysiology data acquisition. Most electrophysiological analysis software is now developed for the IBM PC using the MS-DOS operating system with the range of packages available for the PC now exceeding what was ever available for the PDP11. Software has continued to be produced by research laboratories (Morales *et al.*, 1985; Bevan *et al.*, 1986; Robinson & Giles, 1986; Dempster, 1987, 1989; McCann *et al.*, 1987; Bermejo & Zeigler, 1989; Stromquist *et al.*, 1990). However, there has been a trend towards the commercial distribution of software rather than the free exchange. The original pClamp program produced at CalTech has been substantially enhanced and is now marketed by Axon Instruments (Foster City, CA, USA). Similarly, most suppliers of laboratory interfaces now also supply data acquisition software (e.g. Cambridge Electronic Design

Figure 1.2 A typical laboratory computer system. IBM PC-compatible (16 MHz 80386 CPU) with 40 Mbyte hard disc, CED 1401 laboratory interface, Hewlett Packard Laserjet IIP laser printer, Fylde Electronic Laboratories signal conditioner, Biologic DTR1200 DAT digital tape recorder.

voltage and patch clamp packages). With the development for the IBM PC of a version of the BASIC-23 language, in the form of Axon Instrument's AXOBASIC, many of the programs originally produced for the PDP11 can now be ported to the PC.

A typical laboratory computer and data acquisition system, currently in use in the author's department, is shown in Figure 1.2. The host computer is a 16 MHz 80386-based IBM PC-compatible, with a 40 Mbyte hard disc and a VGA graphics display screen. It is shown with a CED 1401 laboratory interface which is connected to the computer via a ribbon cable and an interface board plugged into one of the slots in the computer expansion bus. Analog signals are fed into the laboratory interface, via a Fylde Electronic Laboratories (Preston, UK) *analog signal conditioning* unit (amplifier and low-pass filter). A Biologic DTR1200 digital tape recorder, used to store analog signals is also shown. The author's SCAN data acquisition program (Dempster, 1987) is running on the computer and a digitized signal is shown on the computer display screen. Hard copies of the signal and tables of results can be printed on the Hewlett Packard laser printer attached to the computer's parallel printer port.

1.10 SOFTWARE AND ANALYSIS METHODS

Laboratory computers have now reached a stage of maturity where neither the cost nor the performance of the hardware proves a significant limitation to its application within the laboratory. Compared with the PDP11 era, the computer analysis system is now probably the least expensive component of the overall electrophysiological laboratory; costing no more than a voltage clamp or microscope. Inexpensive readily available A/D converters can easily digitize signals at sampling rates of 100 kHz or more. However, as is true in other areas of computing, it is becoming clear that the development of appropriate and reliable software is the main impediment to progress in this area.

As the computer becomes the normal means of analysing experimental data, the need to understand the procedures involved, their strengths and limitations and reasons for use, becomes increasingly important. In earlier times, when the user of the program was also the author, this process was largely an automatic consequence of writing the program. However, increasingly much specialist software in laboratories is bought as a commercial product or obtained as a gift. The difficulty in writing such software, let alone the wastefulness of everyone re-inventing the same programs, makes such trends inevitable. One aim of this volume is to redress this balance by presenting a comprehensive account of the methods involved in the practical application of the computer to the analysis of electrophysiological signals.

In spite of the extensive use made of computers in the laboratory over the past 30 years, relatively little has been written in detail on the actual methods used. Discussion is often limited to brief notes in the methods sections of research papers. Such papers that exist on the subject are usually short and give little more than an overview of the particular topic. This is partly due to the difficulties in the publication of computer-based methods. Although the presentation of general principles is useful, much of the value of a computer analysis method lies in the implementation of the program itself. The source code of such programs is often complex, very large, and with features specific to a particular computer or laboratory. They are not particularly suitable for publication via the normal channels such as journal articles. Also many aspects of computer technology fall outside the domain of the life sciences.

The extraordinarily rapid evolution of the computer, from the mainframe, through the mini- to the microcomputer, has also tended to reduce the 'shelf life' of many of the earlier texts on the subject. Nevertheless, much useful information is contained in such texts. For instance, Stacy & Waxman's three volume series *Computers in Biomedical Research* (1965) still contains many useful insights, into basic principles and algorithms. However, it should be borne in mind that technical approaches which were not feasible when a text was written often become commonplace only a few years after publication. For instance, at one time the high cost of the computer demanded that its resources be shared through the use a complex multi-user operating system. However, for the last ten years the emphasis has been on one computer per experimenter and simple single-user operating systems. Other useful texts include: Soucek & Carlson (1976), Cooper (1977), Mayzner & Dolan (1978), Bourne (1981).

Compared to the vast literature on the use of the personal computers for business applications there are relatively few texts discussing the computer in the laboratory, particularly the recording and analysis of analog data. Some texts also focus upon various minutiae of computer operation such as the principles of logic gates and the electronic design of A/D converters (e.g. Gibson & Liu, 1987; Spencer, 1990). Other texts provide only very general overviews (Ouchi, 1987; Beasley, 1990; Thomson & Kuckes, 1989) discussing primarily the use of standard word processing and spreadsheet packages. What discussion of electrophysiological applications of computers that does exist is usually found within more general texts on computers in physiology or the neurosciences (Kerkut, 1985; Mize, 1985; Fraser, 1988).

1.11 THE SCOPE OF THIS VOLUME

The aim of this volume is to discuss the specific computer techniques associated with the analysis of electrophysiological signals. Numerous texts exist on the elementary principles of the operation of the computer and on computer programming, and it is not intended to provide such an introduction here. It is assumed that the reader has a familiarity both with the basic principles of electrophysiology, and with the personal computer and its use for common tasks such as word processing. However, Appendix 1 contains a very brief introduction to digital computer principles. A more complete introduction to the personal computer can be found in Sargent & Shoemaker (1987).

This book is intended to be a practical guide to computer techniques, therefore there are examples of computer programs at the end of each chapter, embodying the concepts discussed. The best way to understand the operation of a computer is to study fragments of the actual program code. Hopefully, this code may also be of use to readers intending to develop their own software. The program listings, however, are not essential to the understanding of the text as a whole. The emphasis will also lie towards the IBM PC family, partly due to its current prominence in the laboratory and partly due to the author's experience. The principles are general and can also be applied to the Macintosh family which although more difficult to programme is nevertheless useable as a data acquisition computer.

To complete this introductory chapter, a brief outline of the general structure of the book will be given. Chapters 2–6 consider the basic principles and practice of the digitization and analysis of signals, while Chapters 7–9 discuss the analysis of some particularly important types of electrophysiological signal.

In Chapter 2 the operation and programming of the laboratory interface is described. This is perhaps the most technically detailed of all the chapters, discussing many specific aspects of computer operation. While it is not important for the electrophysiologist to know how an A/D converter works in exact detail, an understanding of the general principles is essential. In particular, several performance issues and techniques vital to electrophysiological applications will be examined in detail.

Chapter 3 covers the signal conditioning necessary to prepare the analog signal for digitization. This involves amplification of the received signal, in order to match the input requirements of the A/D converter, filtering, event detection, and isolation of the recording system from the experiment to minimize computer-generated electrical interference.

Chapter 4 contains a discussion of procedures for displaying, analysing and enhancing the signals once they have been digitized and stored within computer memory. The automated measurement of signal waveforms is discussed, along with the role of the operator in validating such signals. Signal enhancement techniques such as signal averaging and digital filtering are also introduced.

In Chapter 5 an account is given of some of the standard statistical techniques which can be applied to summarize the waveform measurement results from signal records acquired under a variety of experimental conditions. Some issues of experimental design are discussed, together with techniques for determining the statistical significance of the results. Particular attention is paid to the use of the spreadsheet program as a means of (a) tabulating results and (b) applying statistical tests. Spreadsheet formulae for a variety of simple tests are presented.

Chapter 6 provides a detailed introduction to mathematical modelling and curve-fitting techniques, particularly iterative non-linear least squares. The principles of operation of the commonly used simplex and Levenberg–Marquardt curve-fitting algorithms are discussed along with a comparison of their performance, both in terms of speed and accuracy. Techniques for determining the accuracy of curve-fitting methods are discussed as well as ways of determining which among a series of competing models provides the best fit.

In Chapter 7, procedures are given for the recording and analysis of voltage-clamp data obtained from voltage-activated ionic currents. The use of the computer to generate the voltage clamp command pulse while simultaneously digitizing the resulting membrane current and voltage is treated in detail. The analysis of current–voltage curves and current time course is discussed, together with the fitting of exponential and Hodgkin–Huxley curves. Methods for the digital subtraction of unwanted background currents are presented. Finally, a comparison of the features provided by some of the commonly available voltage clamp software packages is provided.

Chapter 8 discusses the analysis of single-channel currents. The calculation and use of current amplitude histograms and the fitting of gaussian curves are discussed. The half-amplitude threshold and time course fitting methods for detecting the transitions between channel open and closed states are discussed with a comparison of the advantages and disadvantages of each method. Linear and logarithmic dwell time histograms are described with procedures for fitting exponential probability density functions. A comparison of the maximum likelihood and least squares methods is provided. The program listings for this chapter provide the core of a simple single-channel analysis program.

Again, the features of commonly available software are discussed.

In Chapter 9, techniques for recording and analysing the randomly fluctuating 'noise' in whole cell ionic currents are described. Both stationary and non-stationary techniques are covered with current variance and power spectrum analysis. The use of the fast Fourier transform to compute the power spectrum is discussed.

CHAPTER TWO

Digital recording of analog signals

The key procedure which underlies all the computer analysis methods discussed in this book is the conversion of the analog voltage signals derived from the electrophysiological recording system into digital form, suitable for storage and processing within the computer. This chapter discusses the principles of analog to digital conversion and how this can be applied to the recording of electrophysiological signals. To be of any use, this inevitably requires detailed technical discussion of certain aspects of the computer hardware and software. At one level it is important to be able to understand at least the general principles of a procedure which is fundamental to computer-based signal analysis techniques which are rapidly becoming the norm in electrophysiology. It also enables informed choices to be made when choosing laboratory computer systems.

The initial part of the chapter (Sections 2.1–2.5) provides a basic introduction to analog-to-digital conversion and the functions performed by the laboratory interface. A detailed comparison of the particular features of some of the interfaces commonly in use can be found in Sections 2.18–2.24. The remainder of the chapter is probably of more interest to those considering the task of actually programming the interface themselves.

Unfortunately, it is not possible within the space available to provide a really comprehensive treatment of the development of data acquisition software. However, the salient points are discussed here, and this will provide a flavour of some of basic programming styles. The discussion focuses very much on laboratory interfaces and the computer architecture of the IBM PC family. Most of the developments in laboratory data acquisition in recent years have been for the IBM PC and there are many laboratory interfaces available. Little or no development work is now done on earlier types of machine such as the PDP11 minicomputer. As will be discussed later, certain details of the IBM PC design make it eminently suitable for data acquisition work. Given the continued development within this family, this architecture also clearly has many years of life in it yet. To date, much less use has been made of the Apple Macintosh. However, given the elegance and ease of use of that family of computers, this is likely to change in future. Many of the principles discussed here will also apply to the Macintosh, although the details of the implementation might be different.

Practical examples of the digital recording techniques discussed are presented in the listings at the end of the chapter. They provide additional information to the programmer and can be used as the basis of working software.

2.1 DIGITIZATION OF ANALOG SIGNALS

A characteristic feature of the digital computer is that it manipulates integer numbers (i.e. . . . −3, −2, −1, 0, 1, 2, 3, . . .) and only such numbers can be handled within its memory. Analog signals, such as those produced by electrophysiological recording apparatus, are quite different in that they are continuous. The amplitude of an analog voltage signal can take any value within the permissible output range of the device supplying it. Similarly, such a signal varies smoothly with time. A completely accurate copy of an analog signal cannot therefore be represented within a digital computer. A *digitized* approximation to the analog signal can however be made by repeatedly *sampling* the signal voltage level at fixed time intervals, and assigning this amplitude to the nearest integer value that the computer can hold. The continuous analog signal is therefore chopped up and represented inside the computer as a series of integer numbers.

The quality of the digital approximation is dependent on the fineness of both the integer levels used to represent the amplitude and the size of the interval between samples. Figure 2.1 shows two digital recordings of a typical electrophysiological signal, an endplate current. Compare recording (a) consisting of 24 samples collected at a rate of 2 kHz (1 sample/0.5 ms) and the input range split into 32 divisions with (b) at a 20 kHz rate (1 sample/0.05 ms) using 256 amplitude divisions. Record (a) is clearly a

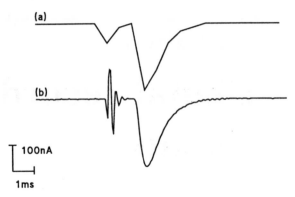

Figure 2.1 Digital recordings of an endplate current: (a) 2 kHz sampling rate (24 samples), 32 level amplitude resolution; (b) 20 kHz sampling rate (240 samples), 256 level amplitude resolution.

poor representation of the original signal, with many of the features of the signal waveform lost compared to (b), which is indistinguishable from a direct oscilloscope record. In practice, digitized electrophysiological signal records usually contain no less than 500, and often 1000–2000 samples, and are quantized in terms of 4096 amplitude intervals.

2.2 THE ANALOG-TO-DIGITAL CONVERTER

The repeated sampling of analog voltage signals is performed by a device known as an *analog-to-digital converter* (ADC). This is a computer-controlled voltmeter, capable of rapidly measuring the analog signal presented to its input and producing an integer number proportional to it, which can then be read into computer memory. The resolution of the ADC is specified as the number of *bits* used to represent the analog voltage. Eight, 12 and 16 bit resolutions, for instance, produce measurements as integer numbers in the range 0–255, 0–4095, and 0–65 535. The wider the range of numbers, the finer the measurement. ADCs are classified in terms of their resolution and the maximum rate at which they can measure the analog signal (sampling rate).

Consideration must be given to resolution and maximum sampling rate when choosing a suitable ADC for digitizing electrophysiological signals. In general, resolution is obtained at the expense of sampling rate and/or cost. A wide variety of ADCs are available fulfilling high resolution, high speed or low cost. The ADC must be capable of sampling at rates high enough to keep up with the time course of the analog signal. Some electrophysiological signals such as voltage-gated Na^+ currents have particularly rapid time courses (1–2 ms) and require high-speed ADCs for digitization.

While 8 bit ADCs are inexpensive and can have very fast sampling rates, a percentage resolution of only 0.4% (and that in the unlikely condition that the signal of interest spans the full voltage range) is not sufficiently fine for research purposes. Sixteen bit ADCs on the other hand, provide a very high resolution at the expense of sampling rate and cost. In addition, noise levels in the typical electrophysiological measurement system are unlikely to be low enough to make effective use of 16 bit resolution. For instance, with a 10 V input range noise levels would have to be less than 300 µV to make use of all 16 bits. Twelve bit ADCs currently provide an adequate compromise, with a resolution more than adequate to match the commonly achievable signal–noise ratios from electrophysiological experiments (100:1) even if the signal does not span the whole input voltage range. Twelve bit ADCs with maximum sampling rates suitable for even the fastest electrophysiological signals (100 kHz) are also available without excessive cost.

2.3 THE LABORATORY INTERFACE

A high-speed high-resolution ADC is a difficult device to construct and most are currently obtained as precision integrated circuits manufactured by a number of specialist suppliers (e.g. Analog Devices, Burr Brown). In addition a number of ancillary support circuits are required to allow the device to be controlled and read from the computer. The complete system is known as a *laboratory interface*. In addition to the ADC the laboratory interface provides:

- high-speed clock(s) for accurate timing
- computer–ADC communications
- multiple analog input channels
- digital-to-analog conversion (DAC)
- digital pulse input and output.

The interface takes the the form of a circuit card (sometimes connected to additional external modules) which is inserted into an expansion slot in the host computer. Most of the computers

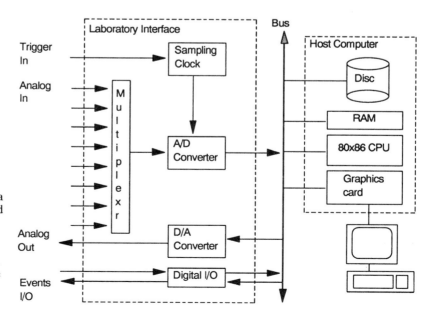

Figure 2.2 Block diagram of a laboratory interface connected to an IBM PC personal computer, showing its A/D, D/A and timing sub-systems. Control information and data is transferred to and from the interface via the expansion bus.

used within the laboratory (e.g. the IBM PC and Apple Macintosh families) have sets of such expansion slots designed to allow the facilities of the computer to be extended by the addition of specialized circuit cards. A detailed description of the IBM PC expansion bus can be found in Eggebrecht (1990).

Figure 2.2 is a generalized block diagram of the functions performed by the laboratory interface which also illustrates how it communicates with its host computer. A program running on the host PC's central processing unit (CPU) can send control codes to, and read data from, the laboratory interface via the computer's expansion slots. The sampling clock can be programmed from the computer to send a pulse at accurately timed intervals to the ADC, instructing it to sample the incoming analog signal. When the conversion is complete the digital number is transferred to the computer and stored somewhere in its memory. If more than one analog channel is to be sampled, then the computer must also instruct the *multiplexer* (a digitally controlled multi-way analog switch), to switch the appropriate analog channel through to the input of the ADC.

2.4 PROGRAMMING THE LABORATORY INTERFACE

Communication between a host computer and the laboratory interface is performed via a set of *input/output (I/O) ports* supplied by the interface card. From the programmer's point of view, I/O ports act in a similar fashion to computer memory except that data written to them is transferred to the laboratory interface rather than stored in RAM. Similarly, data can be transferred from the laboratory interface to the PC by reading the I/O port. I/O ports have unique address numbers, similar to memory addresses, and special machine code instructions (IN, OUT on the IBM PC) for reading and writing to them. Laboratory interface I/O ports can be classified into the following categories, although the exact number of ports and detailed function may vary considerably between manufacturers.

- *Control* ports *written* to in order to command the interface to perform some action, e.g. to initiate an A/D conversion, to start or stop the sampling clock.
- *Status* ports *read* by the computer to determine the result of a previous command, e.g. to determine whether an A/D conversion has been completed, or whether an error has occurred.
- *Data* ports used to transfer data values between the host computer and the interface, e.g. read to obtain the results of an A/D conversion.

A detailed discussion of IBM PC I/O ports in general, and their principles of operation, can be found in Eggebrecht (1990). To provide a specific example, Figure 2.3 shows some of the control and status ports for the National Instruments LAB-PC (National Instruments, Austin, TX, USA) laboratory interface. The A/D conversion process is handled via six I/O ports – two control, one status, and one data port for the A/D converter and two ports which control the sampling clock. Each port can accept or supply an 8 bit number (0–255). It is usually convenient to represent these numbers in binary form, as 8 binary digits, numbered 0–7 from the lowest to the highest, as shown in Figure 2.3. In the control ports, each bit (or cluster of bits) can be considered to act as switches, controlling some operation within the interface, depending on whether they are set (1) or clear (0). Similarly, the bits in the status port act as flags signifying that some event within the interface has occurred.

The LAB-PC interface has eight analog input channels with an amplifier which can be programmed to have gains between ×1 and ×100. Command port 1 is used to select the channel to be digitized and the amplifier gain. A number between 0 and 7 entered into bit positions 0,1 and 2 select the input channel, while a similar 3 bit number entered into bits 5,6,7 selects the amplifier gain. The A/D sampling rate is controlled by Counter A (the LAB-PC has 2 counters A and B) which provides a pulse to the A/D converter to start the conversion. The mode of operation and pulse rate of this counter is programmed via the counter mode and data

Digital Recording of Analog Signals 17

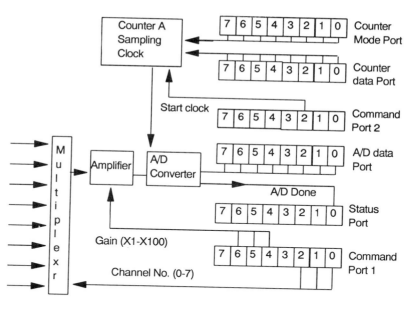

Figure 2.3 I/O ports used to support A/D conversion with the National Instruments LAB-PC laboratory interface for the IBM PC. Command port 1 controls A/D input channel selection and gain. Sampling rate is controlled by a Counter A, set via the Counter A Mode and Data ports. The clock is started using Command port 2. Bit 0 of the Status port indicates when a conversion has been completed.

ports. Once the sampling clock has been programmed, it can be started by setting bit 2 of command port 2. Once started the clock continues to produce A/D start pulses at the programmed intervals. When an A/D conversion has been completed, bit 0 of the status port is set to indicate that a new sample is available for collection by the host computer. The sample can then be read from the A/D converter data port.

2.5 LABORATORY INTERFACE TO HOST COMPUTER DATA TRANSFER

Programming a laboratory interface such as the LAB-PC to digitize analog signals consists of three distinct functions. In the *initialization* phase, the interface must be set up for the required number of A/D conversions, sampling rate and input channels. Second, once sampling has begun, the *data transfer* phase handles the transfer of the stream of samples produced by the A/D converter into the host computer's memory. Finally, after the required number of samples have been collected, the laboratory interface is shut down again during the *termination* phase. Although the details might differ between interfaces, initialization and termination

is usually a straightforward process of setting the appropriate bits in the interface command ports. Data transfer, on the other hand, requires careful attention, particularly when sampling at rates greater than 1–2 kHz. In general, unless the interface has the ability to store A/D samples internally, a sample generated by the ADC, must be transferred into host computer memory before the next conversion occurs and overwrites it. This is an easy task at sampling rates of 1 kHz or less, but becomes increasingly difficult at higher rates. In fact, it usually is the rate at which A/D samples can be transferred from the interface into the host computer memory, known as the *throughput*, which limits the maximum possible sampling rate rather than the conversion time of the ADC itself. There are four commonly used data transfer methods. Not all laboratory interfaces have hardware suitable to support all of the methods.

- programmed
- interrupt driven
- direct memory access
- shared memory.

2.6 PROGRAMMED DATA TRANSFER

Programmed data transfer (also known as programmed I/O) requires the least in terms of hardware sophistication from the laboratory interface. It may be the only method possible with some very low cost interfaces. At its simplest, the method involves requesting the interface to perform an A/D conversion, waiting till it is done, then reading the sample from the interface data port and storing it in memory.

A FORTRAN subroutine (Microsoft FORTRAN V5), making use of the LAB-PC interface, to collect a series of samples from a single A/D channel using the programmed I/O method can be found in Listing 2.1. The program, although quite simple, illustrates many of the basic procedures involved in programming of laboratory interfaces. In general terms, the program implements the following algorithm:

(a) initialize interface card
(b) program the sampling clock
(c) wait until an A/D sample is available
(d) copy the sample from ADC data port to a storage array
(e) repeat (c) and (d) until all samples are collected
(f) stop the sampling clock.

During the initialization phase, the command ports of the interface are first cleared of any settings, ensuring that the interface is in a well-defined starting condition. This avoids the possibility of previous port settings used by other programs interfering with the operations that are about to be set up. The analog input channel and amplifier gain are then written into command port 1.

The clock which controls the A/D sampling in the LAB-PC is the widely used Intel 8253 Counter/Timer. The 8253 is a device providing three digital counter/timer channels. It can be programmed in a variety of modes, to produce either single pulses or streams of pulses at preset intervals. It is used for timing and counting purposes in a wide range of laboratory interfaces and microcomputers. The IBM PC, for instance, uses an 8253 timer to drive its time of day clock. The LAB-PC has two of these devices (Counter A and B) which can be programmed in a number of ways. In the example here, channel 0 of Counter A is used to provide the pulses which start each A/D conversion (see Figure 2.3). The sampling interval is determined by a 16 bit number written into the clock data port, which determines the clock period in units of microseconds. A somewhat terse, but definitive, description of the operation and programming of the 8253 clock can be found in the *Intel Microprocessor and Peripherals Handbook* Vol. 1 (Intel, 1987). A more expansive tutorial on the device can also be found in Spencer (1990).

When the programming of the clock is complete, the A/D data buffer is cleared, and the sampling clock started by setting bit 2 of command port 2. The program then enters the *data transfer* phase. This is a loop which monitors the status port, waiting for bit 0 to be set, indicating that an A/D conversion has completed and a sample is ready to be transferred into the data array in memory. The resulting 12 bit A/D sample is extracted in two parts (low byte, high byte) by reading the data port twice. The two bytes are then recombined, and stored in the data array.

The data transfer process continues until the required number of samples have been collected. Once this occurs the final act of the program is to turn off the sampling clock, shutting down the A/D converter. Just as it was necessary to initialize the interface, it is also good practice to leave the interface in a quiescent condition. In this case, leaving the sampling clock running is not likely to cause problems. However this is not generally true of some of the other data transfer methods to discussed shortly.

2.7 LIMITATIONS OF THE PROGRAM DATA TRANSFER METHODS

The subroutine in Listing 2.1 performs quite well, supporting sampling rates up to 20 kHz without missing any samples. However this is significantly less than the 62.5 kHz maximum conversion rate of the A/D converter in the LAB-PC. The rate limiting step here is the FORTRAN code for testing the A/D status and transferring the sample to the storage buffer which is taking significantly longer than the conversion time. Unfortunately, this is one of the penalties inherent in the use of high level languages such as FORTRAN, C or BASIC. To achieve higher throughput rates the program (or at least the data transfer loop) must be written using the more efficient, but more difficult, assembler language.

A further problem with programmed data transfer, is that the program must constantly monitor the status of the ADC to find out if a new sample is available. Consequently, when recording is in progress it is difficult to use the computer for other operations in case samples are missed. This is important since for many applications it is essential to be able to perform other functions while recording, such as writing data to disc, responding to keys pressed on the keyboard, or updating the display screen. Indeed, at rates above 10 kHz it even becomes necessary to shut down some of the normal functions of the computer. For instance, the IBM PC maintains its time of day information by means of a clock interrupt which occurs every 160 ms. Since this process takes at least 100 µs, samples are likely to be lost at rates much greater than 10 kHz unless this interrupt is turned off. (The 20 kHz performance of the LAB-PC, achieved earlier, is due to the fact that it can temporarily store up to 16 samples in an internal FIFO (First In First Out) buffer, during the delays caused by the clock interrupts.)

However, despite these problems a great deal can be achieved with the programmed I/O method. A good example of this is Axon Instrument's pCLAMP voltage clamp analysis program (Kegel *et al.*, 1985) which uses the Scientific Solutions Labmaster interface (Axon Instruments, Foster City CA, USA). pClamp uses a skillfully coded assembler program loop to both collect A/D samples and generate a voltage stimulus waveform using a D/A converter. Using a PC with a 25 MHz 80386 CPU, pCLAMP can acheive sampling rates as high as 125 kHz. In summary, the programmed data transfer method is easy to implement and understand and can be used with even the simplest laboratory interfaces. However, to achieve the high sampling rates required for most electrophysiological work the A/D sampling program must be carefully written in assembler language. In addition the host computer is fully committed to supervising the sampling process. It is not possible to sample continuously at high rates for long periods since there is no time for the host computer to transfer samples to disc once the RAM memory is full.

2.8 INTERRUPT DRIVEN DATA TRANSFER

In order to improve upon programmed data transfer it is necessary to avoid the need to monitor the ADC status constantly. A mechanism is therefore required to allow the laboratory interface to request the services of the host computer when a conversion has been completed, irrespective of what program is currently running at the time. This is in fact quite a common problem since many of the other peripheral devices used in normal computer operations, such as the keyboard, serial communications ports, floppy or hard drives, all present similar demands. Although it is possible to handle all of these disparate devices by constantly monitoring their status ports one after the other (the original Apple II micro-computer is an example), this is a very inefficient approach which wastes most of the computer's processing power simply waiting for events to happen.

In modern microcomputers like the IBM PC and Apple Macintosh families, a feature known as the *hardware interrupt* has been introduced in order to solve this problem. This is a method by which peripheral devices can signal to the computer's CPU that they require servicing. The currently executing program is then temporarily suspended and control transferred to the appropriate code for dealing with the needs of the peripheral device. Hardware interrupts are implemented in the IBM PC family of computers using the *Intel 8259A Programmable Interrupt Controller* which supports eight independent *interrupt request lines* (IRQ0 . . . IRQ7) on the PC's expansion bus. These lines can be connected to the circuitry of a peripheral device and used to invoke any one of eight different *interrupt service routine* programs. Most of the interrupt lines are already in use with IRQ0 being used to support the MS-DOS time-of-day clock, IRQ1 the keyboard, and others assigned to handle serial and parallel communications ports. However, IRQ5 and IRQ7 are usually available,

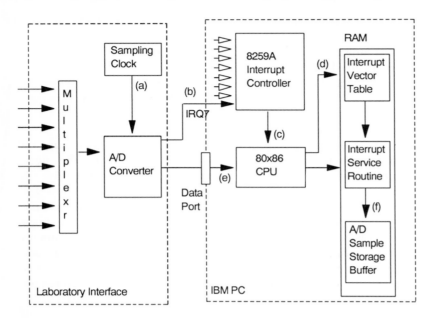

Figure 2.4 IBM PC interrupt-driven transfer of ADC values from the laboratory interface into PC memory. A/D conversions are initiated at regular intervals by sampling clock (a). On completion of a conversion the ADC uses IRQ7 to signal to the 8259A interrupt controller (b) to interrupt the program running on the CPU (c) and transfer control to an interrupt service routine (d) (address stored in the IRQ7 vector table entry). This routine then reads the sample from the A/D converter data port (e) and stores it in memory (f). On completion of the service routine, control returns to the interrupted program.

being associated with unused features of the PC's parallel printer ports, and also IRQ3 if the PC has only one RS232 serial port.

The operation of the IBM PC interrupt system is illustrated in Figure 2.4. The sample timing clock of the laboratory interface is programmed to initiate A/D conversions (a) at regular intervals in the same way as for the programmed I/O example. When a conversion has completed, the ADC produces an interrupt request signal using line IRQ7 of the PC 8259A controller. The controller then interrupts the program currently running on the PC and causes the CPU to execute the interrupt service routine whose address it finds in the interrupt vector table. The service routine then reads the new A/D sample from the ADC data port and stores it in PC memory. It also selects the next analog input channel on the multiplexer if multi-channel sampling is required. Finally, an IRET instruction is issued which returns control of the CPU back to the point in the program when the interrupt occurred.

2.9 PROGRAMMING FOR INTERRUPT-DRIVEN DATA TRANSFER

In order to use interrupts for transferring A/D samples into computer memory the laboratory interface hardware must be attached to the chosen IRQ line, usually by setting a switch or jumper wire located on the interface card. Next an interrupt service routine must be devised which can take the appropriate actions when an A/D conversion has been completed. The address of this routine is then placed into the interrupt vector table located in the first 1 kbyte of computer memory. It is usual to write this code in assembler language since most higher level languages do not have a facility for entering the routine into the interrupt vector table. In any case, it is normally important to make the routine as time-efficient as possible.

An example of an interrupt-driven A/D conversion program, using the Scientific Solutions Labmaster (Scientific Solutions, Solon, OH, USA) interface card, can be seen in Listing 2.2. This was one of the first PC interface cards to come into widespread use in electrophysiology laboratories, and is used with the Axon Instruments pClamp program. The main body of the routine (Listing 2.2 (a)) is written in FORTRAN 77, with only those routines directly associated with interrupt servicing written in assembler language (Listings 2.2 (b)–(d)). This approach avoids having to write large amounts of assembler code while still retaining its advantages where necessary. The FORTRAN language is quite suited for this kind of mixed-language programming since the procedure for executing, and exchanging data with, assembler subroutines is usually well defined. (This is also true of 'C', but not always of the many varieties of BASIC, or PASCAL).

The interrupt-driven routine is similar in many ways to the programmed data transfer routine in Listing 2.1, with an initialization phase setting up the ADC and clock, a data collection phase and a termination phase tidying up after sampling is complete. The Labmaster A/D converter is similar to the LAB-PC with 16 channels, programmable gain amplifier, and a maximum sampling rate of 125 kHz. Sample timing is provided by an AMD 9513 (Advanced Micro Devices Inc., Sunnyvale CA, USA), similar to the Intel 8253, with five cascadable 16 bit timer channels.

The ADC is programmed to send a pulse along its interrupt request line (IRQ7 in this particular case) requesting service from the host computer when a conversion is done. The assembler routine *enable_adc_interrupt* in Listing 2.2 (b) installs an interrupt service routine which responds to these interrupt requests, reading the new sample from the A/D data port and placing into the storage array in memory. This is done by using an MS-DOS function call (37) to place the routine's starting address into the entry for IRQ7 in the interrupt vector table. (Note that the original contents of the IRQ7 entry are first read and retained for later replacement.) The address of the A/D data storage buffer *iadc*, defined in the FORTRAN section of the program, is also passed to the assembler part at this time.

Once sampling has been initiated by starting the Labmaster clock, samples are automatically transferred into the storage buffer as they are acquired. The FORTRAN program can monitor the progress by calling *adc_interrupt_count* which reads the sample counter being updated by *adc_isr*. In this program, the information is used in a simple loop which just displays the value of samples as they are acquired. When the requisite number of samples have been collected, sampling is stopped and the system is tidied up by calling *disable_adc_interrupt* to restore the interrupt vector table to its original condition.

The interrupt-driven data transfer method makes more efficient use of the host computer by freeing it from the need to monitor the status of the A/D converter constantly. It would, for instance, be possible to modify Listing 2.2 to transfer samples from the storage buffer on to a disc file while sampling was in progress, allowing very long digital recordings to be made, limited only by disc space.

2.10 LIMITATIONS OF THE INTERRUPT-DRIVEN METHOD

Unfortunately, there is a price to be paid for using interrupts. When an A/D interrupt occurs, the interrupt service routine must preserve the state of the currently executing program so that it can be restored afterwards. Consequently, the contents of any CPU register used by the service routine must be saved into a storage area and restored before control is returned to the main program. This presents a significant overhead with almost as many instructions involved saving and restoring registers as actually transferring data (see Listings 2.2 (b)). On the IBM PC, this places an upper limit of about 10–20 kHz (depending on CPU speed) on the sampling throughput that the method can support. A detailed discussion of the use of hardware interrupts can be found in Duncan (1988) and a full description of the Intel 8295A device can be found in the *Intel Microprocessor and Peripherals Handbook* Vol. 1 (Intel, 1987).

2.11 DIRECT MEMORY ACCESS DATA TRANSFER

Part of the problem with both the programmed and interrupt-driven techniques is that data transfer is a two-stage process, first a transfer from the ADC to a CPU register and a second transfer from CPU register into memory. The performance of any transfer method that requires the use of the CPU is likely to suffer from the relatively long time taken to execute a series of instructions, each taking 1–2 µs. The RAM memory within the computer, on the other hand, can be read or written within 0.1–0.2 µs. The *direct memory access (DMA)* method avoids these problems by directly transferring data from a peripheral device into the host computer memory. DMA data transfer, however, requires special purpose hardware to be available within the host computer, in the form of a DMA controller device capable of handling the data transfer in the place of the CPU.

The IBM PC family of computers are well suited to making use of DMA data transfer since they have most of the necessary hardware built in as a standard feature. All PCs have at least one *Intel 8237A Programmable DMA Controller* and 80286- and 80386-based PCs have two of them. Each 8237A provides four independent DMA data transfer channels. Peripheral devices request the services of the DMA controller by means of a set of DMA request lines (DREQ0 ... DREQ7) on the PC expansion bus and the DMA controller then takes care of the details of the transfer. The procedure for transferring an A/D sample from a laboratory interface into the host computer memory using DMA is illustrated in Figure 2.5, and proceeds as follows.

(a) The laboratory interface sample clock is programmed to initiate conversions at fixed intervals automatically.

(b) On completion of an A/D conversion, the interface sets the DMA request line *DREQ1* for its DMA channel.

(c) The DMA controller notes the DMA request and (within a few µs) when it is legitimate to do so, freezes the CPU into a *HOLD* state, and signals to the interface via the *DACK1* line that it is ready to receive data.

(d) When the interface receives the DACK1 signal it places the first byte of the A/D sample on to the interface bus. The DMA controller then handles the process of copying the byte into RAM memory. If a second byte is to be transferred (as it usually is for 12 bit A/D samples), the laboratory interface keeps DACK1 set and a second transfer sequence ((c)–(d)) takes place into the next PC memory location.

(e) The DMA controller then releases the CPU which regains control of the interface bus and continues its operations.

Since all the steps ((a)–(e)) in the transfer procedure take place completely in hardware, rather than by the execution of CPU instructions, they are performed very quickly, and can support a maximum data transfer rate of 500 kbyte s^{-1} on a typical computer in the IBM PC family. Also since they do not require any CPU

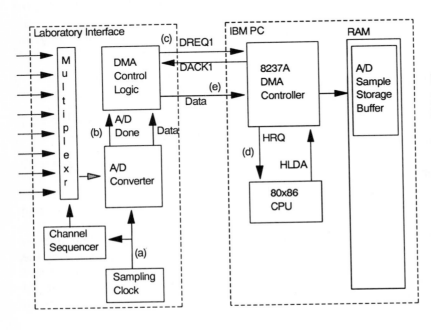

Figure 2.5 DMA transfer of A/D samples from a laboratory interface to an IBM PC. (a) The sampling clock initiates an A/D conversion. (b) When a conversion is done, the DMA logic of the laboratory interface requests, using the DREQ1 line, a data transfer by the 8237A DMA controller. (c) The 8237A halts the CPU, replies to the interface via DACK1 that it is ready to receive data. (d) The A/D sample is copied into PC memory by the DMA controller and the CPU is then released.

instructions or use any CPU data registers, there are no overheads involved in preserving the state of the program before and after the transfer. Just as for interrupt lines, several of the DMA channels are already used by the PC as part of its normal operation; DMA channel 0 is an integral part of the PC's RAM memory system (dynamic memory refresh) and channel 2 is used for data transfer to and from floppy disc drives. DMA channel 1, however, is usually free for use by laboratory interface units and also DMA channel 3 on 80286 and 80386 class PCs.

Laboratory interfaces which support DMA data transfer have to be somewhat more sophisticated in design than those which only support programmed and interrupt-driven modes of operation since they must handle the interactions between the interface and DMA controller, outlined above. In addition, if multi-channel sampling is required, the interface itself must also be able to change the multiplexer setting automatically. These functions can be performed using specially designed hardware logic circuits (the LAB-PC and the Labmaster DMA, for instance, uses this approach) but often an on-board microprocessor is used, such as in the Cambridge Electronic Design 1401 or Data Translation DT2801A interfaces.

2.12 PROGRAMMING FOR DMA DATA TRANSFER

Programming for DMA data transfer requires a detailed understanding of the functioning of the 8237A DMA controller (a complete description of its operation can be found in: *Intel Microprocessor and Peripherals Handbook*, Vol. 1, *Microprocessors*). An example of A/D conversion with DMA transfer is shown in Listing 2.3, using the National Instruments LAB-PC. It can support A/D sampling at rates up to 62.5 kHz without making use of any specially coded assembler language subroutines.

The initialization phase is an extension of that used in the Listing 2.1. The ADC and sampling clock are set up as before, followed by additional steps programming the PC's DMA controller. Before DMA data transfer can take place, the DMA controller must be provided with the following information:

- number of bytes to be transferred
- direction of transfer, i.e. interface-to-memory (WRITE) or memory-to-interface (READ)
- starting address of DMA transfer area in RAM memory.

Programming the number of bytes and transfer direction is a straightforward matter of writing the values to the appropriate ports of the DMA controller as shown in the Listing 2.3. However, defining the starting address of the DMA transfer area in PC memory presents certain problems which stem from differences in how the DMA controller and the Intel 80×86 family of CPUs define the address of locations within the RAM memory.

The Intel 8086 (and the more modern 80286, 80386, or 80486 when using the MS-DOS operating system) has 16 bit CPU registers which can contain numbers between 0 and only 64 575 (64 kbyte). A single register can therefore, by itself, only define that many address locations in memory. Since a total addressable memory of 64 kbyte would have been a severe limitation, the 8086 family of microprocessors make use of two separate 16 bit registers, the segment (DS,ES, CS,SS) and offset registers (SI,DI,BX) combined together to form the memory address according to the following formula.

Linear address = (16 × segment) + offset [2.1]

Using this scheme, a maximum of 1 Mbyte of RAM memory can be directly addressed rather than 64 kbyte. The 8237A DMA controller can also directly access this 1 Mbyte of memory, using two 16 bit registers (designated page and offset) but, unfortunately for the programmer, using a different scheme for calculating the linear address.

Linear address = (64 576 × page) + offset [2.2]

In order to use DMA transfer, the memory address of the start of the data storage array *iadc* must be obtained and written to the DMA controller. This is obtained using the assembler subroutine *VARPTR()* (see Listing 2.1(b)) which provides the segment and offset parts of the address of a FORTRAN variable or array (similar to the VARPTR() and *VARSEG()* command in Microsoft BASIC.) Once the segment:offset address has been obtained in this way, the page:offset address can be calculated by combining and rearranging the above two equations, as can be seen in Listing 2.3.

Since DMA page addresses must start on 64 kbyte memory boundaries (e.g. 0, 64 kbyte, 128 kbyte, etc.) and the address offset register can only count from 0 – 64 kbyte, DMA transfers cannot span page boundaries. Any attempt to do so results in the DMA address offset register overflowing and writing data to the start of the page. For example, if the DMA controller were to be programmed to transfer 2 kbytes into memory starting at address 64512, the data would not be placed into memory locations 64512–66560, but rather into 64512–65535 and 0–1023, usually resulting in corruption of parts of the program or operating system. A simple way of ensuring that the DMA transfer area is contained within a single 64 kbyte memory page is shown in Listing 2.3. The array *iadc*, used for receiving the A/D samples, is defined to be twice as large as necessary. Even if the complete array happens to span two pages, at least one half of the array *must* always be within a single page. The appropriate half can be chosen by verifying that the DMA page values for the first and last elements are the same.

2.13 ADVANTAGES OF DMA DATA TRANSFER

The advantages of using DMA data transfer distinctly outweigh the extra difficulties involved in programming the DMA controller. In comparison with the interrupt-driven method, DMA transfer is much more efficient. Figure 2.6 shows the proportion of CPU processing time (overhead) taken up by the transfer of A/D samples into memory using DMA compared with that using interrupts, for a range of A/D sampling rates from 1–33 kHz. The tests were performed on an IBM PC-compatible with a 16 MHz 80386 CPU, using a Labmaster DMA interface board

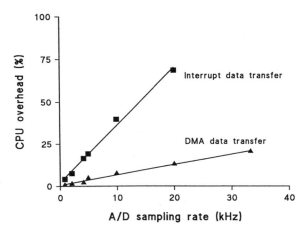

Figure 2.6 A comparison of operating overheads (in terms of % of total available CPU time) incurred in A/D sampling using interrupt-driven and DMA data transfer methods.

which supported both the interrupt and DMA methods. CPU processing time usage was determined by measuring the difference in the number of iterations achieved by a simple computational loop running for 5 s, both with and without A/D sampling in progress. At rates below 1 kHz, neither DMA nor interrupts used more than a few per cent of the available processing time. However with increasing sampling rates the interrupt overhead increases much more rapidly, using 68% of CPU time at 20 kHz while DMA uses only 13%. Interrupt-driven A/D sample throughput is therefore limited to rates less than 30 kHz, while DMA transfer easily extends to 150 kHz. Higher throughputs can be achieved using the 16 bit DMA controller (twice as efficient as the 8 bit controller) available on PCs with 80286, 80386 or 80486 CPUs. Using 16 bit DMA the Data Translation DT2821G interface (Data Translation, Marlboro, MA, USA) can achieve an A/D sampling throughput as high as 250 kHz. Even higher rates can be achieved on the highest performance IBM PC-compatible computers, fitted with the EISA expansion bus (Extended Industry Standard Architecture, an updated 32 bit version of the orginal PC expansion bus, now known as the Industry Standard Architecture bus). The National Instruments EISA-A2000 interface, for instance, can achieve a maximum throughput rate of 1 MHz, using the 'burst mode'

DMA transfer method supported by the EISA bus.

2.14 SHARED MEMORY

The maximum throughput achievable with the DMA method is limited by the speed of data transfer across the computer's expansion bus, which on computers fitted with the standard PC expansion bus is around 250 kHz. To achieve rates above this, without moving to a computer with an enhanced bus such as EISA, it is necessary to avoid using the expansion bus. This can be done by installing a special form of RAM memory known as *dual-port* RAM which, as the name suggests, can have data read or written to it from two separate sources. One side is connected as normal into the host computer while the other is directly connected to the laboratory interface. The interface can therefore write directly into this memory. This technique is known as *shared memory* and sampling rates as high as 1 MHz (e.g. the Computerscope-Phy interface (RC Electronics, Santa Barbara, CA, USA)) can be achieved on a standard ISA bus computer. However, this method will not be discussed further since the performance that it yields is not required for electrophysiological work.

2.15 CONTINUOUS SAMPLING-TO-DISC

The fact that the DMA data transfer method leaves the host computer's CPU free for other tasks means that we can use it as the foundation of the digital recording methods required for electrophysiological signals. Probably the most important method is *continous sampling-to-disc*. There is often a need to acquire very long recording sequences at high sampling rates, a particular example being the recording of single channel currents (see Chapter 8). For such applications, it may be necessary to sustain sampling rates of around 20 kHz for several minutes collecting several million samples. In the programming examples discussed so far, the recordings have been stored in RAM memory. However, such large records exceed the available capacity of the typical computer system's RAM memory (e.g. 640 kbyte under the MS-DOS operating system). The continuous sampling-to-disc method provides a means of writing A/D samples from RAM memory on to disc before the RAM memory limit is exceeded while A/D sampling is still in progress.

A simple method of implementing sampling-to-disc, known as the *double buffer* method, is shown in Listing 2.4. This routine uses the *adc_dma* routine from Listing 2.3 to transfer a stream of A/D samples into a buffer in RAM memory, using DMA data transfer. The DMA controller is programmed to operate in its 'auto-initialize' mode, forcing it to transfer samples into the A/D buffer continuously, cycling back to the start of the buffer rather than stopping when full. The progress of buffer filling is monitored, and when the first half of the buffer is filled, its contents are written on to the disc storage file (a process which may take 10–20 ms) while A/D samples continue to be placed into the second half. The program then waits for the second half to fill and writes it to disc while sampling continues back in the first half again. The process of alternately writing each half-buffer as it is filled continues until the required number of full buffers of samples have been collected. No A/D samples are lost as long as the contents of each half-buffer can be written to disc before the other becomes full.

The maximum continuous sampling-to-disc rate is limited by the rate that samples can be transferred to disc which depends partly on how the procedure has been programmed, the instruction execution speed of the computer, and the performance of the disc drive. The typical PC, currently used in the laboratory, with an average performance hard disc can easily achieve sampling-to-disc rates of around 30 kHz, using standard MS-DOS procedures to write data to file. Rates of 60–70 kHz can be achieved by using special highly efficient routines to write to disc, and rates as high as 250 kHz can be obtained using a fast computer (25 MHz 80386) and high performance SCSI (Small Computer Systems Interface) hard disc drives (Data Translation New Product Handbook, 1991).

Since the nature of the hard disc drive and how it is used is of major importance in determining the achievable sampling-to-disc performance, it is worth discussing the functioning of a hard disc drive in detail. A hard disc drive consists of several rigid metal oxide-coated magnetic discs (platters) stacked on top of each other on a central spindle so that a set of read/write heads (one per side) can be moved radially in and out of the stack. Each disc is divided into a series of several hundred concentric tracks which are themselves radially subdivided into a series of sectors (commonly 17 per track). A sector, holding 512 bytes of data, is the smallest unit that can be read or written to a disc. The procedure for writing to a disc sector involves:

(a) *Moving the read/write head to the correct track*. This can take a quite significant amount of time if the head is located some distance away from the desired location. The disc's *random access time* is the average time taken to move the read/write head between any two randomly chosen tracks on the disc. This may be as long as 60–100 ms on older systems (e.g. IBM PC-XT), but is now commonly less than 30 ms and can be as low as 10 ms.
(b) *Rotational latency*. Once the correct track has been reached, it is then necessary to wait for the correct sector to rotate underneath the read/write head. Since the disc is rotating at 3600 rpm, this may take up to 18 ms.
(c) *Data transfer*. Finally, when the correct sector is lined up, data transfer to the disc can begin at a rate of usually 750 kbyte/s (for the standard ST906 controller used in the PC).

A 750 kbyte/s data transfer rate is clearly very high, but it can only be sustained while a single track is being written. When a track is full, data transfer has to cease until the head is moved to the next track, resulting in access time and rotational latency periods where no data transfer is taking place. Optimum sampling-to-disc performance is achieved by writing data to disc as blocks of sectors rather than as single sectors (ideally a whole track at a time) to minimize rotational latency and also by ensuring that data is stored in adjacent tracks on the disc to minimize the distance the read/write head has to be moved.

It should be noted that data files are not necessarily stored on physically adjacent sectors on the disc surface, a problem which becomes apparent with repeated use of the hard disc. If files are written to an initially empty disc, the data is stored *contiguously* in physically adjacent sectors on the disc. However if some files are subsequently deleted the free space created is mingled among the remaining files in blocks, each block varying in size depending on the size of the file deleted. To make efficient use of this free space most operating systems, including MS-DOS, use a *non-contiguous* file structure which allows files to be split up and stored anywhere on the disc. A map of the blocks of free space throughout the disc is maintained and generally the first available block is used irrespective of where it is on the disc or how large it is. It is not unusual to find that a data file is split over a dozen or more separate areas on the disc. While this method makes efficient use of the available disc space, a high price in performance is paid, with repeated head movements required to access data within heavily fragmented files.

For the highest performance it is essential to ensure that the disc file used to receive sampling-to-disc data has been allocated a contiguous series of sectors of the disc. This can be difficult to achieve since the mechanism for allocating disc space is at a level of the operating system not readily manipulable by the programmer. However a simple, if slightly inflexible, strategy is to start with an newly formatted hard disc and to create a data storage file large enough (e.g. 5–10 Mbyte) for future sampling-to-disc requirements. Since no file deletions have taken place the free disc space is not fragmented and a contiguous storage file will be created. If it is not convenient to reformat the hard disc, an alternative is to use a *disc optimizer* program which shuffles files about on the disc to squeeze out the deleted file space and to create a contiguous block of free disc space. Such programs can often be obtained as part of file/disc management packages (e.g PC-Tools, Central Point Software).

2.16 DETECTION AND RECORDING OF SPONTANEOUS SIGNALS

Continuous sampling-to-disc allows us to make a complete digital recording of the signals under study. While this is an extremely valuable facility it can also be very inefficient when applied to the recording of short-lasting signals occurring only at infrequent intervals. For instance, a typical experiment, on skeletal neuromuscular junction say, might involve the recording of series of miniature endplate currents which last only 2–3 ms and occur spontaneously once or twice per second. It may be necessary to collect large numbers of signals for statistical analysis (e.g. 1000). With continuous recording a record 500 s long would result, taking up 25 Mbyte of disc storage space. However, less than 1% of the total recording would actually contain a signal of interest, and a time-consuming processing of searching through the recording would be required in order to find and analyse the signals.

In order to avoid this 99% waste of time and disc space it is often preferable to only collect A/D samples when a signal is known to be present. This is a relatively simple process when the signals under study are evoked by an external stimulus (e.g. action potentials, nerve-evoked endplate currents) since recording can be linked to an external trigger pulse (see Section 3.10), but spontaneous randomly occurring signals (such as MEPCs) have no such external synchronizing event.

In order to record a spontaneous signal successfully, a means is required to detect its presence in the incoming analog signal. In addition, it is not sufficient to simply start A/D sampling when detection occurs since all pre-trigger signal information would be lost including the leading edge of the signal. In order to retain pre-trigger information, a history of the most recent A/D samples must be retained at all times, in case an event is detected. Finally, since signals are often of small magnitude superimposed on a noisy and often drifting baseline, the signal detection procedure must be able to separate the signal from slow changes in the overall level.

A detection and recording routine is shown in Listing 2.5. The *adc_dma* routine from Listing 2.3 is used in the same way as for sampling-to-disc, with the incoming analog signal being continuously digitized and stored in a memory buffer iadc using DMA transfer in 'autoinitialize' mode to ensure repeated filling of the buffer. With the A/D sampling process devolved to the laboratory interface and the DMA controller, full use of the host computer's CPU is available for running the signal detection program. The signal detection procedure has two distinct phases:

- *Detection phase*, when the incoming A/D sample stream is monitored and the detection criteria applied.
- *Collection phase*, when, once a signal has been detected, the routine waits until the required number of post-trigger samples are acquired. Once all post-trigger samples are available, the sequence of pre- and post-trigger samples containing the signal are transferred to the detected results array.

Signal detection is performed by measuring the difference between incoming A/D samples and a baseline level. When a sample exceeds the required level, detection is deemed to have occurred. The value of the baseline level is made to follow slow shifts in signal level by continuously updating it using a 16 sample running mean derived from the incoming signal. At the moment of signal detection, the A/D buffer already contains the pre-trigger samples. It is then necessary to wait until the required number of post-trigger samples are acquired. For simplicity, Listing 2.5 only detects positive going signals. The fact that it is written in FORTRAN also limits its performance, not in the actual process of sampling, but in the accuracy of detection. The signal, once detected, must be extracted from the A/D buffer before the signal is overwritten by the samples still coming in.

2.17 CHOOSING A LABORATORY INTERFACE

Commercially available laboratory interfaces vary somewhat in features, performance and cost. It is therefore worth summarizing what are

Table 2.1 A comparison of the main features and performance of some laboratory interfaces commonly used in the electrophysiological laboratory.

	CED 1401	DT2801	DT2821	DT2831	DT2841	Labmaster DMA	National Instruments LAB-PC
A/D inputs	16	16	16	16	16	16	8
Maximum rate (kHz)	130	27.5	50–250	50–250	40–750	40–125	64
Automatic channel sequencing	Y	Y	Y	Y	Y	Y	Y
D/A outputs	4	2	2	2	2	2	2
Clocks	5	1	1	5	5	5	6
Digital I/O	16	16	16	16	16	24	8
Program I/O	Y	Y	Y	Y	Y	Y	Y
DMA transfer	Y	Y	Y	Y (2)	Y (2)	Y	Y
Interrupt	N	N	Y	Y	Y	Y	Y
Shared memory	N	N	N	N	Y	N	N
On-board CPU	Y	Y	N	N	N	N	N
On-board memory	Y (8 Mbyte)	N	N	N	Y	N	
Programmable	Y	N	N	N	N	N	N

Notes. Where two numbers are given they indicate a range of optional performances.

the requirements for a suitable laboratory interface for general purpose use within an electrophysiological laboratory. Ideally, it should meet the following specifications.

- 4+ analog input channels (100 kHz sampling rate)
- automatic channel selection for multi-channel recording
- 2+ analog output channels
- 2+ clocks with external trigger capability
- DMA data transfer
- capable of simultaneous A/D sampling and D/A output.

Of the currently available commercially produced laboratory interfaces, some easily meet all of these specifications, others, although deficient in some features, are nevertheless satisfactory for many routine applications. A comparison of the features and performance of a range of the most commonly available of these devices is presented in Table 2.1.

While the discussion here is focused upon the features and performance provided by the laboratory interface hardware, it should be borne in mind that the availability of appropriate software is probably more important. The development of software to control and collect data from a laboratory interface is difficult and time consuming. While most suppliers provide some form of subroutine library for controlling their interface, that may not provide all of the functions necessary. For those without the time or expertise to indulge in an extensive software development program, it is essential to first obtain an appropriate software package and then choose an interface compatible with it. At least one electrophysiological data analysis package is available for each of the interfaces to be discussed.

2.18 SCIENTIFIC SOLUTIONS LABMASTER

The Labmaster, discussed earlier, is possibly the most commonly used laboratory interface at present. There are both practical and historical reasons for this. It has been available for at least 8 years, almost as long as the IBM PC itself, and was one of the first to have useful electrophysiological software developed for it (Kegel et al., 1985). It has been adopted by a number of specialist systems developers in the USA who concentrate on selling to electrophysiogical laboratories such as Axon Instruments and Indec systems. In any case, it is a powerful and flexible device. Although the original Labmaster was somewhat limited by the lack of a DMA data transfer capability, it is now included in the latest Labmaster DMA product. The main thing to be said against is that it is difficult to programme with notoriously obscure documentation, and is physically constructed in an awkward way in terms of the placement of input and output connectors making installation difficult. It is therefore probably better to purchase the Labmaster from one of the systems suppliers mentioned above, as part of a package including input/output connector boxes.

2.19 CAMBRIDGE ELECTRONIC DESIGN (CED) 1401

Cambridge Electronic Design Ltd (Cambridge, UK) has a long history of supplying PDP11 minicomputer-based systems to electrophysiology laboratories within the UK. The CED 1401 interface is widely used within UK research laboratories, more so than the Labmaster. It is a larger and more sophisticated device than the Labmaster with its own programmable on-board microprocessor. It is, in fact, more a self-contained data acquisition computer than a simple laboratory interface and is unique in many ways. It can operate not only with the IBM PC-compatible family of computers but also with the Apple Macintosh, Acorn Archimedes and BBC Micro, and even DEC VAX computers via the standard IEEE 488 interface bus (Marlowe & Mackensie, 1989). It can be fitted with large amounts (8 Mbyte) of on-board RAM for internal high speed data storage. Its on-board processor (a 2 MHz 6502 CPU) runs a multi-tasking interrupt-driven operating system which allows simultaneous A/D sampling, D/A output, and digital pulse sequences. A wide range of

1401 commands (programs) are available for down-loading into the interface from the PC. It is also possible to write customised commands with a cross-compiler. CED produce specialized applications software (notably voltage-clamp and single-channel analysis) for use with the interface.

Recently a new and significantly upgraded version of the CED1401 has been produced, the CED 1401-plus. This device, while being essentially compatible with existing software, is based upon a 20 MHz National Instruments 32GX32 32 bit CPU. The 1401-plus can support simultaneous A/D and D/A sampling at rates of around 100 kHz for each process. The new processor provides a much better match in capabilities to the 80386- and 80486-based PC now in common use in the laboratory.

2.20 DATA TRANSLATION INTERFACES

Data Translation Inc. (Marlboro, MA, USA) is probably the world's largest supplier of high performance laboratory interfaces (and also image capture boards). They produce a wide range of boards from low-cost devices supporting sampling rates of a few kilohertz to devices with 750 kHz rates, pushing the limits of performance that PCs can support. Data Translation boards are also available for the Macintosh II range of computers, VME-bus computers (eg. Sun workstations) and Digital Equipment Corporation VAX computers.

Unlike CED and Axon Instruments, Data Translation's sales seem to be aimed at software developers rather than end-users. They provide large amounts of technical detail concerning their products, but do not produce specialized applications programs for areas like electrophysiology (they do market general-purpose packages, e.g. GLOBAL-LAB). For this reason, they tend to be found in use in laboratories where in-house software development is in progress (e.g. Robinson & Giles, 1986), or where Data Translation-compatible software has been obtained as a gift from another laboratory. Up to the present, the most widely used interface from this range of products has been the DT2801A. It is a relatively inexpensive interface which fully supports DMA data transfer both for A/D sampling and for D/A output. Its maximum sampling rate of 27.5 kHz proves to be quite satisfactory for many applications. The main limitation of the DT2801A is that it cannot perform simultaneous A/D sampling and D/A output which restricts its application to voltage clamp work where it is often desirable to produce a computer-generated voltage command waveform while simultaneously recording current and voltage signals.

Higher performance can be acheived by using the DT2821 series of interfaces (50–250 kHz maximum sampling rate) or the more recent DT2831 which has an unusual facility not found on any of the other interfaces – simultaneous A/D sampling and D/A output using two separate DMA channels, without any use of the host computer's CPU at all. At the very highest end of the range also lies the DT2841 (in combination with a DT7020 co-processor) with a maximum sampling rate of 750 kHz achieved using the shared-memory technique. It is unlikely, however, that an electrophysiologist would require performance of this nature.

2.21 NATIONAL INSTRUMENTS LAB-PC

National Instruments (Austin, TX, USA) is another large supplier of PC interface boards. They are better known as being suppliers of IEEE 488 interface boards for PCs and other computers, being almost the industry standard in that area. IEEE 488 is a interfacing specification which allows measurement devices, such as digital oscilloscopes, frequency counters or multimeters, to communicate with and be controlled by a host computer. This approach is used extensively in the engineering field, but has been less common in the physiological sciences. Recently, however, National Instruments have extended their product range to include laboratory interface boards of the types discussed here. In particular, the National Instrument LAB-PC used in some of the examples in this chapter, is worth mentioning as an inexpensive board which nevertheless provides all of the features required by the electrophysiologist.

Since the LAB-PC is a relatively new product there is less software available for it. At present, of the widely available electrophysiological analysis software packages, only the Strathclyde Electrophysiology Software supports it. However, National Instruments also supply a software package called *LabWindows* which provides a simplified development environment for producing data acquisition and analysis software. This greatly reduces the difficulty of developing software for this interface. While programming skills are still required, the complexities of programming at the assembler level and dealing with the details of hardware I/O ports etc. are avoided.

2.22 LABORATORY INTERFACES FOR THE MACINTOSH

Although the Apple Macintosh computer has proved to be very popular within the academic community, and is found in many laboratories, it has been less frequently used for recording electrophysiological signals. The reasons lie in the software and hardware architecture of the machine. It lacks a DMA controller as a standard feature throughout the complete range of computers. Also the development of laboratory interfaces for the Macintosh started later since early Macintoshes (before the II series) did not have an effective expansion bus. Although the Macintosh graphical user interface is undoubtedly elegant and easy to use, it is considerably more difficult to programme than MS-DOS.

While the above factors have held it back, they have not prevented development of data acquisition systems for the Macintosh. A notable early development was the MacLab system marketed by World Precision Instruments (New Haven, CT, USA). Like the CED 1401, MacLab has an on-board 6502 microprocessor. The system allowed the Macintosh to be used as an oscilloscope or chart recorder. An undoubted limitation was the fact that the interface communicated with its host via a relatively slow serial communications line. More recent versions (MacLab/4) make use of the Mac's SCSI interface bus to provide a more satisfactory communications pathway to the interface.

National Instruments has also devised a solution to the difficulty of Mac programming by providing a graphics-based development system called *Lab-View* (similar, in concept to the *LabWindows* package for the PC) which allows data acquisition programs to be developed without the need to have a detailed knowledge of the Mac operating system. They also supply a number of powerful laboratory interfaces such as the NB-A2000 which provide sampling rates of 1 MHz when combined with their NB-DMA2800 DMA controller board. Specialized electrophysiology software is also becoming available for the Macintosh, such as the *Axo Data* program from Axon Instruments or the M2 software from Instrutech (Mineola, NY, USA). Both of these programs make use of the Instrutech ITC-16 laboratory interface board.

2.23 OTHER LABORATORY INTERFACES

The interfaces discussed so far are only the better known of the many available. A number of others have been promoted for use in the electrophysiology laboratory, including RC Electronic's Computerscope-Phy system and the Intracel S200 interface (Intracel, Cambridge, UK). Keithley (Taunton, MA, USA) also produce a wide range of laboratory interfaces, but without specialized electrophysiology software. Electrophysiological data acquisition system also exists for the Atari ST computer, including the Instrutech M2 system and the EPC-9 patch clamp system (Hera Elecktronik, Germany). Suitable interface boards, such as the Data Translation DT1492, are also becoming available for Sun workstation computers, although there is little software actually to make use of such a combination.

2.24 SUMMARY

Currently, most electrophysiological data analysis systems are based upon the IBM PC family, with a small but increasing use made of Apple Macintoshes. The laboratory interface unit which digitizes the analog signals is a key component in the system. Of these devices, the most commonly used are the Scientific Solutions Labmaster (as supplied by Axon Instruments), the CED 1401, and to a lesser extent the Data Translation DT2801A. Unlike many other areas in computing where an increased standardization has occurred (such as disc drives), each specific manufacturer's interface has its own set of operating conventions. Most software packages for electrophysiogical analysis rarely work with more than one interface (an exception is the Strathclyde Electrophysiology Software) which, in the case of commercial packages, is often proprietary to the supplier of the software. Consequently, the choice of laboratory interface is more determined by the choice of software package than the inherent capability of the interface itself. The availability of appropriate software is a crucial issue and this has often limited the appeal of newer and sometimes better interface cards. These compatibility problems have had the effect of restricting the range of choice of interfaces for use within the laboratory, unless significant program development support is available.

The analog signal digitization techniques discussed so far allow us to make digital recordings of a wide range of electrophysiological signals. Once stored on disc in a numerical form they can then be subjected to the variety of analysis techniques to be discussed in the following chapters. Two important techniques, continuous sampling-to-disc and spontaneous event detection, have been introduced, and examples of their application will be covered in subsequent chapters dealing with the analysis methods which rely upon them.

```fortran
c       Listing 2.1. (a) A/D conversion subroutine using programmed I/O.
c       for National Instruments LAB-PC
        subroutine adc_pio(ichannel,igain,interval,iadc,np)
        integer*2 ichannel                  ! (In) Channel No. (0-7)
        integer*2 igain                     ! (In) Amp. gain (0=x1,7=x100)
        integer*2 interval                  ! (In) Sampling interval (µs)
        integer*2 iadc(np)                  ! (Out) A/D sample storage array
        integer*2 np                        ! (In) No. of samples required
c       National Instruments LAB-PC I/O ports
        parameter(icommand1=16#260)         ! Command port 1
        parameter(icommand2=16#261)         ! Command port 2
        parameter(icommand3=16#262)         ! Command port 3
        parameter(istatus=16#260)           ! Status port
        parameter(iadc_data=16#26A)         ! A/D data port
        parameter(iadc_clear=16#268)        ! Clear A/D data logic
        parameter(icounterA_mode=16#277)    ! Sampling clock mode
        parameter(icounterA0_data=16#274)   ! Sampling clock data

c       Initialization phase
        call outb( icommand1, 0 )
        call outb( icommand2, 0 )           ! Clear all command ports
        call outb( icommand3, 0 )
        ibyte = ichannel + 16*igain         ! Set analogue input channel
        call outb( icommand1, ibyte )       ! and amplifier gain.
c       Program sampling clock (Intel 8253 3-channel counter)
        call outb( icounterA_mode, 16#34 )  ! Put Counter A0 in rate generator
        call outb( icounterA0_data, interval )      ! mode and write sampling interval
        call outb( icounterA0_data, interval/256 )  ! (in µs)
        call outb( icounterA_mode, 16#70 )  ! Force Counter A1 output LOW
                                            ! to enable Counter A0
        call outb( iadc_clear, 0 )          ! Clear LAB-PC A/D data logic
        iadc_done = 0
        do while( iadc_done .eq. 0 )        ! Wait for A/D done bit to be
           call inpb( istatus, iadc_done )  ! to set.
           iadc_done = iadc_done .and. 2#1
        end do
        call inpb( iadc_data, ilo_byte )
        call inpb( iadc_data, ihi_byte )
c       Sample transfer phase
        call outb( icommand2, 2#100 )       ! Start sampling clock
        do i = 1,np
           iadc_done = 0                    ! Wait for A/D done
           do while( iadc_done .eq. 0 )     ! bit in status port
              call inpb( istatus, ibyte )   ! to set.
              iadc_done = ibyte .and. 2#1
           end do
           call inpb( iadc_data, ilo_byte )
           call inpb( iadc_data, ihi_byte )
           iadc(i) = ilo_byte + 256*ihi_byte   ! Store A/D sample in array.
        end do
c       Termination phase
        call outb( icounterA_mode, 16#34 )  ! Stop sampling clock
        call outb( icounterA0_data, 16#a )
        call outb( icounterA0_data, 16#0 )
        return
        end
```

```
        title labio   ; Listing 2.1 (b) Assembler routines to read/write I/O ports.
save_regs       macro                   ; FORTRAN requires DS and BP registers
                push ds                 ; to be saved on stack
                push bp
                mov bp,sp               ; BP points to start of argument list
                endm
restore_regs    macro                   ; Restores the registers saved by
                mov sp,bp               ; save_regs
                pop bp
                pop ds
                endm
unpack          macro argno,nargs       ; Set DS:[SI] =
                mov si,[bp]+(4+4*(nargs-argno+1))   ; address of argument
                mov ds,[bp]+(6+4*(nargs-argno+1))   ; #argno in FORTRAN
                endm                    ; argument list.
labio   segment para public 'code'
        assume cs:labio,ds:nothing
        nargs = 2                       ; Write a byte to an I/O port
        public outb                     ; FORTRAN call outb(iport_no,ivalue)
outb    proc far
        save_regs                       ; Save registers
        unpack 1,nargs                  ; DX = I/O port No.
        mov dx,ds:[si]
        unpack 2,nargs                  ; AL = byte to be written
        mov al,ds:[si]
        out dx,al                       ; Write byte to port
        restore_regs                    ; Tidy up and exit
        ret 4*nargs
outb    endp
        nargs = 2                       ; Read a byte from an I/O port
        public inpb                     ; FORTRAN: call inpb(iport_no,ivalue)
inpb    proc far
        save_regs                       ; Save registers
        unpack 1,nargs                  ; DX = I/O port No.
        mov dx,ds:[si]
        in al,dx                        ; AL = byte from port
        mov ah,0                        ; AH = 0
        unpack 2,nargs                  ; Copy AX to <ivalue>
        mov ds:[si],ax
        restore_regs                    ; Tidy up and exit
        ret 4*nargs
inpb    endp
        nargs = 3                       ; Get segment & offset of address of
        public varptr                   ; FORTRAN variable
varptr  proc far                        ; call varptr( isegment,ioffset,ivar)
        save_regs
        unpack 3,nargs                  ; AX = segment and BX = offset
        mov ax,ds                       ; of argument #3
        mov bx,si
        unpack 1,nargs                  ; Save segment into arg. #1
        mov ds:[si],ax
        unpack 2,nargs                  ; Save segment into arg. #2
        mov ds:[si],bx
        restore_regs
        ret 4*nargs
varptr  endp
labio           ends
        end
```

```
c       Listing 2.2 (a) Interrupt-driven A/D conversion routine.
        subroutine adc_int(ichannel,igain,interval,iadc,np)
            integer*2 ichannel              ! (In) Channel No. (0-7)
            integer*2 interval              ! (In) Sampling interval (µs)
            integer*2 iadc(np)              ! (Out) A/D sample storage array
            integer*2 np                    ! (In) No. of samples required
c       Scientific Solutions Labmaster I/O ports
            parameter(itimer_csr=16#719)    ! Timer control/status port
            parameter(iadc_csr=16#714)      ! A/D converter control/status port
            parameter(iadc_channel=16#715)  ! A/D channel selection port
c       INITIALIZATION PHASE.
            call outb( itimer_csr, 2#11111111 )       ! Reset Labmaster's
            call write_timer_register( 7, 16, 16#4000 ) ! timer system.
            call write_timer_register( 5, 0, 16#0b22 )  ! Set A/D sampling clock
            call write_timer_register( 5, 8, interval/2 ) ! to "repeat mode",
                                                        ! with 0.5MHz clock.
            call enable_adc_interrupt( iadc )         ! Set up A/D interrupt
                                                      ! service routine
            call outb( iadc_csr, 2#01000100 .or. igain) ! Set up A/D converter for
            call outb( iadc_channel, ichannel )       ! selected channel, then start
            call outb( itimer_csr, 2#00110000 )       ! timer running.
c
c       COLLECTION PHASE
c       Monitor the progress of A/D sampling, waiting for the
c       required number of samples to be collected.
            ncollected = 0
            ndisplayed = 0
            do while( ncollected .le. np )
                call adc_interrupt_count( ncollected )  ! No. of samples
                                                        ! collected so far.
                if( ndisplayed .lt. ncollected ) then   ! Display new samples
                    write(*,900) ndisplayed,iadc(ndisplayed)
900                 format(1x,'Sample= ',i5,' Value= ',i5)
                    ndisplayed = ndisplayed + 1
                endif
            end do
c       TERMINATION PHASE
            call outb( itimer_csr, 2#11111111 )       ! Stop sampling clock
            call write_timer_register( 7, 16, 0 )     ! and restore interrupt
            call disable_adc_interrupt()              ! vector table to normal
            return
        end

        subroutine write_timer_register( igroup, ielement, ivalue )
c       Set Labmaster timer channel
            parameter(itimer_data=16#718, itimer_csr=16#719)
            ireg = igroup .or. ielement
            call outb( itimer_csr, ireg )
            call outb( itimer_data, iand(ivalue,16#ff))
            call outb( itimer_data, ivalue/16#100 )
            return
        end
```

```
; Listing 2.2 (b) Assembler A/D interrupt service routine.

  labmaster_code segment para public 'code'
  assume cs:labmaster_code
;
; Labmaster interface card I/O port definitions
;
  adc_csr equ 714h                  ; Labmaster    A/D control/status port
  adc_data_lo equ 715h              ;              A/D data (low byte)
  adc_data_hi equ 716h              ;              A/D data (high byte)
  adc_vector equ 0fh                ;              attached to line IRQ7
  pic_mask_reg equ 21h              ; 8259A mask register (Prog. interrupt controller)
  pic_eoi_reg equ 20h               ;   "     end-of-interrupt reply register
  eoi equ 20h                       ;   "     end-of-interrupt reply code

  buffer_address label dword        ; Memory address in which to store next A/D sample
  buffer_offset dw ?                ; (offset part of address)
  buffer_seg dw ?                   ; (segment part)
  samples_collected dw ?            ; No. of samples collected so far
  old_vector dd ?                   ; Orginal address in IRQ7 entry in vector table

; A/D interrupt service routine
; Responds to Labmaster's A/D done interrupt (on IRQ7) and copies latest A/D
; sample into data buffer at address in <buffer_address>

  adc_isr:
      push ax                       ; Save registers used
      push di
      push dx
      push es
      mov dx,adc_data_lo            ; Get 16 bit value from lo,hi
      in ax,dx                      ; pair of A/D data registers.
      les di,buffer_address         ; Set ES:DI = storage buffer address for sample
      stosw                         ; Copy sample to buffer and increment DI by 2
      mov cs:buffer_offset,di       ; Save new buffer pointer
      inc cs:samples_collected      ; Increment sample counter
      mov al,eoi                    ; Send End-of-Interrupt to 8259A controller
      out pic_eoi_reg,al
      pop es                        ; Restore registers and exit
      pop dx
      pop di
      pop ax
      iret

; .. continued
```

Digital Recording of Analog Signals

```
; Listing 2.2 (b) .. continued ..
  enable_adc_interrupt proc far          ; FORTRAN: call enable_adc_interrupt(iadc)
      public enable_adc_interrupt
      nargs = 1
      save_regs
      unpack nargs,1                      ; Get address A/D data storage array
      mov ax,ds                           ; and put into buffer_seg,buffer_offset
      mov cs:buffer_seg,ax
      mov cs:buffer_offset,si
      mov cs:samples_collected,0          ; Set no. of samples collected to zero
      mov ah,35h                          ; MS-DOS call to get address of IRQ7 vector
      mov al,adc_vector                   ; (AL = vector no., AH =35h )
      int 21h                             ; (Address returned in ES:BX)
      mov cs:old_vector,bx                ; Save it in old_vector
      mov cs:old_vector+2,es
      mov ah,25h                          ; DOS call to place interrupt service routine
      mov al, adc_vector                  ; into IRQ7 vector table entry. (AL= IRQ7)
      mov dx, seg adc_isr                 ; (DS = segment part of <adc_isr> address)
      mov ds,dx
      mov dx, offset adc_isr              ; (DX = offset part of address)
      int 21h
      mov dx,pic_mask_reg                 ; Ensure that IRQ7 is NOT masked off.
      in al,dx                            ; (i.e. that bit 7 = 0 in the
      and al,01111111b                    ; 8259A's interrupt mask register
      out dx,al
      restore_regs
      ret 4*nargs
  enable_adc_interrupt endp
  disable_adc_interrupt proc far         ; FORTRAN: call disable_adc_interrupt()
      public disable_adc_interrupt
      save_regs
      mov dx,pic_mask_reg                 ; Ensure that IRQ7 is masked
      in al,dx                            ; (i.e. that bit 7 = 1 in the
      or al,10000000b                     ; 8259A interrupt mask register
      out dx,al
      mov ah,25h                          ; DOS call to re-attach
      mov al,adc_vector                   ; old vector for
      mov dx, cs:old_vector+2             ; by
      mov ds,dx                           ; writing its starting address
      mov dx, cs:old_vector               ; into vector table.
      int 21h
      restore_regs
      ret
  disable_adc_interrupt endp

  adc_interrupt_count proc far           ; FORTRAN: call adc_interrupt_count( nsamples )
      public adc_interrupt_count
      nargs = 1
      save_regs
      mov ax, cs:samples_collected        ; Get no. of samples
      unpack nargs,1                      ; Return in argument <nsamples>
      mov ds:[si],ax
      restore_regs
      ret 4*nargs
  adc_interrupt_count endp
  labmaster_code ends
  end
```

```fortran
c       Listing 2.3. DMA A/D conversion subroutine
c       for the National Instruments LAB-PC
        subroutine adc_dma(n_channels,igain,interval,iadc,np,imode,itrigger,istart)
        integer*2 n_channels            ! (In) Channel No. (0-7)
        integer*2 igain                 ! (In) Amp. gain (0=x1,7=x100)
        integer*2 interval              ! (In) Sampling interval (µs)
        integer*2 iadc(1)               ! (Out) A/D sample storage array (2*np)
        integer*2 np                    ! (In) No. of samples required
        integer*2 imode                 ! (In) 0=wait till done,1=no wait,2=circular
        integer*2 istart                ! (Out) Starting position of samples in iadc
c       National Instruments LAB-PC I/O ports
        parameter(icommand1=16#260)     ! Command port 1
        parameter(icommand2=16#261)     ! Command port 2
        parameter(icommand3=16#262)     ! Command port 3
        parameter(istatus=16#260)       ! Status port
        parameter(iadc_data=16#26A)     ! A/D data port
        parameter(iadc_clear=16#268)    ! Clear A/D data logic
        parameter(idmatc_clear=16#26A)  ! Clear DMA logic
        parameter(icounterA_mode=16#277)   ! Sampling clock mode
        parameter(icounterA0_data=16#274)  ! Sampling clock data
c       DMA Controller I/O ports (Channel 1)
        parameter( idma_address = 16#02 )     ! Address port
        parameter( idma_count   = 16#03 )     ! Byte counter port
        parameter( idma_mask    = 16#0a )     ! Mask port
        parameter( idma_mode    = 16#0b )     ! Mode of operation port
        parameter( idma_clear_byte = 16#0c )  ! Clear byte
        parameter( idma_page    = 16#83 )     ! Memory page select port
        parameter( idma_status  = 16#8 )      ! Status port
        integer*4 isegment,ioffset,iaddress,ipage_offset,ipage,ipage1

c       Initialization phase
        call outb( icommand1, 0 )             ! Clear all command ports
        call outb( icommand2, 0 )
        call outb( icommand3, 0 )
        call outb( idmatc_clear, 0 )
        ibyte = igain*16 + n_channels - 1
        call outb( icommand1, ibyte )         ! Note how 2 writes
        if( n_channels .gt. 1 ) then          ! are made to command
            ibyte = ibyte + 16#80             ! register 1. This
            call outb( icommand1, ibyte )     ! is required.
        end if
c       Program sampling clock (Intel 8253 3-channel counter)
        call outb( icounterA_mode, 16#34 )    ! Put Counter A0 in rate generator
        call outb( icounterA0_data, interval ) ! mode and write sampling interval
        call outb( icounterA0_data, interval/256 ) ! (in µs)
        call outb( icounterA_mode, 16#70 )    ! Force Counter A1 output LOW
                                              ! to enable Counter A0
        call outb( iadc_clear, 0 )            ! Clear LAB-PC A/D data logic
        iadc_done = 0
        do while( iadc_done .eq. 0 )          ! Wait for A/D done bit to be
            call inpb( istatus, iadc_done )   ! to set.
            iadc_done = iadc_done .and. 2#1
        end do
        call inpb( iadc_data, ibyte )
        call inpb( iadc_data, ibyte )
c ... continued
```

Digital Recording of Analog Signals

```
c Listing 2.3 continued
      call outb( icommand3, 2#1 )                        ! Enable LAB-PC's DMA logic
      call outb( idma_mask, 2#101 )                      ! Disable DMA channel 1
      if( imode .eq. 2 ) then
          call outb( idma_mode, 2#01010101)              ! Continuous circular sweep
      else
          call outb( idma_mode, 2#01000101)              ! Single sweep
      end if
      nbytes = np*2 - 1                                  ! No. bytes to be transferred
      call outb( idma_clear_byte, 1 )                    ! into DMA byte count port
      call outb( idma_count, nbytes )
      call outb( idma_count, nbytes/16#100 )
      call varptr(isegment,ioffset,iadc(1))              ! Segment/offset of data array
      iaddress = ioffset + 16*isegment                   ! Calculate linear address
      ipage = iaddress/16#10000                          ! Calculate 64kbyte memory page
      call varptr(isegment,ioffset,iadc(np))             ! Find the page number of
      iaddress = ioffset + 16*isegment                   ! end of data area in array
      ipage1 = iaddress/16#10000                         ! containing "np" A/D samples
      istart = 1                                         ! If lower half of the array
      if( ipage .ne. ipage1 ) istart = np + 1            ! is not completely contained
      call varptr(isegment,ioffset,iadc(istart))         ! within a single 64K page
      iaddress = ioffset + 16*isegment                   ! use upper half
      ipage = iaddress/16#10000
      ipage_offset = iaddress - ipage*16#10000
      call outb(idma_clear_byte,1)
      call outb(idma_address,ipage_offset)               ! Write starting address of
      call outb(idma_address,ipage_offset/16#100)        ! memory area to hold data
      call outb(idma_page,ipage)                         ! into DMA address port
c     Transfer phase                                     ! and DMA page port
      call outb(idma_mask,2#1)                           ! Enable DMA channel 1
      if( itrigger .ne. 0 ) then
          call outb( icommand2, 2#10 )                   ! Wait for trigger
      else
          call outb( icommand2, 2#100 )                  ! Start sampling clock
      end if
      if( imode .eq. 0 ) then                            ! Mode 0:
          ibyte = 0                                      ! Wait here until
          do while( ibyte .eq. 0 )                       ! A/D sampling is
              call inpb( istatus, ibyte )                ! completed.
              ibyte = ibyte .and. 2#10000
          end do
c         Termination phase
          call adc_stop
      endif
      return
      end
      subroutine adc_stop
      parameter(icommand2=16#261)                        ! Command port 2
      parameter(icommand3=16#262)                        ! Command port 3
      parameter(icounterA_mode=16#277)                   ! Sampling clock mode
      parameter(icounterA0_data=16#274)                  ! Sampling clock data
      parameter( idma_mask    = 16#0a )                  ! Mask port
      call outb( idma_mask, 2#0101 )                     ! Disable DMA channel
      call outb( icommand3, 0 )                          ! Disable LAB-PC DMA
      call outb( icommand2, 0 )
      return
      end
```

```fortran
c       Listing 2.4. Continuous A/D recording-to-disc routine.
c       with DMA transfer routine from Listing 2.3.

        subroutine record_to_disc( iadc, np, nrecords, interval )
        integer*2 iadc(1)              ! A/D temporary storage buffer (np*4)
        integer*2 np                   ! (In) No. of samples in iadc
        integer*2 nrecords             ! (In) No. of records (record=np/2) required
        integer*2 interval             ! (In) Sampling Interval (µs)
        logical lower_buffer_filling
        parameter(iflag=4097)

c       Open binary data file to hold samples
        nbytes = np
        open(unit=1,file='adc.dat',form='binary',recl=nbytes,access='direct')
        irecord = 1                    ! Write a record to move disc
        write(unit=1,rec=irecord) iadc(1)   ! heads to start of file.

c       Start continuous A/D sampling into buffer iadc using DMA in
c       autoinitialise mode.
        call adc_dma(1,0,interval,iadc,np,2,0,istart)  ! Start A/D sampling
        ilower = istart                ! Define start/end points of
        iend_lower = ilower + np/2 - 1 ! upper and lower pairs of
        iupper = istart + np/2         ! sample buffers.
        iend_upper = iupper + np/2 - 1
        lower_buffer_filling = .true.  ! Filling starts in lower buffer.
        iadc(iend_lower) = iflag       ! Insert overwrite flags which
        iadc(iend_upper) = iflag       ! indicate when buffer is full.

c       Transfer samples to file in blocks of size np samples until
c       required number of blocks (nrecords) have been collected.
        do while( irecord .le. nrecords )
          if( lower_buffer_filling .eqv. .true. ) then
            if( iadc(iend_lower) .ne. iflag ) then    ! When flag value
              write(unit=1,rec=irecord) (iadc(j), j=ilower,iend_lower)
              irecord = irecord + 1                   ! at end of lower buffer
              lower_buffer_filling = .false.          ! is overwritten, write
              iadc(iend_lower) = iflag                ! data to file.
            end if
          end if
        else
          if( iadc(iend_upper) .ne. iflag ) then
            write(unit=1,rec=irecord) (iadc(j),j=iupper,iend_upper) ! When flag
            irecord = irecord + 1                     ! for upper buffer
            lower_buffer_filling = .true.             ! is over-written
            iadc(iend_upper) = iflag                  ! write it to file.
          end if
        endif
        end do
        call adc_stop                  ! Stop A/D sampling
        close(unit=1)                  ! Close file
        return
        end
```

Digital Recording of Analog Signals

```fortran
c       Listing 2.5. Spontaneous signal detection and recording routine.
        subroutine detect_signal(iadc,np,interval,ithreshold,np_pre,np_post,iresult)
        integer*2 iadc(1)              ! Temporary buffer of A/D buffer
                                       ! for storing the np most record
                                       ! A/D samples (size=np*2)
        integer*2 np                   ! (In) No. of samples in iadc
        integer*2 interval             ! (In) Sampling interval (µs)
        integer*2 ithreshold           ! (In) Detection trigger level
        integer*2 np_pre,np_post       ! (In) No. pre- & post-trigger samples
                                       ! Note np_pre+np_post<=np/2
        integer*2 iresult(1)           ! (Out) Detected signal array
        integer*4 izero                ! Running mean baseline level.
        parameter( idma_count = 16#03 ) ! Byte counter port
        parameter( idma_clear_byte = 16#0c ) ! Clear byte
        parameter(iflag=4097)

c       INITIALIZATION PHASE: Start continuous sampling into buffer iadc using DMA
c       in autoinitialize mode to fill the buffer iadc(istart .. istart+(np*2)-1)
        nmask = np*2 - 1               ! Byte count mask
        call adc_dma(1,0,interval,iadc,np,2,0,istart)

c       SIGNAL DETECTION PHASE
        izero = 4096
        ip = istart                    ! This avoid premature triggering
        nbytes = nmask
        do while( (iadc(ip) - izero) .le. ithreshold )
           nbytes_old = nbytes
           do while( nbytes .eq. nbytes_old )
10            call outb( idma_clear_byte, 1 )    ! Wait for DMA byte
              call inpb( idma_count, ilo )       ! counter to change
              call inpb( idma_count, ihi )       ! Two bytes transferred
              nbytes = ilo + 256*ihi             ! per sample.
                                                 !
              if( (nbytes .and. 2#1) .eq. 0 ) goto 10  ! Result must be odd number
           end do
           ip = ((nmask - nbytes-2).and.nmask)/2 + istart  ! Location of latest
                                                 ! A/D sample in buffer
           izero = 16*izero                      ! Adjust signal zero
           izero = izero + iadc(ip)              ! using a 16 point
           izero = izero / 17                    ! running mean.
        end do

c       COLLECTION PHASE
c       When a signal is detected - wait until the required number of post-trigger
c       samples (np_post) have been collected. (A marker flag is written into the
c       buffer location for the last sample. When that flag is overwritten, the
c       required number of samples have been acquired.)
        ilast = ((nmask - nbytes + 2*np_post) .and. nmask)/2 + istart
        iadc(ilast) = iflag
        do while( iadc(ilast) .eq. iflag )
        end do
        call adc_stop                            ! Stop A/D sampling
        do i = np_pre+np_post,1,-1               ! Copy the record
           iresult(i) = iadc(ilast)              ! of np samples from
           ilast = (ilast - 1)                   ! the A/D buffer in
           if( ilast .le. istart ) ilast = istart+ np - 1  ! to results array.
        end do
        return
        end
```

CHAPTER THREE
Analog signal conditioning

While introducing the topic of analog to digital conversion, the assumption has been made that the analog voltage signals produced by the electrophysiological recording system were appropriately matched to the input requirements of the laboratory interface. This is not always the case and there is often a need to *condition* (i.e. adjust) the analog voltage signals to make the best recording or, indeed, make a recording at all. Signal conditioning may involve amplification of the signal level, removal of constant DC levels, signal filtering, and the detection and processing of synchronization pulses.

A typical signal conditioning system required for the digitization of intracellular potentials is shown in Figure 3.1. The raw voltage signal from the microelectrode amplifier is passed through an amplification stage and then to an electronic filter before being fed into the A/D converter of the laboratory interface. Many intracellular signals last only a fraction of a second and require some form of stimulus to evoke them, a typical example being nerve evoked endplate potentials.

It is necessary to synchronize digital recording with the stimulus event, using the synchronization output of the stimulator to trigger A/D sampling.

3.1 SIGNAL AMPLIFICATION

Signal amplification is often necessary because the A/D converter in the laboratory interface has a relatively low voltage sensitivity compared to more familiar devices such as the oscilloscope. Some ADCs have only fixed input ranges (e.g. ±5 V for the CED 1401) or a limited programmable range (±10 V, ±5 V, ±2.5 V, ±1.25 V, Data Translation DT2801A). Even low-cost oscilloscopes usually have a range of 10 or so input sensitivities, from 5 mV/div. (±100 mV) to 5 V/div. (±50 V). This lack of sensitivity can create problems when working with electrophysiological instruments which have restricted output levels. Microelectrode amplifiers for

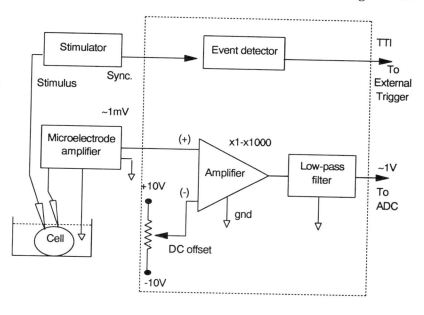

Figure 3.1 Analog signal conditioning system. A small signal (~1 mV) from a microelectrode amplifier is amplified 1000 times by a differential amplifier, resulting in a ~1 V signal suitable for A/D conversion. DC offsets are removed from the signal before amplification by adjusting the DC level into the (−) input of the amplifier. The signal is also low-pass filtered.

instance (e.g. the World Precision Instruments 705), may not amplify the measured cellular signal voltage at all, or at most by a factor of 10. While the oscilloscope voltage sensitivity can usually be adjusted to match the signal level being provided, the same is not always true for the laboratory interface. The problem becomes particularly acute when studying small signals, such as miniature endplate potentials (MEPPs) with amplitudes of only 1–2 mV. Since a 12 bit ADC with an input range of ±1.25V has a sensitivity of 0.61 mV/bit, the digitized MEPP would only be a few bits in amplitude.

To make effective use of the full 12 bits of ADC resolution, small signals must be amplified to be several orders of magnitude larger than the ADC sensitivity. In the case of the MEPP, to achieve a 0.1% resolution, the signal must be amplified 1000 times, spanning 25% of the ADC input voltage range. To be versatile, and allow matching to a variety of different signal types, the amplifier should have a variable *gain* or amplification factor between 1 and 1000. The gain may be varied continuously using a precision multiturn potentiometer or switchable between a series of fixed levels as in an oscilloscope. It should be noted that if a switched gain amplifier is chosen a sufficient number of gain levels should be available, e.g. 1–2–5–10–20–50–100–200–500–1000×.

3.2 DC OFFSET REMOVAL

It may also be necessary to remove any constant DC level that might exist in the analog signal before amplification takes place. Again taking MEPPs as an example, although they are small signals, they are superimposed on top of a large −90 mV cell resting potential. If the original signal were to be amplified 1000 times it would, in theory, result in a signal with a DC level of −90 V. This kind of signal level is not, of course, obtainable by most laboratory instrumentation which has output voltage limits of ±10 V. In any case, levels of more than 20–30 V will often damage the input stages of the laboratory interface.

To avoid this problem the −90 mV resting potential must be subtracted from the signal before amplification. This can be achieved using a *differential amplifier* as shown in the signal conditioning system in Figure 3.1. A differential amplifier, as the name suggests, amplifies the difference between two signals ((+) and (−) in Figure 3.1). DC levels are subtracted from the signal by feeding a constant voltage level from the potentiometer into the (−) input which is subtracted from the signal from the microelectrode amplifier, fed into the (+) input, before being multiplied by the amplifier. This method is also known as *input DC offset*.

Another method for eliminating DC levels which has been occasionally used in the past is the use of *AC coupling*. By inserting a capacitor in series with the signal, it is possible to block the passage of the constant DC level while allowing through transient waveforms. Unfortunately, it is not usually possible to do this without also causing some distortion of the signal. In general, in situations where the details of the signal waveform are of interest it is preferable to use the DC offset method.

3.3 SIGNAL FILTERING

According to Fourier theory an analog signal can be described as the sum of a series of sine wave components of various amplitudes spread over a range of frequencies, from DC (0 Hz) extending to as high a frequency as necessary. Slowly changing aspects of the signal are represented with low frequency components and rapidly changing parts with high frequency components. Filtering procedures selectively remove particular frequency components and thereby modify the time course of the analog signal. *Low-pass* filtering removes high frequency signal components above a defined filter *cut-off* frequency, effectively smoothing the signal. Conversely, *high-pass* filtering removes DC and low frequency components, leaving the transient rapidly changing parts of the signal (AC coupling is a form of high-pass filtering). The combination of the two is *band-pass* filtering. The filtering most generally applied as a signal conditioning process is of the low-pass variety and it is done for two reasons.

- *Anti-aliasing* filtering to eliminate potential artifacts in the signal induced by the digital sampling process.
- Smoothing of the signal by removing high frequency components of background noise to improve the signal–noise ratio.

3.4 ALIASING

As discussed in Chapter 2, while an analog signal is a continuous quantity, the digital recording of

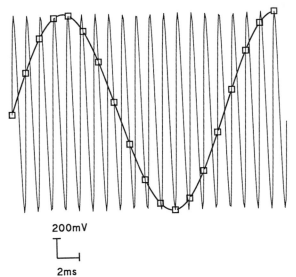

Figure 3.2 Aliasing. Sampling a 1 kHz sine wave at 800 Hz intervals, a rate less than the Nyquist rate of 2 kHz (samples shown as squares), produces a highly misleading digital recording with a false frequency of 57 Hz. Shown superimposed on an unaliased recording sampled at 10 kHz.

the same signal is restricted to a series of samples at a fixed intervals. If the sampling interval is not small, compared with the time course of the signal, the digital version will not be a faithful representation of the original signal. This problem is particularly acute in the case of periodic signals where the phenomenon of *aliasing* occurs, producing a highly misleading digital record. Figure 3.2 shows two digital records of the same 1 kHz sine wave signal, a faithful record sampled at a 10 kHz rate and a record sampled at 800 Hz. The apparent frequency of the sine wave after digitization at the lower sampling appears to be 57 Hz rather than the true 1 kHz as obtained from the high sampling rate record.

When sampling at rates that are high relative to the signal time course, aliasing is not a problem but it is useful to know how high the sampling rate need be to avoid aliasing. Nyquist's theorem (Nyquist, 1928) provides this information, with the minimum valid sampling rate (the *Nyquist rate* f_{nyq}) given by

$$f_{nyq} = 2 f_{max}$$ [3.1]

where f_{max} is the highest frequency component present in the analog signal. Since frequencies above the Nyquist limit are not faithfully recorded by a digital recording system, and can only act to distort the remainder of the signal, it is essential to filter these frequency components out of the analog signal before it is digitized – a process known as *anti-alias filtering*. While the Nyquist theorem defines the minimum valid sampling rate there is no reason why substantially higher sampling rates cannot be used. It is often preferable to err upon the safe side and choose a sampling rate substantially higher than the Nyquist rate. This makes the digitized records more presentable when displayed on the PC's graphics screen. In our laboratory, for instance, EPP and MEPP signals which are filtered to have few frequency components in excess of 5 kHz are routinely digitized at sampling rates of 25 kHz. On the other hand, signals recorded for spectral analysis (see Chapter 9) are often sampled using the Nyquist or only slightly higher sampling rates.

3.5 SIGNAL SMOOTHING AND NOISE

Electrophysiological signals are generally recorded in the presence of background noise inherent to the recording procedures. Such noise (defined here as unwanted signals obscuring the signals of interest) can be split into two main types.

- *Interference* from external signal sources such as electromagnetic waves from 50/60 Hz AC mains power lines or the switching on and off of electrical device.
- *Random noise* from sources such as Johnson (thermal) noise in high resistance microelectrodes or amplifier input stages.

Interference signals can usually be completely removed by the use of electrical shielding of the recording apparatus and appropriate attention to the electrical ground connections of the apparatus. Details of interference reduction methods can be found in Morrison (1986).

Random noise, on the other hand, cannot be so easily removed since it is often inherent to the recording apparatus. One of the most commonly encountered random noise sources is *Johnson noise*, a property of all electrical conductors. It manifests itself as random fluctuations in voltage measurements, due to the thermally induced movement of the charge carriers, e.g. electrons in a metal conductor or ions in a salt solution. Johnson noise is a form of *white noise* since its random fluctuations are spread evenly over an infinite range of frequencies. The magnitude of this noise is proportional to the frequency bandwidth of the recording apparatus and also to the resistance of the conductor, given by the equation

$$V_j = \sqrt{4kTRf_c} \qquad [3.2]$$

where k is Boltzmann's constant (1.38×10^{-23} J/K), T is the absolute temperature (K), R is the resistance (Ω), and f_c is the upper frequency limit (Hz) of the recording bandwidth set by the cut-off frequency of the low pass signal conditioning filter. Johnson noise turns out to be the factor which limits the minimum recordable signal using intracellular microelectrodes. It is a particular problem when dealing with small cells which require sharp, and therefore high resistance, microelectrodes. At a temperature of 20°C (293K) a 20 MΩ microelectrode has a Johnson noise level of 8.04 nV/Hz. In a recording system with a bandwidth of 5 kHz, this results in a random background noise signal with an amplitude of 40.2 μV (rms). The rms (root mean square) value, commonly used to express the amplitude of random signals, is equivalent to the signal standard deviation. Such a signal will therefore have a peak–peak value of $6V_j = 241$ μV. Further discussion of Johnson noise and similar noise sources such as shot noise can be found in DeFelice (1981).

As can be seen from Equation 3.2, the wider the bandwidth set by f_c the larger the background noise. It is therefore useful when recording signals of small amplitude to use a low-pass filter to remove as much of the higher frequency components of the random noise as possible without affecting the signals of interest. This can be particularly important when recording current signals from some voltage clamp systems where there may be substantial amounts of high frequency background noise. However, if the frequency components of the signal itself extends

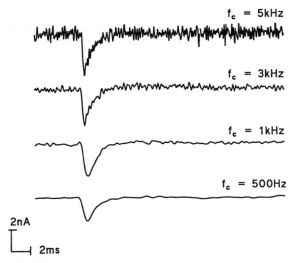

Figure 3.3 Effects of low-pass filter cut-off frequency on background noise. Four MEPC signals filtered at 5 kHz, 3 kHz, 1 kHz, 500 Hz (8-pole Bessel filter). (Records courtesy of Dr T. Searl, University of Strathclyde.)

Table 3.1 Effect of low-pass filter cut-off frequency (f_c) on MEPC peak amplitude, rate of rise, exponential decay time constant (τ) and signal–noise ratio.

f_c	Peak (nA)	Rate of rise (nA/ms)	τ (ms)	S/N ratio (Peak/rms)
5 kHz	−3.9	26.7	0.99	11.5
3 kHz	−3.1	16.9	1.04	12.9
1 kHz	−2.8	8.2	0.97	30.6
500 Hz	−2.2	3.9	0.91	30.6

beyond the filter cut-off then some of its components are going to be removed as well, resulting in some distortion of the signal time course. An optimal choice of filter f_c is therefore one which minimizes the background noise without inducing significant signal distortion.

Figure 3.3 shows the effect of low-pass filtering with cut-off frequencies of 5 kHz, 3 kHz, 1 kHz and 500 Hz on voltage-clamped rat neuromuscular junction miniature endplate currents (MEPCs). Such signals typically require the careful choice of low-pass filter cut-off frequency, being of small amplitude (approximately 4 nA in this case) superimposed on a relatively high background noise. As the filter f_c is shifted to lower frequencies, a clear reduction in the background noise can be seen. However, at the lower frequencies (1 kHz and 500 Hz) the MEPC signal itself begins to be noticeably changed.

Table 3.1 shows the quantitative changes introduced by the cut-off frequency in some MEPC signal parameters of experimental interest; peak current, maximum rate of rise, decay time constant and signal/noise (S/N) ratio (computed as the ratio of the MEPC peak value divided by the rms amplitude of the background noise). The results were derived as the average of measurements on a set of 15 MEPCs repeatedly analysed with the filter set at each cut-off frequency. As f_c is reduced, the MEPC peak height and, most markedly, the rate of rise of the signal is reduced. On the other hand, the time constant of decay is hardly affected. It should also be noted that at the lowest frequency (500 Hz) the S/N ratio ceases to increase as the signal begins to be filtered as much as the noise. The choice of the optimum cut-off frequency will therefore depend on what aspect of the signal is of prime interest. If rate of rise was important 5 kHz would have to be used to avoid distortion in the signal. On the other hand if only the decay time constant was important then 500 Hz could be used.

3.6 FILTER TYPES AND CHARACTERISTICS

So far, we have discussed filters as if they acted perfectly, completely removing frequencies above the cut-off without affecting those below and without distorting the remaining signal in any way. In reality, such perfect filter characteristics cannot be obtained and the practical use of filters requires certain compromises to be made. The performance of a filter can be described by its *frequency response* curve, essentially the fraction of the signal passed over a range of frequencies extending from DC upwards. The frequency response curve can be measured by feeding a range of sine wave signals into the filter and measuring the output. If the input signal V_i is described by

$$V_i(t) = V \sin(2\pi f t) \qquad [3.3]$$

then the output signal V_o is given by

$$V_o(t) = A(f) V \sin(2\pi ft + \theta(f)) \quad [3.4]$$

where $A(f)$ is the *attenuation* response of the filter and $\theta(f)$ is the *phase* response, the relative displacement in time between the input sine wave and the output.

The filter cut-off frequency is commonly defined as the *half-power* or -3 dB (decibel) point on the attenuation–frequency curve. Although the choice of the signal power level (power is defined as V^2/R) may seem a strange choice, it is chosen because it is more consistently meaningful when applied to signals of widely varying shape compared to a simple half-amplitude level. The unit of the *decibel* is used as a convenient measure of signal gain or attenuation ratios. It is a logarithmic value allowing very large ratios to be neatly expressed. The ratio of the amplitude of the input and output signal power levels is in dB:

$$dB = 10 \log_{10}\left(\frac{P_o}{P_i}\right) \quad [3.5]$$

Taking into account that $P = V^2/R$, the decibel can be re-expressed in terms of signal amplitude as

$$dB = 20 \log_{10}\left(\frac{V_o}{V_i}\right) \quad [3.6]$$

A value of -3 dB therefore corresponds to a power ratio of 0.5 and an amplitude ratio of 0.707.

3.7 THE RC FILTER

The simplest implementation of a low-pass filter consists of the combination of a resistor and capacitor as shown in Figure 3.4 (a). This filter's attenuation response (in terms of signal power) is given by the equation:

$$|H(f)|^2 = \frac{1}{1 + (f/f_c)^2} \quad [3.7]$$

where

$$f_c = \frac{2\pi}{RC} \quad [3.8]$$

with R the resistance (Ω) and C the capacitance (farads). Figure 3.4 (a) shows the effect of feeding a 1 kHz sine wave signal through the RC low-pass filter ($f_c = 950$ Hz). The amplitude has been reduced by a factor of 0.7 and the output signal delayed by 0.12 ms (a phase shift of 0.79 radians). Figure 3.4 (b) shows the filter attenuation-frequency response plotted as log power versus log frequency. The response starts off flat at low frequencies (i.e. no attenuation) and starts to diminish at frequencies approaching f_c. At high frequencies ($\gg f_c$), the log–log response falls off linearly. The slope of this region, in units of dB/decade (10 × change in frequency), is known as the *roll-off* and is a measure of the sharpness of the filter cut-off.

Figure 3.4 (c) shows the phase reponse of the filter. As frequency increases the amount of phase delay also increases. The phase response is important in that it provides a measure of the

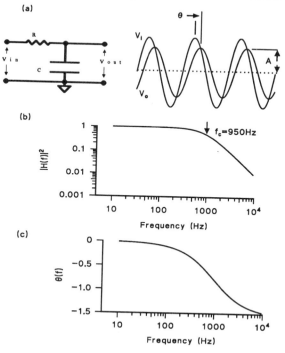

Figure 3.4 (a) Simple passive low-pass RC filter, $f_c = 950$ Hz. Effect of RC filter on a 1 kHz sine wave. (b) Filter attenuation-frequency response (log–log). (c) Filter phase–frequency response.

time delay applied to the frequency components of the filtered signal. If the same time delay is not applied to all components equally then the process of filtering will distort the shape of the signal, even for frequencies within the pass-band of the filter. A filter design which has this equal time delay property is known as a *linear phase* filter. As will become apparent, not all filters have linear phase properties.

In comparison with an ideal 'brick wall' filter which removes all frequencies above its cut-off, the RC filter has a rather poor frequency response. There is significant attenuation of signal frequencies below f_c and a relatively shallow roll-off above. In fact, 42% of the total signal power passed by the filter is at frequencies above f_c.

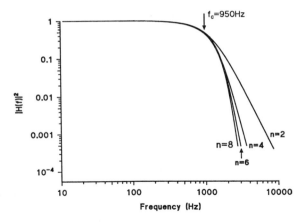

Figure 3.5 Improvements in sharpness of cut-off obtained using multiples stage of active filters. Attenuation–frequency response curves for Bessel filters (f_c = 950 Hz) with 1, 2, 3 and 4 stages (2nd, 4th, 6th and 8th order).

3.8 ACTIVE FILTERS

While simple filters can be constructed using combinations of passive components such as resistors and capacitors, it is more practical to construct *active* filter circuits using operational amplifiers (very high gain differential amplifiers in the form of integrated circuits, e.g. the National Semiconductor LM741, Op-07). A major advantage of active filter designs is that they allow complex filters to be built up by connecting several individual simple stages in series. Multi-stage active filters can be designed with much improved frequency response performances compared to the RC filter.

The properties of such filters can be most concisely described in the frequency domain using Laplace transforms which relate the output signal in terms of the input signal and the filter transfer function $H(s)$

$$V_o(s) = H(s) V_i(s) \quad [3.9]$$

where s is the complex frequency ($x + iy$). The transfer function for the RC filter is, for instance

$$H(s) = \frac{1}{(s + RC)} \quad [3.10]$$

The transfer function in a single equation, contains all the information previously expressed by the filter attenuation and phase response functions. Details of the use of Laplace transforms and the transfer function in the analysis of signals can be found in Martin (1991). A key feature of a filter design is the number of *poles* or points in the frequency domain where the denominator of the transfer function becomes zero. These determine the characteristic cut-off points of the filter frequency response. The larger the number of poles in the transfer function, the sharper the filter roll-off that can be achieved.

In general a single operational amplifier stage within a multi-stage filter design can be made to yield a two-pole transfer function. In order to add more poles and improve the filter response further, extra stages must be added in series. Figure 3.5 shows the improvements in sharpness of cut-off obtained as extra stages are added to a filter (Bessel-type, see below). An introduction to operational amplifier techniques and simple filter design can be found in Derenzo (1990) or Horowitz & Hill (1980). In practice a compromise must be reached between the attainable filter response and the cost of large numbers of stages. Typical commercially available filters such as the Frequency Devices LPF902 (Haverhill, USA) often consist of four stages and are therefore 8 pole (or 8th order) filters.

3.9 BESSEL AND BUTTERWORTH FILTERS

A variety of circuit designs can be used for the individual stages in the filter, conferring different attenuation and phase frequency response characteristics. There is, however, no single ideal filter design with a choice having to be made between the sharpness of the filter's roll-off after f_c and degree of distortion (due to non-linear phase response) produced in the signal.

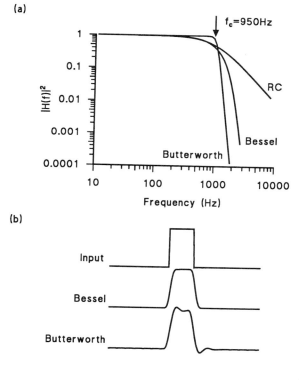

Figure 3.6 Bessel and Butterworth frequency response and time domain characteristics. (a) Attenuation vs frequency response for eight pole Bessel and Butterworth filters, compared with a two pole RC filter (f_c = 950 Hz). (b) Effect of filtering on signal waveform.

Figure 3.6 (a) shows the frequency response for two opposing types of filter design, Bessel and Butterworth (both 8th order), with the RC filter (1st order) for comparison. The *Butterworth* filter design produces a filter with a very sharp roll-off of 160 dB/decade, compared to 20 dB/decade for the RC filter. Only 3% of the signal passing through the filter is at frequencies exceeding f_c. The Butterworth filter, however, does not have a linear phase response resulting in distortion of the passed signal. This manifests itself as ringing at the leading and trailing edges of a Butterworth filtered rectangular voltage pulse, as can be seen in Figure 3.6 (b). Such signal distortion prevents the use of Butterworth filters in situations where the shape of the filtered waveform must be faithfully preserved. Unfortunately, this includes most forms of electrophysiological signal analysis. (Noise analysis is an exception, see Chapter 9.)

The *Bessel* filter, on the other hand, is an example of a filter design with a linear phase response which does not add ringing distortion to the filtered signal waveform. While the Bessel filter attenuation response in Figure 3.6 (a) (135 dB/decade roll-off, 18% > f_c) is not as good that of the Butterworth, it is much improved compared with the RC filter. For general purpose use, therefore, Bessel filters or similar linear phase response filter designs should be used with 8th order filters being a reasonable compromise between performance and complexity.

Bessel and Butterworth (or Chebyshev, another very sharp cut-off, non-linear phase filter) filter designs can be obtained from a variety of manufacturers, the most noteworthy being the Frequency Devices LPF901 and LPF902, Kemo VBF8 (Beckenham, UK) and Barr & Stroud (Glasgow, UK).

3.10 EVENT DETECTION AND TRIGGERING

As discussed earlier, the start of digital recording must often be synchronized with an external event such the stimulation of a nerve or the onset of a voltage jump command to a voltage-clamp. Most laboratory interfaces provide a means for such synchronization, usually by linking the start of the A/D sampling clock to a pulse on an *external trigger* input. This input is usually designed to respond to a standard TTL (transistor–transistor logic) pulse. TTL is the most common of a number of interfacing standards, allowing the interconnection of digital logic circuits. TTL signals are defined nominally as two levels, LOW = 0 V (<1.4 V) and HIGH = 5 V (>3.5 V). Generally, a TTL transition rather than a specific level (e.g. HIGH-to-LOW or LOW-to-HIGH) is the trigger signal, with the HIGH-to-LOW transition being used most often, since it provides the best immunity from noise accidentally triggering the system.

In order to ensure that the initial phase of an event is recorded, the synchronization pulse must be provided *before* the main stimulation pulse. Most stimulators can provide such a synchronization pulse but it is not always electrically compatible with the laboratory interface's TTL input. Often a signal is provided which is the wrong polarity, perhaps a negative voltage pulse, or too large (e.g. +12 V such as provided by the Grass Instruments S44 or S88 stimulators). A similar problem occurs when a synchronization pulse has been recorded on an FM tape recorder (e.g. Racal Store 4, Racal Ltd, Southampton, UK) which often has an output voltage range of no more than ±1 V, too small to trigger the TTL circuit directly.

A circuit is therefore required with a variable trigger threshold which can be set to fire on a wider range voltage transitions than the laboratory interface's own TTL trigger. Such circuits can be constructed using an integrated circuit device known as a *comparator*, a typical example being the National Semiconductor LM311. This device is like an operational amplifier, in that it accepts two input voltage signals (V+, V−), but provides a TTL digital output. When V+ > V− the comparator output is set to 0 V (TTL LOW) and conversely, if V− < V+ the output is 5 V (TTL HIGH). A variable threshold trigger circuit can be constructed by feeding the desired trigger voltage level into one input and the trigger signal into the other. A TTL transition then occurs when the trigger signal exceeds the threshold. Details of design and construction of comparators can be found in Horowitz & Hill (1980), or Coughlin & Villanucci (1990).

3.11 SOURCES OF ELECTRICAL INTERFERENCE

Electrophysiological recordings are well known to be prone to interference from external sources. This is not surprising since the amplitudes of the signals of interest are many orders of magnitude smaller the AC mains voltages powering the apparatus within the laboratory. The most common problem is the coupling of 50/60 Hz electromagnetic radiation from the AC mains supply into the inputs of the sensitive recording headstage. A variety of coupling modes are possible, capacitive, inductive, and conductive (Wolf & Smith, 1990). Capacitively coupled interference is transmitted via the small but significant mutual capacity that exists between any two conductors. Within the laboratory the main source of this is the unshielded AC mains power wiring.

Inductive and conductive coupling occurs primarily as a consequence of multiple connections to the electrical ground. Electrical signals must be made with reference to some standard potential, usually the potential of the ground (or earth) underneath the building. Two signal connections must therefore be made to the tissue, a signal measurement connection and a ground connection. Equally there must be a signal and ground connection whenever an electrical signal is transmitted between two pieces of apparatus. In the normal electrophysiological recording system there are several devices all with their own ground connections. Conductive interference arises because the connections between the ground of a piece of apparatus and the master ground is not perfect (i.e. of zero resistance) and therefore it is possible to find small voltages produced there by the AC power supply. When the grounds of two devices are connected together a current may flow along the ground connection due to these differences in ground potential. If such a current flows into the sensitive headstage of the microelectrode amplifier it can often overwhelm the electrophysiological signal being studied. Inductive interference is a similar current due to the

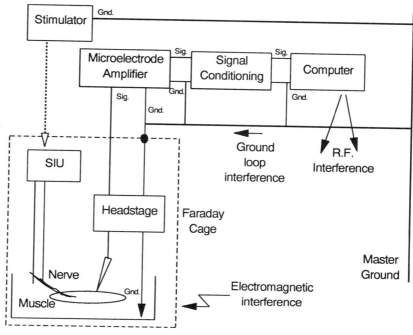

Figure 3.7 Typical sources of electrical interference, and preventative measures. Faraday shield prevents capacitively coupled 50/60 Hz AC from mains wiring. A single ground line, via headstage prevents ground loops passing through tissue. Stimulus isolation unit allows a stimulus pulse to be applied without a direct electrical connection.

multiple ground connections acting as a loop antenna picking up electromagnetic radiation from the AC mains power.

Computers in the laboratory have the potential to contribute additional interference from the high-speed digital pulses associated with their operation which create large amounts of radio frequency noise. Problems may occur if this noise is coupled into the sensitive recording headstage, either through the ground circuits or electromagnetically.

The interference problems discussed above may be eliminated (or at least minimized) by the appropriate layout of the ground connections within the recording system, as shown in Figure 3.7. The cell or tissue under study is surrounded by a grounded metal cage (Faraday cage) which effectively screens the capacitively coupled 50/60 Hz and most of the radio frequency (R.F.) radiation. Each device in the recording system (microelectrode amplifier, signal conditioner, computer stimulator) is grounded via its power cable. Consequently, there are likely to be multiple ground paths and grounds loops to some degree.

It is worth noting that it was once common practice to avoid such ground loops by disconnecting the ground wire from all devices except one and grounding the whole setup via external connections to a single point. This is to be discouraged, unless the work is carried out to appropriate standards by a qualified electrican, since it potentially compromises the electrical safety of the apparatus. When ground loops do occur, they can often be traced to poor connections on distribution boards or particular pieces of apparatus and remedied without such measures.

It is the ground connection to the tissue bath which is of particular importance and, in modern designs, this is often made via a single connection on the headstage of the recording amplifier. It is important to avoid additional ground connections to the bath from other apparatus, a ground loop of the type discussed above may be formed inducing large amounts of interference. A detailed discussion of the procedures for eliminating electrical interference can be found in Wolf & Smith (1990) or Morrison (1986).

In relation to the computer, ground loop problems are of much less concern than they used to be. The modern personal computer is a small device, powered from the same circuits as the rest of the apparatus. It is often well shielded to meet US FCC (Federal Communications Commission) regulations. In comparison, the older minicomputers were larger, used much more power, and might be actually located in another room. In a well-designed PC-based system, the computer need not generate a significant amount of interference. Modern microelectrode recording amplifiers and similar devices prevent significant ground loop interference from being coupled into the recording. The output of these devices can generally be directly connected to the A/D converter inputs of the laboratory interface without experiencing significant problems. Similarly, the computer is often used to apply stimuli to the tissue under study. Often the recording device itself has a facility for this purpose, e.g. the command voltage input of a voltage clamp or the current injection control of a bridge amplifier such as the Axoclamp. Again, in general, few problems are encountered when attaching the laboratory interface D/A output to such inputs, because no additional connection is made to the tissue bath.

3.12 STIMULUS ISOLATION

One exception to this rule occurs when it is required to stimulate a nerve directly – to produce a nerve-evoked endplate current, for instance. Usually a voltage pulse of 5–20 V amplitude, lasting 100–200 μs is required to stimulate the nerve. Such pulses are often produced by stimulators designed for the purpose (e.g. Grass S88) but it can also be done using a computer. However, in either case the stimulus must be applied without adding any additional ground connections to the bath. This can be done using a *stimulus isolation unit (SIU)* which permits a stimulus pulse to be coupled to the nerve without a direct electrical connection. An SIU consists of two halves, in complete electrical isolation from each other. One half accepts an input pulse from a stimulator or D/A converter, the other half applies the pulse to the nerve. Most SIUs use either optical or transformer (electromagnetic) coupling to achieve the isolation between the input and output.

The transformer-coupled SIU uses the principle that there is no direct circuit connection between the input and output windings of a transformer. However, a transformer can only pass AC signals while it is often desirable to produce DC stimulus pulses. A transformer-coupled SIU circumvents this problem by creating a high frequency AC carrier signal which is modulated by the stimulus pulse. The combined signal passes across the transformer and is decoded at the other side, and applied to the tissue. An advantage of the transformer-coupled approach is that the carrier signal can also be used to provide a power source for the isolated end of the SIU. A typical example of a transformer-coupled isolator is the Grass SIU5, used with the Grass range of stimulators. This device is interesting in that it can produce an isolated stimulus pulse of significant amplitude (100 V maximum). Unfortunately, the SIU5 cannot be directly connected to a D/A output since it draws a large amount of current from the stimulator during operation. It is also possible to construct SIUs, using transformer coupled isolation amplifiers ICs such as the Burr-Brown 3656 (Derenzo, 1990).

The optically coupled SIU uses a light emitting diode/photodiode pair to transmit the pulse amplitude and duration in terms of light intensity. Unlike the transformer-coupled SIU, the isolated side requires a power source. Mains power cannot be used since this would undermine the isolation therefore a stack of batteries are used. This can be inconvenient since a large number of batteries are needed to achieve high output voltages. Optical SIUs, like the World Precision Instruments A360R, use a set of rechargable batteries which can be recharged overnight *in situ*. On the other hand, the optical SIU draws much less current on the input side and can therefore be directly connected to a computer D/A output.

3.13 MODULAR SIGNAL CONDITIONING SYSTEMS

A practical signal conditioning system suitable for recording voltage clamp signals would consist of the following:

Analog signal conditioning

Table 3.2. Modular signal conditioning systems suitable for use in the electrophysiological laboratory. Typical examples of amplifier, low-pass filter and trigger circuit are shown with their key performance features.

Product	No. of modules	Amplifier	Filter	Trigger
Neurolog (Digitimer Ltd)	NL900 13r	NL106 X0–10, X0–100$_v$ 10 V offset	NL125 0.5 Hz-5 kHz$_v$ −40 dB/decade	NL515 ±10 V trigger AC/DC
Axon Instruments	5b/10r	AI2130 X1–1000 10 V offset	AI2040 1 Hz-99 kHz −160 dB/decade	AI2020A +10 V AC/DC
Fylde Electronic Laboratories	8b/16r	FE-2940A X1–1000 ±6 V	FE-301AF 500 Hz–5 kHz −160 dB/decade	FE294TU ±10 V AC/DC

b = Bench mount case; r = 19" rack mount case.

Two differential DC amplifiers (current and voltage)

    Gain: X1–X1000
    Bandwidth: DC–10 kHz
    Input DC offset: ±10 V

One low-pass Bessel filter (8th order) (current)

    Cut-off: 10 Hz-10 kHz

One variable threshold trigger unit.

    Input range: ±10 V
    Output: TTL

Such circuits can be constructed using published designs using standard, easily available, operational amplifiers. This can be an economical route to take if you have access to the services of an electronics laboratory in-house or have the expertise and facilities for the construction of electronic circuits. On the other hand, there can be significant amounts of construction work involved, particularly in the case of a multi-stage filter unit with a wide range of switchable cut-off frequencies.

Much effort can be avoided by purchasing commercially available signal conditioning systems. Generally such systems consist of a chassis or mainframe (often mountable in a standard 19" rack) containing a power supply into which a variety of interchangeable signal conditioning modules can be inserted. Some manufacturers, particularly Axon Instruments Inc. and Digitimer (Welwyn, UK) produce such systems specifically for the neurosciences market. Other companies, such as Fylde Electronics Laboratories Ltd (Preston, UK), with products developed primarily for the engineering/aerospace industry should not be neglected, since they are nevertheless suitable for electrophysiological applications. Some of these companies, Fylde, for instance, will also produce modules customized to suit individual needs when requested. A comparison of some of the features of these modular conditioning systems is shown in Table 2.3.

3.14 COMPUTER-CONTROLLED SIGNAL CONDITIONING

In order to obtain an accurate digitized record of the amplitude of the experimental signals it is necessary for the exact amplification applied during signal conditioning to be known to the computer recording system. This is often done by manually entering the settings of the signal conditioning instrumentation into the computer program either before or after recording. Although quite workable this system is prone to operator error and somewhat inflexible, especially if it is necessary to change the gain settings several times during a recording session.

Recently, this problem has begun to be addressed by the development of signal conditioning instrumentation whose settings can be directly controlled from the computer system, notable examples being the Axon Instruments CyberAmp system and the Cambridge Electronic Design 1902 filter/amplifier. Both of these signal conditioners allow amplifier gain and low-pass filter cut-off frequency to be set by the computer system. In both cases the signal conditioner is connected to the computer system via the standard RS232 serial port. Allowing the computer program to control the signal conditioning directly has great potential to simplify the operation of the computer recording system.

3.15 SUMMARY

This chapter has discussed some of the issues involved in making analog signals suitable for digitization by the laboratory interface; amplification, DC offset, filtering, trigger synchronization, and electrical interference. Appropriate signal conditioning can make a considerable difference to the quality of digital recordings and should not be neglected. Many suppliers of computer systems for the electrophysiological laboratory are beginning to recognize the need to provide suitable signal conditioning systems as well as the laboratory interface and the applications software. Ideally, the computer software should be able to control, or at least read, amplifier/filter settings. This is beginning to be possible with computer-controlled signal conditioners such as Axon Instruments CyberAmps and CED's 1902 unit, and this is a trend which is likely to continue.

CHAPTER FOUR

Signal analysis: measurement of waveform characteristics

One of the main advantages of recording signals on a computer system is the ability to perform a wide range of quantitative measurements faster, and to a higher degree of accuracy, than could be attained using earlier manual methods. The details of the signal analysis and the exact measurements made depend to some extent on the kind of signal under study. The procedures applied to EPCs will, for instance, differ markedly from those applied to the analysis single ion channel currents. Even for the same type of signal only simple measures such as the signal peak amplitude may be required whereas in other cases detailed analysis of the signal time course must be performed. There are however common elements to many procedures. This chapter and the following chapter cover these general aspects of the signal analysis procedure. Later chapters will cover the procedures appropriate to particularly important methods such as for single-channel analysis and methods such as non-linear curve fitting which require a detailed treatment.

4.1 PRIMARY AND SECONDARY SIGNAL ANALYSIS

The analysis of analog signal waveforms can be thought of as a process of data refining where the essential features of a waveform are extracted and represented in a more refined form. As shown in Figure 4.1, there may be several distinct stages to the process, each stage operating on the results of the previous, producing a higher level product. The electrophysiological signals, originally in analog form, are converted into digital form by the process of digitization, resulting in a series of digitized records. The *primary analysis* stage is focused upon the individual digitized record and performs the extraction of the interesting features within the waveforms stored within the records. This consists of the following activities:

- *Inspection and validation* of the signal record(s) as being acceptable for quantitative analysis.

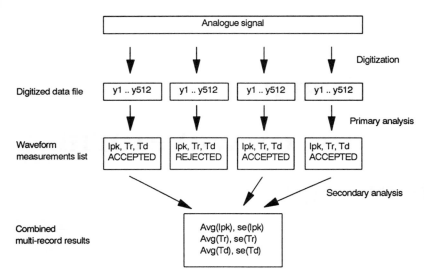

Figure 4.1 Stages of signal analysis. Analog signals are digitized and stored on disc as records of 512 A/D samples. Characteristic waveform features (I_{pk}, T_r, T_d) for each acceptable record are extracted in the primary analysis stage and stored in the measurements list. Means and standard deviations for each parameter within the set of records are computed during the secondary analysis stage.

- Quantitative signal *measurement* procedures for deriving the *characteristic features* of signal waveforms.
- Procedures used to *enhance* the quality of signals such as signal averaging.

Primary analysis results in a list containing the characteristic waveform measurements for each valid record within the data file. This effects a significant amount of data reduction, by condensing the information within each 512 sample record into a small set of characteristic parameters.

Secondary analysis procedures operate on the measurements list and are focused on the set of records as a whole with the aim of assessing the variability between records, either with simple repetition or under different experimental conditions. Typically, statistical procedures come into use at the secondary analysis stage, a simple example being the computation of the mean and standard deviation of the data within the measurement list. Secondary analysis procedures will be discussed in detail in Chapter 5.

4.2 INSPECTION OF SIGNAL RECORDS

One of the most important aspects of the primary analysis stage is the visual inspection and validation of the digitized signal records. It will be well known to anyone who is involved in electrophysiological experimentation (or in almost any branch of the life sciences) that the results of experiments are often highly variable. There are many reasons why recordings may vary between one experiment and the next – simple quantitative biological variations between animals or cells within a single animal, flaws in the experimental procedure such as damage to the cell due to poor microelectrode impalements, blocking of electrodes, instability in the voltage-clamp apparatus. Electrophysiological recordings cannot therefore be automatically assumed to be perfect representations of the true behaviour of the cell or tissue under study.

Before any meaningful quantitative analysis of such signals can be made, a qualitative assessment of the validity of the signals must be performed by the experimenter. In the days before computer analysis, this kind of assessment was inherent in the analysis procedures used. Signals would be photographed from oscilloscope screens and measured with a ruler or calipers from the enlarged film image. Although this manual procedure was exceedingly time consuming, it did have the advantage of forcing the experimenter to look at each and every record which was being included in the analysis.

A provision for the inspection and validation of signals is not, however, inherent in analysis

Table 4.1 PC display screen graphics resolution.

Display adapter	Horizontal	Vertical	Colours
IBM colour graphics adapter (CGA)	640	200	2
Hercules monochrome display	720	348	2
IBM enhanced graphics adapter (EGA)	640	350	16
IBM video graphics array (VGA)	640	480	16
IBM extended VGA (XGA)	1024	768	256
Apple Macintosh IIsi	640	480	256

by computer. Firstly, the records are not immediately observable, being stored as series of numbers on magnetic discs. Secondly, the nature of the analysis procedure is dependent on the computer software which could, in theory, be capable of measuring and analysing recordings without ever displaying the actual signal waveforms to the experimenter. A computer program will usually produce some kind of result no matter how absurd the data fed into it and it is not always possible to tell that an error has occurred from the final results. *A provision for the inspection and validation of digital recordings is therefore an essential feature of good signal analysis software.*

4.3 COMPUTER GRAPHIC DISPLAYS

Computers, when they first began to be used in the laboratory in the form of the minicomputer (DEC PDP8, PDP11), had a very limited capability of displaying data graphically since the video display units (VDU) or teletypewriters could only display alphanumeric characters. It was therefore common practice at that time to display digitized signals on oscilloscope screens driven by the laboratory interface digital-to-analog converters (Cooper, 1977; Bourne, 1981). This was quite satisfactory but generally resulted in the graphical and alphanumerical data being displayed on different screens.

The oscilloscope display method has now been made largely unnecessary by the rapid improvement of the graphical display capabilities of the personal computers which have replaced the minicomputer in the laboratory. A PC display screen operates in much the same way as a television screen. A screen of alphanumeric characters and graphics is generated by rapidly moving the electron beam in a series of horizontal sweeps from the top to the bottom of the screen while turning the beam intensity on and off to form a pattern of illuminated dots on the screen. The number of lines per screen and number of dots per line determine the vertical and horizontal resolution of the displayed image. Display resolution and the number of displayable colours have constantly improved since the initial development of the PC, as illustrated in Table 4.1 which shows the commonly available display screens for the IBM PC and Apple Macintosh family of computers.

The current standard for resolution is that of the IBM video graphics array (VGA) display which has a resolution of 640 (horizontal) by 480 (vertical) and can display in 16 colours. This is a quality of display resolution which proves to be quite satisfactory for general purpose use. Higher resolution displays can certainly be obtained (e.g. 1024 × 1024) but at increased cost and size of display. Due to the resolving power of the human eye, 19″ or larger display screens rather than the common 12″ or 14″ are needed to make effective use of the enhanced resolution.

4.4 DISPLAYING DIGITIZED SIGNAL RECORDS

Figure 4.2 shows a display screen from the signal inspection section of the author's synaptic current analysis program SCAN, as it would be seen by the operator. A series of endplate current records are in the process of being inspected. The screen is split into three windows containing the signal record, a menu of options, and status

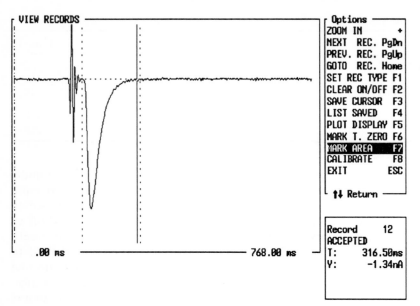

Figure 4.2 Display screen from a synaptic current analysis program (SCAN) used for visual inspection and validation of digitized recordings. The operator selects options from the menu on the right. A 512 sample record of an EPC is shown.

information. The graphical image on the screen consists of a 640 × 480 rectangular set of dots (pixels) each of which can be illuminated individually in one of 16 colours. The image of a digitized signal record is displayed on the screen by finding and turning on the appropriate set of pixels based on the values of the digitized samples. For instance, the record in Figure 4.1 consists of 512 samples each of which contains a value (0–4095, 12 bit ADC resolution) specifying the amplitude of the signal. To plot this record to lie within a rectangle bounded by screen coordinates (X_{min}, Y_{min}) and (X_{max}, Y_{max}), a set of (X, Y) screen coordinates are obtained for each ADC sample y_i in the record using the following equations:

$$X_i = \frac{(X_{max} - X_{min})(i - 1)}{n} + X_{min} \quad [4.1]$$

$$Y_i = \frac{(Y_{max} - Y_{min}) y_i}{4096} + Y_{min} \quad [4.2]$$

where n is the number of samples in the record (512).

It should be noted that the vertical resolution of the display screen (480) is not sufficient to represent all the possible values of the 12 bit ADC samples (4096), causing a significant deterioration in amplitude resolution of the displayed image compared to the original digitized record. For the purpose of inspecting digitized signals this does not prove to be a great problem as long as the signal is displayed over the full area of the screen as in Figure 4.1. Measurements, on the other hand, should always be taken from the orginal digitized record rather than from the screen coordinates (X, Y). Another way round the screen resolution problem is to display only a magnified part of the record. The following scaling equations can be used to display a magnified image of the part of the digitized between samples $i_{min} \ldots i_{max}$ and between amplitude levels $y_{min} \ldots y_{max}$

$$X_i = \frac{(X_{max} - X_{min})(i - i_{min})}{(i_{max} - i_{min} + 1)} + X_{min} \quad [4.3]$$

$$Y_i = \frac{(Y_{max} - Y_{min})(y_i - y_{min})}{(y_{max} - y_{min})} + Y_{min} \quad [4.4]$$

4.5 HARD COPIES OF SIGNAL RECORDS

While the screen display of signal records is satisfactory for inspection and measurement purposes, there is almost always a need to produce a paper copy of at least some of the records. A variety of *hard copy* devices can be used for this purpose, such as the dot matrix printer, laser printer and digital pen plotter. Each of these devices use quite different kinds of technology which determines their strengths and limitations.

The dot matrix printer is an inexpensive low to moderate quality printing device which was extensively developed and popularized by the rise of the personal computer. Characters are formed on the page by the impact of a line of print pins pressing through an inked ribbon as the print head moves across the paper. Such printers have two distinct modes of operation. In *text* mode the printer produces alphanumeric characters from a small variety of built-in typefaces. In *graphics* mode, each pin in the print head is directly controllable allowing complete images to be printed. Dot matrix printers are raster graphics devices like the display screen, with the image expressed as a pattern of dots rather than lines (vectors).

In order to plot a graph on a dot matrix printer it is necessary to generate a *bit map* describing the dots to be printed and then send that to the printer line by line. An industry standard exists for this purpose in the form of the printer codes, originally developed by Epson for their range of dot matrix printers, and which have been adopted by the majority of printer manufacturers in this market area. A simple way to produce a bit map is to use the one that already exists for the display screen and transfer that to the printer, a process known as a *screen dump*. The MS-DOS operating system provides such a screen dump program (GRAPHICS.COM). A distinct limitation of screen dumps, however, is that they produce images with a resolution no better than that available on the display screen (640 × 480). The dot matrix printer itself can produce images with a resolution of 120 DPI (dots per inch) which on an A4 page is equivalent to 1020 × 1400.

To exploit the available printer resolution fully, a specific printer bit image has to be created directly from the digitized signal values, just as was done for the display screen. This *rasterization* process is a somewhat complex task, since there is no simple way of drawing lines, symbols, or even alphanumeric characters when the printer is in graphics mode.

Higher resolution and better formed lines can be achieved using a *digital plotter*, such as the Hewlett Packard HP7470A or HP7475. These devices draw images on single sheets of paper using fine fibre-tipped pens, under servo-mechanical control, by the computer. Plotting instructions are issued (usually via the RS232 communications line) to the plotter, in the form of a graph plotting language which specifies the pen movements required to create the plot. Hewlett Packard plotters use a code called the *Hewlett Packard Graphics Language (HPGL)*. HPGL has become a standard in this area and is widely supported. A particular advantage of plotters is that they accept instructions in *vector* format, with simple commands for pen movement, line drawing and character generation, making them much easier to programme than dot matrix printers. Digital plotters are also capable of higher resolution with the pen position controllable to a fraction of a millimetre. The HP7475, for instance, has a theoretical resolution of 10 900 × 7650 over an A4 page.

Laser printers apply the kind of technology developed for photocopying to high quality computer printing. A raster image is written by a laser beam on to a photosensitive drum which picks up toner where the image is and transfers it to the paper. Laser printers are capable of producing very high quality print in a variety of type styles and sizes and high resolution graphics (300 DPI). They vary in terms of cost and printing speed but there are two main standards in the area; the Hewlett Packard Laserjet family (Laserjet II, IIP, III, IIIP) and Postscript printers.

The basic Laserjet printer resembles a dot matrix printer in terms of functionality, being a raster printing device with text and bit map graphics modes. Often only a restricted range of typefaces and sizes is available. Graphs can be plotted on these printers but the same kind of bit map generation problems exist as for the dot matrix printers. However, some of the latest Laserjet printers (Laserjet III, IIIP) now also support a version of the HPGL language, allowing them to emulate a digital plotter. This greatly enhances their usefulness in the laboratory.

Adobe Postscript is a *page description language* in the same sense that HPGL is a graphics language. It can be used to specify in very exact terms the shape, size and position of characters and lines on the page. Laser printers, such as the Apple Laserwriters, which can interpret the Postscript language, can usually produce a much wider range of typefaces and sizes than the simpler Laserjet printers. Postscript is also a very effective vector-based graph plotting language. Overall, the Postscript laser printer currently is the best single choice for both printing and plotting within the laboratory, with the ability to print both high quality text and graphics.

4.6 MEASUREMENT OF SIGNAL CHARACTERISTICS

A typical experiment produces a set of signal records obtained under varying experimental conditions, in the presence and absence of drugs for instance, or at different cell membrane potentials. There is a wide variety of measurements that can be made depending on the specific nature of the electrophysiological signals under study and on the reasons for performing the analysis. However, a very large category of measurements can be classed as quantification of a transient signal's waveform characteristics and require similar types of procedures. Synaptic currents and potentials, nerve and cardiac action potentials, voltage-activated ionic currents and single ion channels all fall into this category. In general terms they can be said to be signals which are transient in nature, occur in response to some stimulus (not necessarily under the control of the experimenter), rise to a peak or a steady state, and then terminate, perhaps with some characteristic decay waveform. For such signals one or more of the following measurement parameters are often required for each individual signal record:

- peak amplitude (positive, negative, absolute)
- steady-state amplitude
- time to rise to peak
- rate of rise to peak
- duration
- frequency of occurrence.

Figure 4.3 shows sets of typical waveform characteristic measurements for different types of electrophysiogical signals.

4.7 CARDIAC ACTION POTENTIAL

Figure 4.3 (a) shows a set of measurements made on a cardiac muscle action potential (CAP). The CAP has a rapid rise to a sharp peak within a few milliseconds, followed by a decay to a long lasting plateau phase and finally a decline which overshoots then returns to the resting potential. The CAP is a complex waveform generated by the interaction of Na^+, Ca^{2+} and several different K^+ conductances with different time courses (Noble, 1975). Characterization of its waveform requires several parameters; maximum rate of rise dV/dt_{max}, peak amplitude AP_{pk}, duration in terms of time taken to decay to 50% and 90% of the peak (APD_{50}, APD_{90}), and hyperpolarizing overshoot, AP_{pk-}. Each of these measurements provide information on different aspects of the functioning of the ion conductances in the cell membrane. Variations in dV/dt_{max} and AP_{pk} primarily provide information on the effects on the Na^+ conductance. Changes in APD_{50} and APD_{90} indicate actions on Ca^{2+} or K^+ conductances.

4.8 VOLTAGE-ACTIVATED CURRENTS

Measurements made on action potentials provide much useful information on the potential physiological effects of agents acting on ion channels. However, it is often necessary to use the voltage clamp technique to obtain more detailed information on precisely which species of ion channel is being affected by the drug in such experiments. Figure 4.3 (b) shows a voltage-activated Ca^{2+} current from a voltage clamped

Signal analysis: measurement of waveform characteristics

of the ion channel. The analysis of voltage-activated currents is discussed in detail in Chapter 7.

4.9 ENDPLATE CURRENTS

The final example, Figure 4.3 (c), is a synaptic current, a voltage-clamped nerve-evoked endplate current (EPC) from rat skeletal muscle. The EPC waveform rises rapidly to a peak value then decays to zero again with an exponential time course. It is characterized by its peak amplitude I_{pk}, 10–90% rise time T_{rise} and the time constant τ of the exponential decay. Although often simpler in form than the action potential or voltage-activated currents the EPC, and particularly its peak amplitude I_{pk} can be modified by a wide range of factors of both pre- and post-synaptic origin.

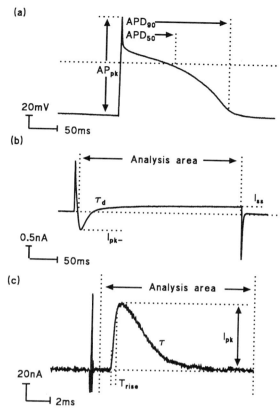

Figure 4.3 Characteristic parameters of typical electrophysiological signals. (a) Peak amplitude (AP_{pk}) and duration (APD_{50}, APD_{90}) of a cardiac muscle action potential. (b) Peak (I_{pk}), steady-state (I_{ss}) amplitude and decay time constant (τ_d) of a voltage-clamped cardiac muscle Ca^{2+} current. (c) Peak amplitude (I_{pk}), rise time (T_{rise}) and decay time constant (τ) of a nerve-evoked endplate current. (Records courtesy of Drs G. Boachie-Ansah and T. Searl, Strathclyde Univsersity.)

cardiac muscle cell. In this case the currents rise to a peak following a sigmoidal path and then decay exponentially to a steady-state value. The current can be characterized by the peak current I_{pk}, rise time (time taken to rise from 10% to 90% of the peak) T_{rise}, steady-state current I_{ss}, and the time constant of the exponential decay from peak to steady-state, τ_d. The dependence of these parameters on the magnitude of the voltage-step provides information on the kinetic behaviour of the ion channels supporting the current and allows the formulation and testing of varieties of hypothetical models for the gating

4.10 SEMI-MANUAL MEASUREMENT USING SCREEN CURSORS

The simplest form of computer-based measurement tool for deriving waveform characteristics is the movable cursor readout of the screen display of digitized signal records. A typical implementation of the technique can be seen in the display screen that was shown in Figure 4.2. A vertical cursor is placed on the screen over the displayed signal record. It can be shifted back and forth along the record by pressing appropriate functions keys, in this case the left and right arrow keys on the cursor key pad. For any given cursor position, the numerical value of the corresponding ADC sample in the digitized record is displayed (bottom right in Figure 4.2). The user can therefore measure the value of each individual sample within the displayed record. Characteristic parameters such as I_{pk} can be measured for each record simply by moving the cursor along the signal until the maximum value is found.

4.11 CALIBRATION OF MEASUREMENTS

It is usually preferable to display the cursor readout values in terms of the signal being studied (e.g. mV for membrane potentials, pA, nA or µA for currents) rather than in the integer ADC units in which the data is actually stored. In addition, the signal level must also be related to the zero level of the electrophysiological signal. It is difficult to avoid introducing spurious offset voltages (e.g. microelectrode tip potentials) into measured electrophysiological signals and, consequently, the measured absolute zero voltage level bears no direct relationship to the zero level of the quantity being measured. Three values are therefore required for a completely calibrated cursor measurement:

- ADC sample value at the cursor position
- ADC value for the zero level
- scaling factor relating ADC values to real signal units

The signal level y is related to the binary integer level adc for each sample by the following equation

$$y_i = g \frac{V_{max} - V_{min}}{adc_{max} - adc_{min} + 1} (adc_i - adc_z) \quad [4.5]$$

where V_{max} and V_{min} are the maximum and minimum limits of the the input voltage range, adc_{max} and adc_{min} are the limits of the corresponding integer ADC levels, adc_z is the zero level relative to which measurements are being made, g is the conversion factor which determines the measured signal unit in terms of volts. A typical 12 bit ADC might quantize analog voltages over a range of ±5 V into 4096 integer levels producing a integer result over the range of 0–4095. The conversion factor g is dependent on the recording and signal conditioning instrumentation. It would normally be entered into the program by the user as part of the setting up procedure for recording. In Figure 4.1, for instance, the EPC is a current signal with units of nanoamperes and a conversion factor of 10 mV/nA is in use. Some software, notably pCLAMP, when used in combination with AxoClamp or AxoPatch amplifiers, has the facility to acquire gain information directly by reading voltage levels on the 'gain sender' outputs of these devices.

4.12 DEFINING THE ZERO REFERENCE LEVEL

Two quite different approaches can be taken to the definition of the zero level adc_z, depending on what kind of measurements are required. If measurements are to be made relative to a defined point within the signal record then a value of adc_z can be calculated directly from each individual record, i.e. a *record-relative* zero level. Alternatively, if measurements of, say, the absolute cell membrane potential are required, each individual record may not contain this information itself and an external *absolute* zero level is required. The difference between the two types of zero reference level can be illustrated by practical examples.

In some types of recordings, particularly of synaptic currents like the EPC in Figure 4.3 (c), the signals of interest are transient waveforms superimposed on an otherwise unrelated steady baseline level. In such cases, measurements of signal amplitudes are made relative to this baseline rather than the actual zero current level. This is done by setting adc_z to the average of a series of ADC samples selected from a portion of the record preceding the actual signal. For instance, in Figure 4.3 (c) adc_z has been calculated from the average of the first 20 ADC samples at the beginning of the record. It is not unusual to find that the baseline level may drift during the recording of a long series of records. Use of a record-relative zero level conveniently compensates for such baseline drift.

There are however other situations where the absolute value of the signal level is the primary quantity of interest. The cardiac action potential in Figure 4.3 (a) is treated this way. The membrane potential is measured in terms of the absolute cell membrane potential rather than relative to a baseline within the record. A zero level cannot be obtained directly from the signal record since there is no clearly definable region at which the signal is known to be zero. In such cases, it would be necessary to collect a zero level calibration record (perhaps recorded before the microelectrodes were inserted into the cell). A single value of adc_z would then be derived from that single record and applied to all the rest.

4.13 AUTOMATED PARAMETER MEASUREMENT

While the cursor-based approach to signal measurement has the merit of simplicity, it involves the operator in many repetitive operations on each individual record. This may be quite satisfactory for the analysis of a small number of records but can be very time consuming when large numbers (e.g. 1000–2000) of records are involved. For such cases, it is preferable to automate the process of measuring waveform characteristics, using software to seek out and measure, for instance, the signal peak amplitude without any user involvement. However, automated waveform measurement raises the possibility of flawed records corrupting the results without the user's awareness. It is still necessary to inspect each record visually before allowing it to be passed for automatic analysis. In the SCAN program (see Figure 4.2 again) the operator displays each record individually on the screen and, using the 'Set Rec. Type' menu option, marks valid records as ACCEPTED and faulty records as REJECTED. This information is permanently stored in the file along with the actual recording.

It will be apparent that the criteria for acceptance or rejection of a signal record is a key issue. Some might argue that the facility to exclude records from analysis is a dangerous thing, likely to lead to biased results with the operator (perhaps even unconsciously) choosing only the records that fit some preconceived theory. This is always a possibility that must be guarded against, but the complexity and imperfections of electrophysiological recordings make it difficult to avoid having to make some kind of selection.

4.14 ACCEPTANCE/REJECTION PROCEDURES

To take the analysis of MEPPs as an example, Figure 4.4 shows a series of records all obtained from the same frog neuromuscular junction. Figure 4.4 (a) and (b) show what are generally regarded to be 'normal' MEPP signals, in that they have fairly rapid rise times (<1 ms) and decay exponentially. Certain rejection criteria are straightforward and uncontroversial, such as the elimination of records corrupted by external interference or 'glitches', like that in Figure 4.4 (c). Similarly, some records may be 'normal' but unsuitable for the proposed analysis procedure. Figure 4.4 (d) shows a record containing two MEPPs which have occurred almost simultaneously and are therefore superimposed (a not unexpected event for signals occurring randomly at a high enough rate). Measuring the peak height of such an event is scarcely representative

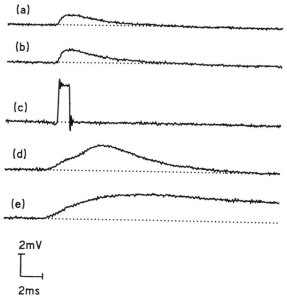

Figure 4.4 Acceptance/rejection of MEPPs from frog skeletal muscle neuromuscular junction. (a) and (b) are typical 'normal' MEPPs which would be accepted for analysis. (c), (d) and (e) would be rejected; (c) is a 'glitch' due to external interference, (d) is the summation of two near-simultaneous MEPPs, and (e) is a large slow MEPP of unknown origin. (Data courtesy of Professor I.G. Marshall, Strathclyde University.)

of the height of a single MEPP. Although it might certainly be possible to devise a measurement procedure which could unscramble the heights of the two signals, if such multiple events occur relatively infrequently it is convenient simply to exclude them from the analysis.

On the other hand, there are signals which, although differing from the 'normal' (i.e. most common) MEPPs, cannot be rejected on any simple criteria. Figure 4.4 (e) shows an example of a signal from the same recording which is more than twice as large, has a slower rise, and is much longer lasting than the MEPPs in (a) and (b). In this recording such signals comprised about 4–10% of the total records. The origin of such 'abnormal' signals is not clear, but they have often been noted in the past (Gage & McBurney, 1975). Such records create a dilemma, in that there is no reason to believe them to be artifacts, but their profound difference from the 'normal' MEPPs makes it difficult to combine them in a single analysis meaningfully. In such circumstances it is important to define exactly what aspects of the recording are being analysed. If it is the 'normal' MEPPs that are being studied it is correct to reject the large 'abnormal' MEPPs. However, when the results of such an analysis are presented later, it is essential that the basis of the rejection criteria is made explicit and the existence of the 'abnormal' category admitted.

Clearly, 'normal' and 'abnormal' are value judgements and are expressions of what the experimenter is interested in studying and what question the experiment was designed to answer. Acceptance/rejection decisions are therefore unavoidable and are a central part of the analysis procedure and interpretation of experimental results. The rejection criteria for the MEPP experiment can be summarized:

- reject all records known to be artefacts
- reject records technically unsuitable for analysis
- reject records containing events outside the remit of the experiment
- abandon the analysis if an excessive number of rejections have to be made
- state rejection criteria when presenting results.

4.15 SOFTWARE FOR AUTOMATED MEASUREMENT

Once each record has been marked as ACCEPTED or REJECTED by the qualitative acceptance criteria, the automated analysis procedures can be safely applied. A typical set of automated measurement algorithms are shown in Listing 4.1. The routine fills a set of arrays with lists of peak positive amplitude, rise time, area and average level for each ACCEPTED record.

Even though bad records have been excluded, care must still be taken to ensure that the computer is measuring the right thing within the signal record. It is often necessary to exclude certain portions of the range of samples within the signal record from the analysis. Many electrophysiological signal records contain large stimulus artifacts as a consequence of the electrical stimulation involved in producing the signal. An example of such an artifact can be seen in Figure 4.3 (c) in the large spike immediately preceding the EPC. An automated peak-finding procedure would have no means of distinguishing between the peak of the artifact (usually of no intrinsic interest) and the peak of the signal with the larger of the two being returned. Consequently, the range of samples within the record which are to be analysed must be restricted to exclude the stimulus artifact. Such a feature is implemented in Listing 4.1.

Listing 4.1 searches through the selected subseries of records within a data file. For each record, the set of integer ADC samples (512 in this case) is read into an array. The same zero level considerations apply to automatic analysis as for the cursor measurements. In this case a record-relative zero is used, with adc_z derived from the average of the first 20 ADC samples within the record. All signal amplitudes are measured relative to this point. The location of the signal peak amplitude is then found by searching the array for the maximum value (within the analysis limits $i\_start \ldots i\_end$). The rise time is measured as the time taken for the signal to rise from 10% to 90% of the peak value and is computed by stepping backwards starting at the peak and counting the number of samples

Figure 4.5 Different approaches to data file structures for storing waveform measurements. (a) SCAN data file integrating digitized signal record and waveform measurements in a single file. Each file consists of a header block (H) (defining record sizes, number of records in file, etc.) and a set of records containing an analysis block (A) with waveform measurements and a sampled data block (S) containing the original digitized signal. (b) pCLAMP file structure where the digitized signal records in the data file produced by CLAMPEX are analysed by CLAMPAN or CLAMPFIT. Waveform measurements are stored in a variety of separate files.

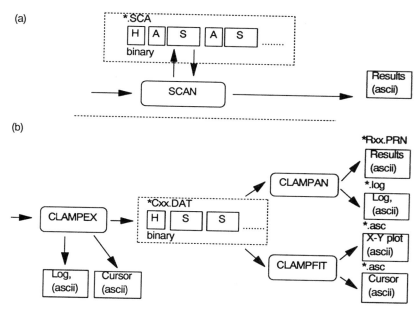

between when the level drops below 90% peak until it reaches 10%.

These analysis routines have been simplified to better illustrate the principles involved. They are therefore limited in that they are restricted to the analysis of positive-going signals and the assumption is made that first 20 ADC samples in each record can always be used as the zero level. It is, however, a quite straightforward process to add a negative peak search and rise time procedure, and to allow a more flexible definition of the zero level.

4.16 THE MEASUREMENTS LIST

The end product of the primary signal analysis stage is a set of characteristic measurements for each valid signal record within the digitized data file. These results may now be subjected to further stages of analysis, particularly statistical analysis, to investigate the differences between records. The measurements list may be stored in a variety of formats depending on the design of the signal analysis program. It is possible to integrate the measurement data into the original digitized data file if space has been set aside for this purpose. This is the approach taken within the SCAN program as shown in Figure 4.5 (a). Each record within the digitized data file consists of a 512 byte analysis block (A) in addition to the data block (S) containing the ADC samples. The analysis block is used to contain the ACCEPTED/REJECTED status of each record, and the range of waveform measurements performed by SCAN. Integrating the measurement list has the advantage of keeping all the information together within an single file, but it inevitably leads to a more complicated data file structure.

A quite different approach is taken by the pCLAMP software as shown in Figure 4.5 (b). Unlike SCAN, signal recording and analysis functions are performed separately by different programs. Records are digitized by the CLAMPEX program and stored in its binary format data file (*cxx.DAT). Signal analysis is performed using either the CLAMPAN or CLAMPFIT programs which operate upon the binary data file. The waveform measurements are stored in a variety of different results files. This approach allows simpler programs and file structures to be used at the expense of the proliferation of large numbers of separate files.

Each approach has its own merits and limitations. Electrophysiological data analysis often involves the processing of large amounts of data, often running into thousands of individual signal records. Even small improvements in the efficiency and ease of use of the software can be worthwhile. Integrating recording, primary waveform measurement and secondary statistical analysis procedures into a single program allows these operations to be carried out with a minimum of steps for the user without constantly having to change programs. On the other hand, it is not practical to include all possible analysis procedures in a single program. Such a program would become large and unwieldy, even if the time were available to write the software. SCAN is a highly customized program intended to handle large numbers of synaptic current records efficiently while the pCLAMP package is intended to be a general-purpose electrophysiological analysis program.

4.17 EXPORTING DATA AND RESULTS

Most signal analysis programs such as SCAN or pCLAMP are designed mainly for primary analysis of the signal waveform and often have limited secondary statistical analysis and may also lack the ability to produce high quality graphs of results. It is therefore necessary to be able to pass results generated by the primary analysis program to other programs for further analysis. This may involve the use of spreadsheets, statistical analysis, or scientific graph plotting programs. While it is clearly possible to print out the waveform measurement results and then manually re-enter the data into such software via the keyboard this is a slow and error prone process. It is much preferable to avoid having to retype the results by directly *importing* the measurement list into the statistics program. However, in order to do this, the program to receive the data must be able to interpret the file containing the list of measurements. In this respect complex highly integrated file structures, such as the SCAN data file, which are unique to a particular program, may prove a disadvantage since they cannot be readily interpreted by other packages. It is therefore essential that programs using data files with unusual structures be able to *export* results in a more commonly understood format.

Table 4.2. An ASCII code text file containing a list of current and voltage measurements for four records.

'Record' <sp> 'Current nA' <sp> 'Volts mV' <cr> <lf>
1 <sp> −3.2 <sp> −40.1 <cr> <lf>
2 <sp> −4.1 <sp> −50.3 <cr> <lf>
3 <sp> −5.2 <sp> −60.1 <cr> <lf>
4 <sp> −6.2 <sp> −70.3 <cr> <lf>
5 <sp> −7.0 <sp> −79.8 <cr> <lf>
6 <sp> −8.2 <sp> −80.1 <cr> <lf>
<eof>

Special ASCII characters	*Name*	*ASCII code (decimal)*
<cr>	carriage return	13
<lf>	line feed	10
<sp>	space	20
<tab>	tab	9
<eof> =	end of file	26

Data can be represented within a computer system in a variety for forms or codings which vary between different computer systems and different programs. The need for a common standard to allow different systems to communicate has been long recognized and resulted in the ASCII code (American Standard Code for Information Interchange) which assigns a number to each letter of the alphabet, number and some punctuation marks. The ASCII code was initially developed to allow computers to provide a common standard for displaying text on VDUs. It was adopted for use on the personal computer and now forms the basis of the coding for almost all computer-based alphanumeric text, including most word processors. A table of ASCII code numbers can be found in the operations manual of most computer systems. For this reason almost all data analysis programs have the capability to import ASCII-coded data files.

A simple example of an ASCII-coded file containing a short measurements list is shown in Table 4.2. A series of current and voltage measurements have been obtained from six digitized signal records. There are three items of information per record, record number, current and voltage. Most data analysis programs such as spreadsheets expect such data to be stored in a tabular format with each row corresponding to a record and each column to a measurement. A row is generally defined as a series of ASCII characters terminated by the pair of carriage return <cr> and line feed <lf> characters. Columns are separated by one or more *white space* characters (<sp> or <tab>) or punctuation characters (comma, semi-colon, etc.). The ASCII <eof> (end of file) code is used to indicate the end of the table. If textual information such as the column labels shown in the example are to be included in the file, it is important to distinguish between the white space and punctuation which might occur within the text and that indicating column boundaries. This is commonly done by placing the text within a pair of quotation marks, as shown.

The pCLAMP software makes extensive use of ASCII format files with all data files, with the exception of the files containing the digitized signal records, stored in this format. These files can thus be readily imported into a wide range of secondary analysis programs. The tabular format of such ASCII files corresponds closely to the row/column format of spreadsheet programs. Almost all of such programs (e.g. Lotus 1-2-3, Borland Quattro) can easily import data into their own internal data file formats. Similarly, scientific graph plotting programs such as Biosoft Fig.P or Jandel Sigmaplot, can often selectively extract pairs of columns from the ASCII file and import them into the graph file. The ASCII format is also sufficiently general to allow data to be exported to computer systems other than the IBM PC family, such as the Apple Macintosh or Sun workstations. However, some additional care may be required in defining the file format to ensure compatibility. Macintosh software, for instance, often requires the columns to be separated by <tab> characters rather than any white space character. Similarly, rows of ASCII text under the UNIX operating system are terminated by a single <lf> character rather than the <cr> <lf> pair.

Given the advantages of ASCII format files, it might be asked why any other file format should be used at all. There is however, a penalty to be paid for its use. ASCII-coded files do not store data as efficiently as binary code files. For instance, a 12 bit ADC sample is a number between 0–4095. Such a number requires 2 bytes of storage space in binary form, but 5 bytes in ASCII form (four digits and a white space character). ASCII coding therefore requires at least 100% more storage space than binary coding. ASCII-coded files are also much slower to read and write than binary format. Computations can only be performed in binary format, therefore ASCII-coded data must be transformed into internal format and back again whenever the file is read or written. This can take 10 times as long as directly reading a binary format file. For these reasons it is usually preferable to store digitized signal records in binary format.

4.18 SIGNAL AVERAGING

The digital storage of the electrophysiological signal provides the opportunity to perform a wide range of operations to enhance the signal and to reduce background noise. Probably, the most common technique is *signal averaging*, which can be applied whenever it is possible to obtain a repeated series of essentially similar signals. The data in a digitized record can be considered as consisting of two components, the signal itself and a randomly fluctuating backgound noise. An average record can be computed by summating the sample values at corresponding positions within each record and dividing by the number of records, n:

$$y_{\text{avg}}(i) = \frac{1}{n} \sum_{j=1}^{n} y_j(i) \qquad [4.6]$$
$$[i = 1 \ldots 512]$$

A FORTRAN subroutine for averaging a series of signal records, using this algorithm, is shown in Listing 4.2. If the background noise is truely random there will be no correlation between the noise component of one record and the next. In the process of summation, noise components are likely to cancel out because at each sample point the noise may be of positive sign in one record and negative in the next. On the other hand, the signal component *is* correlated between records, therefore sample points will either be all positive or all negative and no cancelling occurs. Averaging therefore preserves the signal while reducing random background noise.

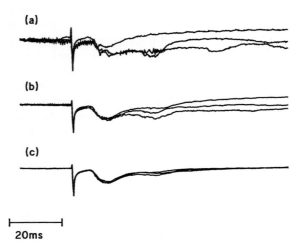

Figure 4.6 Averages of stimulus-evoked potentials. (a) Individual evoked potentials recorded from rat cortex. (b) Point-by-point averages of sets of four EVPs. (c) Averages of sets of 16 EVPs. (Courtesy Dr O. Holmes, University of Glasgow.)

Signal averaging can be shown to its most dramatic advantage in situations where a signal is of low amplitude relative to background noise, but can be repeatedly evoked by some stimulus. In such cases averaging of repeated stimuli can recover a clear image of a signal which is often barely recognizable in each individual sweep. This is a technique often used in neurophysiological recording of extracellular evoked potentials (EVPs) within the central nervous system. Figure 4.6 shows a series of EVPs recorded from an electrode inserted into the cerebral cortex of a rat, evoked by stimulation of the rat's forepaw. The individual stimuli in Figure 4.6 (a) produce quite small responses on top of a noisy and fluctuating baseline. Looking at individual records such as these it is difficult to distinguish the general shape of the evoked response following the stimulus. The response varies markedly between records preventing any meaningful measurement. Figure 4.6 (b) shows the result of point-by-point averaging individual responses in sets of four. A clear stimulus-induced negative-going potential now becomes apparent, but there is still a high degree of variability later in the record. Raising the average to 16 records, as shown in (c), finally produces a clear image of the evoked potential with little variation between averages. It can be shown that the background noise is reduced in proportion to the square root of the number of records averaged.

It might be noted that signal averaging effects a smoothing of the signal similar to that obtained by the low-pass filtering described in Chapter 3. However, unlike filtering, averaging has the advantage that the reduction in background noise is obtained without adversely affecting the time course of the signal itself. Signal averaging has been used in many areas of electrophysiology as a means of extracting otherwise unanalysably small signals from background noise. It has proved particularly useful in the recording and enhancement of the very small 100–300 µV postsynaptic potentials (PSP) encountered in recordings from central nervous system neurones where many hundreds of PSPs may be averaged at a time (Jack *et al.* 1981; Redman & Walmsley, 1983).

4.19 AVERAGING OF DETECTED SPONTANEOUS SIGNAL

In order to perform averaging, a means must exist to accurately superimpose the samples of corresponding points on each signal record. With stimulus-evoked signals, this is straightforward since the acquisition of each record can be accurately synchronized with the stimulus pulse (assuming that the latency between stimulus and response does not vary excessively between records). Not all electrophysiological signals to which averaging might be applied have a convenient external synchronizing point, spontaneously occurring MEPCs being a typical example. Similarly, some stimulus-evoked signals can have a large and significantly varying latency between stimulus and response (e.g. Brock & Cunnane, 1988). In order to apply averaging to such signals successfully without distorting the signal shape, a synchronizing point on the record itself must be used. Records are therefore adjusted to superimpose their alignment points before being added to the average.

The choice of the alignment points depends on the shape of the signal and care must be taken to find a stable and easily detectable point on the waveform. This is usually on the leading edge since most commonly encountered signals rise more rapidly than they decay. In practice, the mid-point of the rising phase rather than the peak should be used. Although the peak is the seemingly obvious choice, it proves to be a poor alignment point since it creates an unintended artifact in the average correlated with the background noise. This is illustrated in Figure 4.7 which shows the averages of a series of MEPC records (collected using the event detection procedure described in Section 2.16), aligned using the signal peak level (b) and the mid-point of the rising phase (c). It can be seen that peak alignment results in a small rapidly decaying exponential phase at the peak of the signal not present in the mid-point alignment, or in individual records (a). This effect occurs because the act of selecting the peak level of the record removes one of the criteria for noise reduction by averaging; that the background noise signal should not be correlated with the signal. Therefore at the peak of the average, the background noise cancellation is less effective, overemphasizing the height of the peak. Since the background noise is random, this effect decays quite rapidly away from the alignment point, resulting in the extra exponential decay component seen in Figure 4.7 (b). Selection of the mid-point of the rising phase is less sensitive to background noise and avoids this effect.

Listing 4.3 shows an averaging subroutine modified to align signal records by the mid-point of their rising phases. The mid-point is found by first determining the size and location of the signal peak level within each record then stepping backwards sample by sample until the signal level falls below 50% of the peak. Each record is added to the summation array, with its sample points slightly offset to cancel out the difference between the new record's mid-point and the mid-point of the average (defined as the mid-point of the first record processed). Unlike the simple averaging routine in Listing 4.2, Listing 4.3 is limited to the averaging of positive-going signals by the design of the mid-point detection. If negative-going signals are to be averaged, then it is necessary to invert the signal record before processing, then re-invert the resulting average.

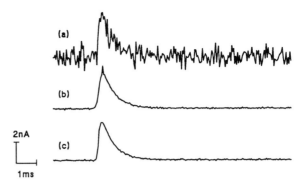

Figure 4.7 Detection and averaging of spontaneous signals (MEPCs). (a) Single detected MEPC. (b) Average of 200 MEPCs aligned by peak amplitude (note spike artefact at peak). (c) Average of the same MEPCs as (b) but aligned by the mid-point of the rising phase. (Data courtesy of Dr T. Searl, University of Strathclyde.)

4.20 DIGITAL FILTERS

The digital averaging techniques so far discussed can only be applied to signals which can be evoked repeatedly. However, if only a single record exists, it is possible to obtain some smoothing of the signal by averaging adjacent sample points using the *n-point moving average* algorithm,

$$y(i) = \frac{1}{n} \sum_{j=i-n+1}^{i} x_j \qquad [4.7]$$

where x_i is a point from the original signal record and n is the number of points to be averaged. The results of applying Equation 4.7 with $n = 10$ to a 255-point signal record containing an MEPC is shown in Figure 4.8. The moving average is essentially a low pass *digital filter* algorithm and has many features in common with the analog low pass filters discussed in Chapter 3. It has similar limitations, as can be seen from Figure 4.8. The digital filter has greatly reduced the background noise on the signal record but it has also distorted the signal (cf. Figure 3.3).

The moving average is one of the simplest of a wide range of possible digital filter algorithms based upon the linear equation

$$y(i) = \sum_{j=1}^{n} \alpha_j x_{i-j} + \sum_{k=1}^{m} \beta_k y_{i-k} \qquad [4.8]$$

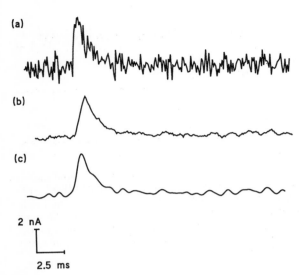

Figure 4.8 Digital low-pass filters. (a) Raw unfiltered 255 sample digitized record of an MEPC. (b) Results of a 10-point moving average filter (Equation 4.7). (c) Results of a gaussian filter with $f_c = 0.05$ (Equation 4.9).

where the output y_i is obtained as the weighted sum of not only the n most recent input samples $x_{i-1} \ldots x_{i-n}$ but also the m most recent output values $y_{i-1} \ldots y_{i-m}$. The filter response characteristics are defined by the values of the α_j and β_k weighting factors. Filters can be implemented using α or β factors alone or combinations of both. Filters containing only β factors are described as *recursive* since they operate upon previous instances of the actual filter output. Filters with only α factors, such as the moving average, are described as *non-recursive*. Generally, recursive digital filter designs require fewer terms in the summation than non-recursive designs with similar performances. However, the process of feeding back the output signal into the filter algorithm has a potential for instability which the non-recursive design does not. The tendency of recursive filters to produce prolonged oscillations when supplied with sharp pulse input leads them to be described as *infinite impulse response (IIR)* filters. In comparison, non-recursive filters are described as *finite impulse response (FIR)* filters.

4.21 THE GAUSSIAN FILTER

Essentially the same issues of sharpness of frequency response cut-off and signal distortion apply to digital filter algorithms as to analog filters and they are described in much the same terms: amplitude, phase response, and transfer functions. Digital filter algorithms with contrasting characteristics, like the analog Bessel and Butterworth designs, can be found. As before, it is usually more important to avoid distortion of the signal than to achieve the sharpest cut-off. This is generally easier to achieve using non-recursive designs. In practice, the moving average filter, although simple to implement, is neither efficient in operation nor provides the best frequency response. A better alternative is the *gaussian* filter, so called because the filter coefficients are derived from a symmetrical set of $2n + 1$ coeffcents based upon the gaussian function, i.e.

$$y(i) = \frac{1}{\sqrt{(2\pi\sigma^2)}} \sum_{j=-n}^{n} \exp\left(\frac{-j^2}{2\sigma^2}\right) x_{i+j} \quad [4.9]$$

The frequency response characteristics of the filter are determined by the standard deviation σ of the gaussian function which can be related to the filter cut-off frequency by

$$\sigma = \frac{0.1325}{f_c} \quad [4.10]$$

where f_c is expressed in units of the sampling rate. The gaussian filter has similar characteristics to the Bessel analog filter. The results of the gaussian filter with $f_c = 0.05$ applied to the MEPC can be seen in Figure 4.8 (c). It achieves the same improvement in background noise with less distortion of the MEPC than the moving average. Further discussion of the gaussian filter, particularly as applied to the analysis of single-channel currents can be found in Colquhoun & Sigworth (1983) and Sachs (1983). A gaussian filter subroutine based upon one described there by Sigworth is presented in Listing 4.4. The uses of digital filters applied to other electrophysiological signals are also discussed by Wilkison (1991). A basic introduction to the design and use of digital filters can be found in Lynn & Fuerst (1989).

4.22 DIGITAL SIGNAL PROCESSORS

Digital techniques are gaining popularity in many areas of signal processing due to a much greater flexibility in design possibilities. Increasingly, filters are being implemented using *digital signal processors (DSP)* such as the Texas Instruments TMS32000 series or the Motorola DSP56000. The DSP is a special purpose microprocessor adapted for the processing of analog signals and the implementation of digital filter algorithms. When fitted with appropriate A/D and D/A converters the DSP can perform all of the functions of a complex multi-stage analog filter. Some general-purpose computer systems such as the Next or the Silicon Graphics Indigo workstations have digital signal processors fitted as standard and several DSPs are available as plug-in cards for the IBM PC (Ariel PC-56 or DSP-16; Amplicon DAP 2400) or Apple Macintosh (National Instruments NB-DSP2300). DSPs are not limited to digital filtering and are beginning to be used for tasks such as neural event recognition (Willming & Wheeler, 1990). It is likely that, as they become increasingly common components of a computer system, many more uses will be found for them.

4.23 SIGNAL ANALYSIS SOFTWARE PACKAGES

Most laboratory interface suppliers now recognize the need also to provide software to operate them. The nature and scope of this software varies greatly, from simple digital oscilloscope programs to complex signal analysis packages. A distinction also has to be made between general packages (usually with a bias towards engineering) and those specifically designed for the analysis of electrophysiological signals.

The Axon Instruments pCLAMP package is probably the most commonly used of the commercially available packages designed specifically for the analysis of electrophysiological signals. Software produced by Cambridge Electronic Design for their CED1401 interface is also in widespread use, particularly in the UK. Similar packages are produced by RC Electronics (ComputerScope-Phy.) and Intracel (ADCAD). All of the above programs run on the IBM PC family of computers. Packages also exist for the Apple Macintosh, notably the MacLab system distributed by World Precision Instruments and AXOLAB from Axon Instruments.

Software is often offered on a non-commercial basis (sometimes free of charge) by a number of research laboratories (Robinson & Giles, 1986). This software may often be of high quality but, as might be expected, limited support and documentation may be provided. An exception to this is the author's Strathclyde Electrophysiology Software project at the University of Strathclyde (Glasgow, UK) which was set up with the intention of supplying a range of electrophysiological analysis software on a non-commercial basis.

Most packages provide features such as digital averaging, sometimes digital filtering, but vary greatly in signal measurement facilities. Both SCAN and pCLAMP, discussed earlier in this chapter, provide automated signal measurement. Other software, such as CED's SIGAVG (SIGnal AVeraGer) does not. It should be noted that there is no single 'best' signal analysis program. Programs, even those purporting to be of general use, are often designed with specific kinds of task in mind. Consequently, some programs excel in particular areas and lack features for others. For instance, SCAN, although usable for a number of purposes, was designed to analyse large numbers of endplate currents. It has an efficient automated measurement system and a means of making ACCEPT/REJECT classifications of each record. pCLAMP, on the other hand, has a less effective measurement system, being designed more with voltage-activated currents in mind. However, pCLAMP can record up to four channels while SCAN only accepts one.

4.24 SOFTWARE DEVELOPMENT ENVIRONMENTS

The software discussed so far have all been applications programs. It is now more common to use such programs than to develop software of one's own. This has been an almost inevitable trend. Software development is difficult and time consuming. If suitable analysis software is readily available there is little merit in re-inventing yet more similar programs. However, unusual experimental methods and protocols cannot always be handled by existing software packages. It is in such circumstances that there is still a need to develop software within the laboratory.

The applications programs discussed earlier were all developed from scratch, using standard computer languages, such as FORTRAN, PASCAL or Microsoft QuickBasic, and 8086 assembler code. It has taken many years for such packages to reach their current state of sophistication. Much of this time is usually spent in learning to program the laboratory interface directly and developing a library of subroutines for various purposes; e.g. data acquisition, graphical display, signal analysis. This 'from the ground up' approach is not feasible for the general researcher who wants to quickly develop an occasional very specific application.

A number of suppliers have recognized this problem and now provide *software development environments* for data acquisition and analysis which greatly simplify the development of custom software. These environments are often based upon an existing language such as QuickBasic or C, extending it by including a library of pre-written data acquisition and analysis functions. Currently, the three best known environments are Axon Instrument's AxoBasic, National Instrument's LabWindows and Keithly Asyst.

AxoBasic is essentially a library of routines which adds data acquisition, display and array arithmetic functions to Microsoft QuickBasic. Its strength lies in its similarity to the BASIC-23 language which was widely used by electrophysiologists to develop software for the PDP11, and for which many programs still exist. LabWindows is similar in concept to AxoBasic, but considerably more sophisticated and general purpose. While AxoBasic simply provides an extended set of commands for QuickBasic, LabWindows includes a menu-driven program generator which greatly simplifies the creation of programs even by novices. It provides support for a pop-up menu system, graphics, data analysis and data acquisition facilities such as continuous sampling-to-disc. It can generate program in either the QuickBasic or C languages. Asyst was one of the first data acquisition programming environments. Unlike the other two it uses its own specific language, not unlike the FORTH computer language, rather than extending a standard one.

National Instruments produce a similar development environment for the Macintosh called LabView, which exploits the graphical user interface of that computer to further extend and simplify the program generator concept. LabView programs are created by connecting together various program elements graphically on the screen. 'Virtual Instruments' can be created in this way with a front panel shown on the screen with buttons, meters and oscilloscope displays.

4.25 SUMMARY

This chapter has outlined some of the basic analysis techniques applied to electrophysiological signals: characteristic waveform measurements, signal averaging and digital filtering. The important role of the operator in making qualitative decisions on the validity of signals has also been emphasized. Some possible sources for obtaining software have been discussed. Commercially available software has improved greatly in quality and sophistication over the past 5 years and is likely to continue doing so. The appearance of good program development environments such as LabWindows or AxoBasic has also made it easier to develop 'in-house' software within the laboratory.

Listing 4.1. Automatic waveform measurement routine.

```fortran
      subroutine measure(ir_start,ir_end,i_start,i_end,scale,dt,
     & record,peak,avg,trise,area,nresults)
      integer*2 ir_start,ir_end          ! (In) Range of records to be measured
      integer*2 i_start,i_end            ! (In) Range of ADC points to be analysed
      real*4 scale                       ! (In) ADC-calibration factor
      real*4 dt                          ! (In) ADC sampling interval (ms)
      real*4 record(1),peak(1),trise(1),avg(1),area(1)  ! Results arrays (Out)
      integer*2 nresults                 ! No. of records analysed (Out)
      parameter(nsectors_per_record=3,nsectors_per_header=1,nbytes_per_sector=512)
      integer*2 istatus(256),iadc(512)   ! Record status and data buffers
      character*8 status                 ! (First 8 bytes of istatus buffer
      equivalence( status,istatus)       !    contains 'ACCEPTED' or 'REJECTED'
      open(unit=1,file='data.sca',access='direct',form='binary',recl=512)
      do irecord = ir_start,ir_end
       isector = (irecord-1)*nsectors_per_record + nsectors_per_header + 1
       read(unit=1,rec=isector) (istatus(i),i=1,256)
       if( status .eq. 'ACCEPTED' ) then         ! Only analyse "accepted" records
         read( unit=1, rec=isector+1 ) (iadc(i),i=1,512)
         nresults = nresults + 1
         record(nresults) = float(irecord)
         sum = 0.                                ! Find zero level for
         do i = 1,20                             ! record.
            sum = sum + float( iadc(i) )         ! Computed as average
         end do                                  ! of first 20 ADC points
         izero = int( sum/20. )
         ipeak = -huge( iadc(1) )
         do i = i_start,i_end                    ! Find most positive value
            if( ipeak .lt. iadc(i) ) then        ! of the ADC points
                ipeak = iadc(i)                  ! within the analysis
                ipeaki = i                       ! range defined by
            endif                                ! <i_start> & <i_end>
         end do
         ipeak = ipeak - izero
         peak(nresults) = float(ipeak)*scale
         sum = 0.                                ! Find the average value
         do i = i_start,i_end                    ! and total area
            sum = sum + float(iadc(i)-izero)     ! of the ADC points within
         end do                                  ! the analysis area
         avg(nresults) = scale*sum/float(i_end-i_start+1)
         area(nresults) = scale*sum*dt
         i10 = ipeak/10                          ! Calculate time
         i90 = ipeak - i10                       ! taken to rise from
         i = ipeaki                              ! 10% to 90% of peak
         irise = 0
         do while( (iadc(i)-izero.ge.i10) .and. (i.ge.i_start) )
             if( iadc(i)-izero .le. i90 ) then
              irise = irise + 1
             endif
             i = i - 1
         end do
         trise(nresults) = float(irise)*dt
       endif
      end do
      return
      close(unit=1)
      end
```

```fortran
c      Listing 4.2. Signal averaging routine.
c      Averages a series of records from a data file produced by the SCAN program.
c      File format; HSDDSDDSDDSDD... where H= 512 byte file header block,
c      S= 512 byte record status block, D= 512 byte A/D data block.
c
       subroutine average_record( iaverage, ir_start, ir_end )
       integer*2 ir_start,ir_end       ! (In) Range of records to be averaged
       integer*2 iaverage(1)           ! (Out) average to main program
       integer*2 navg                  ! (Out) No. of record averaged

       integer*4 isum(512)             ! Summation array (Note 32 bit integer)
       parameter(nsectors_per_record=3,nsectors_per_header=1, nbytes_per_sector=512)
       integer*2 istatus(256),iadc(512) ! Record status and data buffers
       character*8 status              ! (First 8 bytes of istatus buffer
       equivalence( status,istatus )   !  contains 'ACCEPTED' or 'REJECTED'

       open(unit=1,file='data.sca',access='direct',form='binary', recl=512)
       np = nbytes_per_sector*(nsectors_per_record-1)/2

       do i = 1,np                     ! Clear summation array
         isum(i) = 0                   ! and number averaged
       end do
       navg = 0
c
c      Add each record that is marked as ACCEPTED to summation array
c
       do irecord = ir_start,ir_end
         isector = (irecord-1)*nsectors_per_record + nsectors_per_header + 1
         read(unit=1,rec=isector) (istatus(i),i=1,256)
         if( status .eq. 'ACCEPTED' ) then        ! Only analyse "accepted" records
           read( unit=1, rec=isector+1 ) (iadc(i),i=1,np)
           do i = 1,np
              isum(i) = isum(i) + iadc(i)
           end do
           navg = navg + 1
         endif
       end do
c
c      Calculate average (Note test to avoid divide by zero)
c
       if( navg .gt. 0 ) then
         do i = 1,np
            iaverage(i) = isum(i) / navg
         end do
       endif
       close(unit=1)
       return
       end
```

Signal analysis: measurement of waveform characteristics

```fortran
c       Listing 4.3. Signal average with alignment of rising phases.
c       (Same file structure as Listing 4.2)
        subroutine average_record_by_midpoint( iaverage, ir_start, ir_end, navg )
        integer*2 ir_start,ir_end            ! (In) Range of records to be averaged
        integer*2 iaverage(1)                ! (Out) average to main program
        integer*2 navg                       ! (Out) No. of record averaged
        parameter(nsectors_per_record=3,nsectors_per_header=1, nbytes_per_sector=512)
        integer*2 istatus(256),iadc(512)     ! Record status and data buffers
        character*8 status                   ! (First 8 bytes of istatus buffer
        equivalence( status,istatus)         !  contains 'ACCEPTED' or 'REJECTED')
        integer*4 isum(512)                  ! Summation array (Note 32 bit integer)
        integer*4 izero                      ! Zero level
        open(unit=1,file='data.sca',access='direct',form='binary', recl=512)
        np = nbytes_per_sector*(nsectors_per_record-1)/2
        do i = 1,np                          ! Clear summation array
           isum(i) = 0                       ! and number averaged
        end do
        navg = 0
        imidj = 0
        do irecord = ir_start,ir_end
           isector = (irecord-1)*nsectors_per_record + nsectors_per_header + 1
           read(unit=1,rec=isector) (istatus(i),i=1,256)
           if( status .eq. 'ACCEPTED' ) then  ! Only analyse "accepted" records
             read( unit=1, rec=isector+1 ) (iadc(i),i=1,np)
             izero = 0
             nzero = 10
             do i = 1,nzero                   ! Calculate the zero level
                izero = izero + iadc(i)       ! for this record as
             end do                           ! the average of the first
             izero = izero / nzero            ! 10 A/D samples.
             ipeak = -32767
             do i = 1,np                      ! Find the peak value
                if( iadc(i) .gt. ipeak ) then ! and its location within
                    ipeak = iadc(i)           ! the record.
                    ipeaki = i
                endif
             end do
             imid = (ipeak - izero)/2 + izero ! Find the mid-point of the rising phase
             imidi = ipeaki
             do while( (iadc(imidi).gt.imid) .and. imidi.gt.1 )
                 imidi = imidi - 1
             end do
             if( imidj .eq. 0 ) imidj = imidi
             do j= 1,np                       ! Add new record
                i = max(min(j-imidj+imidi,np),1) ! to summation array
                isum(j) = isum(j) + iadc(i)   ! with shift to align
             end do                           ! mid-points.
             navg = navg + 1                  ! ( Note checks to ensure
           endif                              !   that summation keeps
        end do                                !   within array bounds)
        if( navg .gt. 0 ) then                ! Calculate average
           do i = 1,np
              iaverage(i) = isum(i) / navg
           end do
        end if
        close(unit=1)
        return
        end
```

```fortran
c       Listing 4.4. Gaussian digital filter.
c       (based on Sigworth, 1983)
        subroutine gaussian_filter(ix,iy,np,fc)
        integer*2 ix(np)                    ! (In)  Input signal array
        integer*2 iy(np)                    ! (Out) Filtered output array
        integer*2 np                        ! (In)  No. of sample points
        real*4 fc                           ! (In)  Filter cut-off frequency (samples-1)
        parameter(max_nc=54)
        real*4 a(max_nc)                    ! Filter coefficients array

        sigma = .132505/fc
        if( sigma .ge. .62 ) then
c
c       Generate positive half of filter coefficients
c                    1              j2
c       a(j) =   -------  exp( - ------ )     j=0,nc
c                √(2πσ2)           2σ2
c
        b = -1./(2.*sigma*sigma)             ! Create coefficients until
        nc = 1                               ! limits of numerical precision
        a(nc) = 1.                           ! is reached.
        sum = .5
        temp = 1.
        do while( (temp.ge.2.*epsilon(temp)) .and. (nc.lt.max_nc) )
           nc = nc + 1
           temp = exp( float(nc-1)*float(nc-1)*b )
           a(nc) = temp
           sum = sum + temp
        end do

        sum = sum*2.                         ! Normalize coefficients
        do i = 1,nc                          ! so that they summate to 1.
           a(i) = a(i)/sum
        end do
        nc = nc - 1
      else
c
c       Special case for very light filtering (See Colquhoun & Sigworth, 1983)
c
        a(2) = (sigma*sigma)/2.
        a(1) = 1. - 2.*a(2)
        nc = 1
      end if

        do i = 1,np                          ! Apply filter
           sum = 0.                          ! to data in ix(1..np)
           do j = i-nc,i+nc                  ! and place in iy(1..np)
              l = max(min(j,np),1)           ! Note how filter
              k = iabs(j-i)+1                ! summation is truncated
              sum = sum + float(ix(l))*a(k)  ! at each end of the array
           end do
           iy(i) = int(sum)
        end do
        return
        end
```

CHAPTER FIVE

Statistical analysis and presentation of results

The analysis procedures to be discussed in this chapter begin at the point where the previous signal analysis chapter leaves off – the lists of characteristic waveform measurements which were the end result of the primary signal analysis process. Here, we discuss the basic methods of presentation of experimental results and the statistical procedures used to summarize them in quantitative terms and draw conclusions. This can be considered to be the final stage of the process of information refining started by the digitization of the analog signal and signal waveform analysis.

The manner of analysis and presentation of data has considerable bearing on what kind of information can be extracted from experimental results. Procedures vary greatly, but it is useful to group them together into the following loose classification:

- analysis of effects of a specific treatment on the cell
- analysis of time-dependent trends
- analysis of the distribution of waveform measurements.

Any or all of these approaches may be applied during a series of experiments. The word 'treatment' is used in a particularly wide sense here. It may be the application of a drug, changes to the cell membrane potential, bathing solution composition or temperature, even the application of a particular stimulus pattern to the cells. In any case, the common factor is that a systematic modification is applied to the preparation under study with the aim of observing and quantifying its effects on cell behaviour. Such designs have wide application, particularly in pharmacological studies to determine the effects of a drug quantitatively. An example might be the effects of curare on neuromuscular transmission, measured by recording nerve-evoked endplate currents before and after the application of the drug.

Treatment-type experiments of the type just discussed are based upon the assumption that the properties of the recorded waveforms (e.g. amplitude, time course, etc.) remain constant unless the experimental conditions are changed. However, it is not uncommon to find that an electrophysiological signal may change on repetition in a systematic way, decreasing in amplitude for instance. Such trends in a waveform measurement may yield further insight into the mechanisms underlying the electrophysiological signal under study. Similarly, even when a systematic record-to-record trend is not apparent, there may still be useful information inherent in the apparently random variations in a waveform measurement that can be revealed by the analysis of the frequency distribution of the measurements.

At this stage, statistics play a central role in the analysis process by allowing the development and testing of specific quantitative hypotheses against the experimental observations. An overview of the common statistical procedures as applied to typical results from electrophysiological experiments is therefore presented. Attention is focused upon the computational procedures without detailed derivation. A more complete introduction to the theory and practice of statistics can be found in many textbooks (Colquhoun, 1971; Lee & Lee, 1982; Daniel, 1987).

5.1 SPREADSHEETS

Before proceeding to the actual statistical techniques, it is worth discussing the basic procedures for storing and manipulating the numerical data. These are simple tables of numbers but can often be quite extensive, running to hundreds or thousands of items. A class of programs known as *spreadsheets* prove to be well suited to the handling of this kind of data. Spreadsheet programs were initially developed as a means of automating accountancy procedures (Schuyler, 1985), the term spreadsheet referring to a large sheet of paper divided into rows and columns into which figures were entered and manually computed. The development of spreadsheet programs contributed to the early success of the microcomputer, particularly the Apple II with the VISICALC spreadsheet and the IBM PC with Lotus 1-2-3. The spreadsheet has proved to be a powerful tool for the storage and manipulation of numerical data, applicable to many areas in addition to accountancy, particularly the analysis of scientific results (Matheny, 1984; Hewtt, 1986). Numerous programs are available commercially with Lotus 1-2-3, Borland Quattro and Microsoft Excel being among the most widely used.

The spreadsheet program provides a computer-based equivalent of the paper spreadsheet in the form of a table into which data can be entered. The table is divided into columns (A–Z) and rows (1–99999) so that each table entry or *cell* has its own unique address (A1, B10, Z999, etc.). Each entry can contain a number, a textual label, or a *formula* defining a mathematical operation. The contents of a cell or column of cells can be made a function of the contents of other cells. In addition built-in mathematical *functions* are available which can be made to operate upon defined blocks of cells. For instance, the function @ avg (a1 . . . a100) computes the average of the block of cells from a1 to a100. A list of some of the functions useful for statistical analysis are shown in Table 5.1. It is possible to produce very large spreadsheets containing hundreds of columns and thousands of rows, providing a means of tabulating all the results from a complete series of experiments, automatically calculating means, standard deviations, etc. Spreadsheets also have the capability of displaying tabulated information graphically

Table 5.1 Some of the statistical and mathematical spreadsheet functions from the Borland Quattro Pro spreadsheet, operating upon the block of cells A1 . . . A100.

@count(a1 . . . a100)	Number of entries in block a1 . . . a100
@sum(a1 . . . a100)	Sum total of a1 . . . a100
@avg(a1 . . . a100)	Arithmetic mean of a1 . . . a100
@stds(a1 . . . a100)	Sample standard deviation of a1 . . . a100
@vars(a1 . . . a100)	Sample variance of a1 . . . a100
@sqrt(a1)	Square root of a1
@exp(a1)	Exponential of a1
@ln(a1)	Natural logarithm of a1

Figure 5.1 A PC screen display of the Borland Quattro Pro spreadsheet containing a simple set of experimental data. The program's ability to show data in numeric and graphical form simultaneously is illustrated.

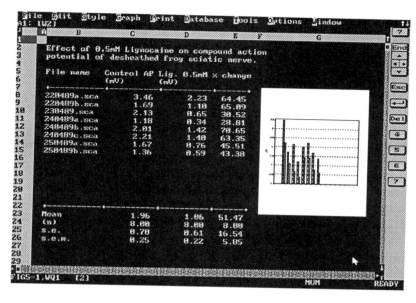

in a variety of different formats such line graphs, bar charts and pie charts.

One of the greatest advantages of the spreadsheet is its ability to combine numeric, textual and graphical information within the same format. This allows it to be used as an electronic laboratory notebook. Figure 5.1 shows a display screen from the Borland Quattro Pro V2.0 spreadsheet being used to store and analyse results from a series of eight experiments to investigate the effects of the local anaesthetic lignocaine on the extracellular action potential of frog sciatic nerve. Series of action potentials have been recorded both in the presence and absence of 0.5 mM lignocaine. The peak amplitudes of the action potentials have been computed using the signal analysis techniques discussed in Chapter 4, resulting in a series of eight control and eight drug-treated amplitudes. Control amplitudes are stored in column C and drug-treated in D. Note how a large amount of explanatory information stored with the numerical results makes it possible to work backwards to the raw results. In particular, column B contains the names of the data files containing the digitized recording. Similarly, a description of the experiment has been placed in cells B1 and B2. Quattro also has the useful facility to embed graphical data into the spreadsheet display. A bar chart of the 16 control/drug data pairs is shown adjacent to the table of values.

5.2 DESCRIPTIVE STATISTICS

The data in the spreadsheet in Figure 5.1 illustrates the usual variability of results from biological experiments. The amplitudes of the eight control action potentials show a marked variation between cells (1.18–3.46 mV). In such circumstances, relatively little can be inferred from comparisons between single instances of control and treated groups, since it is not possible to distinguish drug-induced effects from inherent variability. However, by observing the results from a repeated series of experiment, the magnitude of the variabilty becomes apparent and reliable estimation of the control and drug-treated amplitudes becomes possible. Both the variability and the 'best' estimates for experimental measurements are summarized using *descriptive statistics*, the most common of these being the

- arithmetic mean
- median
- mode

- standard deviation
- standard error of the mean.

The mean, median and mode are measures of *central tendency*, intended to provide a single value representative of all the values within a group of observations. The mean is probably the most commonly used

$$m_x = \frac{1}{n} \sum_{i=1}^{n} x_i \qquad [5.1]$$

where n is the number of observations in the group. The median is the mid-value when the observations are ranked into order of size and the mode is the most commonly occurring value within the group.

The variability of the measurements within each group can be quantified by the *variance* – the average of the squared differences between the individual measurements and the mean calculated from,

$$Var = \frac{1}{n-1} \sum_{i=1}^{n} (x_i - m_x)^2 \qquad [5.2]$$

Note that the sum of the squared differences is divided by $n-1$ to form the variance rather than simply the number of observations n. This accounts for the fact that the mean value m_x used in Equation 5.2 is itself computed from the sample group, effectively reducing the number of independent measurements by one. In this respect, the variance is said to have $n-1$ *degrees of freedom*. Without this correction for degrees of freedom, the variance computed from small samples would tend to underestimate the true variance of the underlying population. The variance is expressed in terms of squared units of the quantity being measured. A more convenient expression of the variability, in the same units as the mean, is the square root of the variance, *the standard deviation*

$$s_x = \sqrt{Var} \qquad [5.3]$$

It is useful to calculate the expected variability of the mean itself, provided by the standard error of the mean,

$$s.e.m. = \frac{s_x}{\sqrt{n}} \qquad [5.4]$$

The mean, standard deviation, and standard error computed for the control and treated groups can be seen in Figure 5.1. The theory underlying these equations and a more detailed introduction to statistics using spreadsheets can be found in Mezei (1990) or Soper & Lee (1990). Similar discussions but with examples using conventional programming languages can be found in (Lee & Lee, 1982, BASIC; Davies, 1971, FORTRAN).

5.3 THE NORMAL DISTRIBUTION

Descriptive statistics, as their name suggests, describe the nature of the observations within the data sets. The eight control measurements in our example have a mean value of 1.96 mV and span the range 1.18–3.46 mV. However, clearly it would be naive to think that these values have determined the range of all possible values. Another sample of eight measurements would, most likely, yield a different set of values for the descriptive statistics. To be of practical use, we need to know to what extent we can rely upon the results obtained from this sample, or put another way, a means of predicting from the sample statistics the likely distribution of all possible measurements that, hypothetically, one could make. In order to do this it becomes

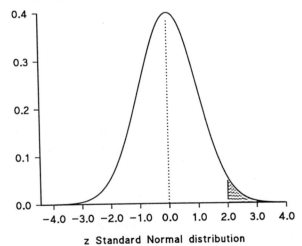

Figure 5.2 Normal probability density function, generated from Equation 5.5 with $\mu = 0$, $\sigma = 1$. Tail probability, $p\ (2 \leq z)$ shown as shaded area.

necessary to make certain assumptions about this underlying distribution.

If an experiment, such as the measurement of AP amplitude discussed here, is repeated many thousands of times under the same experimental conditions it is not unusual (although not universal) to find that the measurements are distributed in the fashion similar to that shown in Figure 5.2. The majority of measurements are clustered about the mean, becoming increasingly rare away from that value, with the bell-shaped distribution shown. The shape of distributions such as this can be represented mathematically by a *probability density function*, which defines the probability of obtaining any particular experimental value. The distribution in Figure 5.2 is the *normal* or *gaussian* probability density function.

$$y(x) = \frac{1}{\sqrt{(2\pi\sigma^2)}} \exp\left[\frac{-(x-\mu)^2}{2\sigma^2}\right] \quad [5.5]$$

where μ is the mean value and σ the standard deviation for the population of measurements. If the probability density function is known for a population it is possible to determine the probability that a measurement will lie within a particular range of values by integrating the pdf within these values

$$p(x_l \leq x \leq x_h) = \int_{x_l}^{x_h} y(x)\,dx \quad [5.6]$$

Using this approach it can be shown that about 68% of the measurements lie within one standard deviation of the mean value ($\mu \pm \sigma$) and that 99.7% lie within three standard deviations. Such limits constitute *confidence intervals* within which it can be said with a certain probability that all measurements will lie, the 95% confidence interval being approximately two standard deviations. Similarly, the *tail probability* of obtaining a measurement which exceeds a particular value can be computed from the integral

$$p(\geq x) = \int_{x_l}^{\infty} y(x)\,dx \quad [5.7]$$

For instance, $p(\geq 2)$ is shown by the shaded area underneath the pdf curve in Figure 5.2. Tables of the area underneath the normal distribution curve can be found in most statistical text books (Daniel, 1987; Lee & Lee, 1982). The tables are normally presented as cumulative distributions of the standardized normal curve ($p(\leq z)$, $\mu = 0$, $\sigma = 1$) where z is

$$z = \frac{(x-\mu)}{\sigma} \quad [5.8]$$

The initial reason for the development and publishing of statistical tables was to avoid the lengthy integral computations involved in the calculation of the normal probability. With the use of computers this is no longer an issue. In fact, it is inconvenient to have to look up a table manually when the rest of the computations have been done automatically. The normal area can be computed by direct numerical integration of Equation 5.5 or by one of a number of approximation formulae (Thistead, 1988; Press et al., 1986). The following formula to find the probability $p(\geq z)$ is based upon the Applied Statistics Algorithm AS66 by Hill (1985).

$$p(\geq z) = \cfrac{b_1 \exp(-(z^2)/2)}{z - b_2 + \cfrac{b_3}{z + b_4 + \cfrac{b_5}{z - b_6 + \cfrac{b_7}{z + b_8 - \cfrac{b_9}{z + b_{10} + \cfrac{b_{11}}{z + b_{12}}}}}}} \quad [5.9]$$

The values of b_1–b_{12} are shown in Table 5.2. Equation 5.9 can be easily implemented using spreadsheet formulae with the constants b_1–b_{12} stored within a column. For values of z greater than 1.28, an accuracy of at least the arithmetic precision of the PC (six decimal figures) is attainable.

5.4 STUDENT'S *T*-DISTRIBUTION

It is difficult, in practice, to develop confidence intervals based upon Equations 5.5–5.9 because

Table 5.2 Constants for AS66 approximation to the normal tail probability.

b_1	0.398 942 280 385	b_2	3.805 2E-8	b_3	1.000 006 153 02		
b_4	3.980 647 94E-4	b_5	1.986 153 813 64	b_6	0.151 679 116 635		
b_7	5.293 303 249 26	b_8	4.838 591 280 8	b_9	15.150 897 245 1		
b_{10}	0.742 380 924 027	b_{11}	30.789 933 034	b_{12}	3.990 194 170 11		

they properly apply only to the situation where the parameters (μ, σ) of the underlying population distribution are known. This is not normally the case with only the mean and standard deviation of a limited sample being available. However, if it can be assumed that the underlying distribution is normal, it is possible to derive a pdf for the quantity

$$t = \frac{x - \mu}{\frac{s}{\sqrt{n}}} \qquad [5.10]$$

Student's t-distribution is a family of related distributions, with a similar bell shape to the normal distribution, but a width dependent on the number of degrees of freedom (i.e. independent observation, $n-1$, see Equation 5.2) used to compute the standard deviation s. When the number of degrees of freedom is large (e.g. >30) the t-distribution is indistinguishable from the normal distribution. Distributions with small degrees of freedom, however, tend to be more spread out with a narrower middle and wider tails. In essence, the t-distribution accounts for the extra uncertainty incurred by having to calculate the standard deviation from the sample groups.

Tail probabilities and confidence intervals for the t-distribution can be computed by integration in the same fashion as for the normal distribution. Most statistics textbooks contain tables of the critical t-values associated with particular tail probability areas (e.g. 0.01, 0.05, 0.1, etc.) for a variety of degrees of freedom (1–100).

For example, the t-value for a set of 95% confidence limits corresponds to tail areas of 0.025 at either end of the distribution. For a sample of eight measurements (i.e. seven degrees of freedom) the critical value can be obtained by looking up the value $t_{0.025,7}$ in the tables. The 95% confidence interval for the estimation of the mean is then given by

$$x \pm t_{0.025,7} \left(\frac{s}{\sqrt{n}}\right) \qquad [5.11]$$

which in the case of the example is 1.96 ± 0.59. A more detailed discussion of the theoretical basis and applications of the normal and the t-distribution can be found in Daniel (1987).

Confidence limits allow us to make definite statements about experimental results. In this case it can be asserted that there is a 95% probability that the mean value of the AP amplitude lies within the range 1.43–2.5 – given, of course, the assumption that the underlying distribution is normal. It is not unusual to find data to be normally distributed, but with small sample sizes (<100) it is difficult to verify normality rigorously and this is rarely done in normal experimental practice. The purist might say that this then completely undermines the use of the test. However, in practice, there is often little evidence to lead one to believe that the underlying population distribution is not normal. On the other hand, if the distribution of measurements within a sample shows a marked asymmetry, or if there is evidence of there being several peaks within the distribution, clearly some doubt must be cast upon the normality assumption. In such circumstances the confidence limits based upon the normal distribution are likely to be inaccurate.

Just as for the normal distribution, it is inconvenient to have to look up tables of t-distribution values when using a computer. Given the ability to find normal tail probabilities such as by using Equation 5.9, it is possible to find t or other distribution probabilities by transforming the t-value into an equivalent normal variate. Thistead (1988) suggests Peizer & Pratt's transform for this purpose

$$z = \left(n_f - \frac{2}{3} + \frac{1}{10n_f}\right) \sqrt{\frac{1}{n_f - 5/6} \log_e \left(1 + \frac{t^2}{n_f}\right)}$$

$$[5.12]$$

where n_f is the number of degrees of freedom. This transformation/normal approximation method can be seen applied in several of the statistical test spreadsheets to be discussed shortly.

5.5 STATISTICAL SIGNIFICANCE

In computing mean values and confidence intervals, we have considered the control and drug-treated groups within the example as separate entities. However, one of the prime reasons for doing the experiment was to determine what, if any, effect the application of 0.5 mM lignocaine had on AP amplitude. Examining the mean values for each group there does appear to be a difference between control amplitude (1.96 mV) and the amplitude in the presence of the drug (1.06 mV). The question naturally arises, bearing in mind the variability of the results, whether this difference is *significant*, i.e. actually due to the effects of the drug.

Significance, in statistical terms, has a very precise meaning. It is the probability that the observed results can be explained solely in terms of the inherent variability within the data without supposing any treatment effect at all. This can be expressed in terms of two mutually exclusive hypotheses:

- the *null hypothesis* that the observed difference can be wholly accounted for by the inherent variability of the measurements, and not due to any effect of the treatment; and
- the *alternate hypothesis* that the observed difference is due to a treatment effect.

It is not possible to prove the alternative hypothesis to be true directly, but it is possible to determine the plausibility of the null hypothesis. The competing hypotheses are tested by computing a *test statistic* from the observed measurements, for which the probability of the null hypothesis being true can be determined. Numerous different test statistics have been developed each with their own strengths and limitations.

In constructing a hypothesis test statisticians consider two types of possible error. A *Type 1* error is defined as concluding that a difference is significant when it is not, i.e. falsely rejecting the null hypothesis. A *Type 2* error, on the other hand, is failing to detect a significant difference when it actually occurs, i.e. falsely accepting the null hypothesis. If a test is to be of practical use, it is necessary to be able to define the probability of committing Type 1 or 2 errors. The Type 1 error probability is determined from the tail area of the probability density function for the test statistic and is also described as the significance probability. A typical criterion for significance is that the probability of a Type 1 error must be less than 0.05. The Type 2 error probability determines the *power* of a statistical test to detect differences in results, and is dependent on both the chosen Type 1 significance level and the number of measurements in the sample group.

5.6 STUDENT'S T-TEST

In the case of the lignocaine experiment, the null hypothesis is that the observed difference between the mean values of the control and treated groups is due only to random variations in AP amplitude. In such circumstances, it is common to use *Student's t-test* to determine the significance of the observed difference. The test statistic t is computed from the formula

$$t = \frac{m_t - m_c}{\sqrt{(s_g^2/n_t + s_g^2/n_c)}} \quad [5.13]$$

where s_g is a combined standard deviation derived from both the control and treated groups,

$$s_g = \sqrt{\frac{(n_t - 1)s_t^2 + (n_c - 1)s_c^2}{n_t + n_c - 2}} \quad [5.14]$$

Large differences between the group means relative to the group standard deviation produce large t-values, irrespective of the absolute magnitude of the group mean values. To determine the significance of t it is necessary to determine the probability of a value as large as that observed occurring by chance, assuming no real difference between groups. This can be obtained from the tail area

$$p(\geq t, n_f = n_c + n_t - 2) \quad [5.15]$$

from the t-distribution with n_f degrees of freedom.

The t-test can be easily implemented within a spreadsheet as shown in Figure 5.3 which contains the example data set from Figure 5.1. The control and drug-treated sample groups have been entered into columns A and B. Group means, pooled standard deviation, etc. and t-value are computed in column D. The t-distribution tail probability is computed using the transformation/approximation methods discussed earlier. The observed t-value in cell E13 is transformed into an equivalent z-value using Equation 5.12 (E14) which is then used to calculate the normal tail probability using Equation 5.9 implemented in E15. The details of the spreadsheet formulae can be found in Listing 5.1. A t-value of 2.738 has been computed for the difference between the mean values of control and drug sample groups. With 14 degrees of freedom, the probability of encountering a value as large or greater than this ($p(t \geq 2.738)$) is 0.0081. It is therefore safe to reject the null hypothesis and conclude that lignocaine has significantly depressed the AP amplitude.

There are certain conditions that have to be satisfied before the t-test can be used. First, the overall population of measurements from which the two samples are derived must adhere to a *normal* distribution. Second, the control and treated groups must not have markedly different standard deviations. The requirement for the control and treated groups to have the same standard deviation is at least as important as the normality condition, perhaps more so since it is more easily violated. In essence, for the t-test to be valid the treatment must only change the average level of the quantity being measured, not its variability. Whether the variability of a signal is preserved after a drug treatment depends very much on the source of the variation. If it is due to a source such as instrumentation noise it is not likely to be altered by a drug treatment of the cell. On the other hand if the variation is inherent in the cellular system under study then it may scale in proportion to the average signal

	A	B	C	D	E
1					
2			Student's T-test (unpaired)		
3					
4	Control	Drug	Results	Control	Drug
5	+-------+-------+-----------+-------+----+	
6	3.46	2.23	Mean	1.964	1.061
7	1.69	1.1	s.d.	0.704	0.611
8	2.13	0.65	s.e.m.	0.249	0.216
9	1.18	0.34	(n)	8.000	8.000
10	2.01	1.42	Difference (1-2)		0.902
11	2.21	1.4	Pooled s.d.		0.659
12	1.67	0.76	Degrees of freedom		14.000
13	1.36	0.59	Student's T		2.738
14			z (equivalent)		2.40792
15			p(>=t)		0.00802
16					
17				0.39894228	2.40792
18				3.8052E-08	2.89903
19				1.000006153	0.00802
20				0.000398065	
21				1.986153814	
22				0.151679117	
23				5.293303249	
24				4.838591281	
25				15.15089725	
26				0.742380924	
27				30.78993303	
28				3.99019417	

Figure 5.3 Unpaired t-test spreadsheet to determine the significance of the difference between two groups of measurements. t-distribution tail probability is computed within the sheet using the transformation and approximation formulae discussed in the text. Formulae in Listing 5.1.

amplitude. It is quite possible to have a treatment effect which both decreases the average and increases the variability. In such circumstances, a real effect of treatment might be classed as non-significant by the t-test.

In the case of unequal group standard deviations a modification to the t-test (Daniel, 1987) can be used. The t-value is computed as before, except that a pooled standard deviation is not used

$$t' = \frac{m_t - m_c}{\sqrt{(s_t^2/n_t + s_c^2/n_c)}} \quad [5.16]$$

The number of degrees of freedom is computed from (Mendenhall & Sincich, 1989)

$$n_f = \frac{(s_c^2/n_c + s_t^2/n_t)^2}{\dfrac{(s_c^2/n_c)^2}{n_c - 1} + \dfrac{(s_t^2/n_t)^2}{n_t - 1}} \quad [5.17]$$

Equation 5.16 should be used in preference to Equation 5.13 if group standard deviations differ by more than a factor of two.

5.7 THE PAIRED T-TEST

The t-test as applied to the AP amplitude data has so far ignored the fact that the data are paired. Each control/treated pair of measurements is the result of an experiment on a single nerve. As a general principle, it is preferable, where possible, to relate treatment effects to a paired control within the same cell since this makes it possible to reduce or eliminate the effects of inter-cell variation. This of course makes the experiment more difficult since the cell under study must be maintained for a longer period in order to record both control and treated data. However, due to the reduction in variation of the results, fewer numbers of paired experiments need to be performed to achieve a statistically significant result.

With paired data, instead of formulating the t-value from the difference between group means, it is possible to use the mean of the control-treated pair differences

$$t = \frac{m_{t-c}}{\sqrt{(s_{t-c}^2/n)}} \quad [5.18]$$

where m_{t-c} is average of the paired treated–control differences, s_{t-c} is their standard deviation and n the number of pairs. In the case of data transformed as a percentage of control data such as that in Figure 5.1, the formula is slightly amended to

$$t = \frac{100 - m_\%}{\sqrt{(s_\%^2/n)}} \quad [5.19]$$

where $m_\%$ and $s_\%$ are the average and standard deviation of the set of paired treated measurements expressed a percentage of the control.

In either case, the tail probability $p(\geq t)$ is computed from the t-distribution with $n - 1$ degrees of freedom (since we are dealing with a set of n independent observations and one computed average rather than $n_t + n_c$ and the two averages for the unpaired case). Applying the paired t-test to the percentage depression in AP amplitude induced by drug treatment in the example data set (column D in Figure 5.1) yields a value of $t = 8.78$ with $n_f = 7$, $p(t \geq 8.78) = 0.00003$. Note the much higher level of significance than for the unpaired test. A spreadsheet for the paired t-test can be found in Listing 5.2.

5.8 NON-PARAMETRIC TESTS: THE WILCOXON RANK-SUM TEST

While the t-tests are useful, they undoubtedly require some assumptions about the underlying (and unknown) population distribution of measurements. t-Tests are classed as *parametric* tests since they are based on the known properties of an assumed underlying normal population distribution. Such tests are *sensitive*, in the sense that they require smaller sample sizes to discriminate differences than other tests. On the other hand, they are not *robust* tests since if the basic assumptions about the population distribution are not true then they produce incorrect results.

There are often circumstances where it is more important to be assured that the test is correct than to have a high sensitivity. Such robustness is gained by using *non-parametric* tests which reduce the possibility of error by making as few assumptions about the data as possible. The *Wilcoxon rank-sum test* is a non-parametric equivalent to the t-test which computes the significance of the difference between two groups of data without making any assumptions about the underlying population distribution.

The essence of the test is to sort the combined set of observations from both the control and treated groups in order of increasing size and assign a rank to each observation from the lowest (1) to the highest ($n_c + n_t$). Observations with equal values are assigned the average of their

	A	B	C	D	E	F	G	H	I	J
1								{/ Sort;Block}a10..b10~		
2		Wilcoxon Rank-sum test.						{/ Sort;Block}{?}~		
3								{/ Sort;Key1}A10~A~		
4	Group 1 rank sum			89	n=		8	{/ Sort;Go}		
5	Group 2 rank sum			47	n=		8	{quit}		
6										
7	Data		Grp.	Raw	Ties	Tie	Ranks	Group 1		Group 2
8				rank		sums	+ties	n ranks		n ranks
9	+--------+----+-----+---+------+-----+--+-----+--+-------									
10	0.34		2	1	1	1	1	0 0		1 1
11	0.59		2	2	1	2	2	0 0		1 2
12	0.65		2	3	1	3	3	0 0		1 3
13	0.76		2	4	1	4	4	0 0		1 4
14	1.1		2	5	1	5	5	0 0		1 5
15	1.18		1	6	1	6	6	1 6		0 0
16	1.36		1	7	1	7	7	1 7		0 0
17	1.4		2	8	1	8	8	0 0		1 8
18	1.42		2	9	1	9	9	0 0		1 9
19	1.67		1	10	1	10	10	1 10		0 0
20	1.69		1	11	1	11	11	1 11		0 0
21	2.01		1	12	1	12	12	1 12		0 0
22	2.13		1	13	1	13	13	1 13		0 0
23	2.21		1	14	1	14	14	1 14		0 0
24	2.23		2	15	1	15	15	0 0		1 15
25	3.46		1	16	1	16	16	1 16		0 0

Figure 5.4 Spreadsheet implementation of the Wilcoxon rank-sum (or Mann–Whitney) test, a non-parametric equivalent to the t-test.

combined ranks. For each group the sum total of its ranks (R_c, R_t) can then be calculated. Either sum can be used as the test statistic, R_t being used here. If there is no real difference between control and treated groups, there should be an approximately equal distribution of high and low ranks within each of the groups resulting in a value of R_t close to

$$R_t = \frac{(n_c + n_t)(n_c + n_t + 1)}{4} \quad [5.20]$$

If the treatment has increased the parameter being measured, there will be a preponderance of large ranks in the treated group resulting in a large value for R_t. Conversely, if the treatment has caused a reduction, R_t will be small. Therefore both large and small values of R_t indicate significant difference between groups. Critical high and low values for determining the significance of R_t can be obtained from statistical tables as a function of n_c, n_t and the required tail probability (Daniel 1987; Mendenhall & Sincich, 1989).

It is often more difficult to implement non-parametric tests using spreadsheets since the range of functions offered by the spreadsheet are not well matched to the operations required. Parametric tests only require arithmetic operations, such as summation and multiplication, and mathematical functions such as exponentiation whereas the key operations in many non-parametric tests require the sorting of the data into ascending order. Arithmetic functions, implemented as formulae, are carried out automatically whereas sorting operations normally require the user to make a series of menu selection steps.

A spreadsheet implementation of the rank-sum test is shown in Figure 5.4. Due to the limitations in the facilities provided by the spreadsheet, it is formatted slightly differently from the previous examples. Data values from both sample groups are entered into column A. Column B contains the values 1 or 2 to indicate which group each entry, in the corresponding row in A, belongs to. The data must be sorted into ascending order in order to find their rank positions from column C. This is done using the Sort operation from the spreadsheet's menu which requires a series of operations to define the block of entries to be sorted and the sort criteria.

The relatively complex sorting operation is partially automated by using the spreadsheet's *macro* language. Macros are short programs stored within a series of cells within the spreadsheet, allowing operations to be linked together and performed automatically at a single key press. Any operation that can be performed by pressing a function key or selecting a menu option has a corresponding macro command. Spreadsheet macro languages have developed to the extent that they have become simple

Table 5.3 Results from the parametric and non-parametric tests when applied to the nerve action potential data set from Figure 5.1.

Test	Statistic		Significance
t-test (unpaired)	$t = 2.74$	$n_{df} = 14$	$p(t \geqslant 2.74) = 0.008$
t-test (paired)	$t = 8.27$	$n_{df} = 7$	$p(t \geqslant 8.27) < 0.0004$
Wilcoxon rank-sum	$t_1 = 47$	$n_c = 8, n_t = 8$	$p(t_1 \leqslant 47) = 0.025$
Wilcoxon paired	$t_0 = 0$	$n = 8$	$p(t_0 \leqslant 0) = 0.005$

programming languages. In Figure 5.4 the macro program shown in column H selects the block to be sorted from columns A and B, then the column to be used as the sorting key (A) and initiates the operation. The rest of the test is carried out using arithmetic operations. Columns E, G and F perform operations which correct the rank value for ties.

If R_t exceeds the expected value from Equation 5.20, then the significance is determined by the probability of a rank sum greater than or equal to R_t, $p(\geqslant R_t, n_c, n_t)$ is computed. Conversely, if R_t is less than the expected value, $p(\leqslant R_t, n_c, n_t)$ is computed. It is difficult to implement a spreadsheet-based direct calculation procedure for tail probability as was done for the parametric tests. It is possible to compute such probabilities for small sample sizes using a direct simulation procedure to enumerate all possible rankings for a given pair of sample group sizes and determine the proportion which fulfill the probability criterion (Colquhoun, 1971). Unfortunately, this kind of procedure requires a more powerful programming language than provided by the spreadsheet. A discussion of the use of a spreadsheet macro language can be found in Mezei (1990).

Note that in order to test for a significant difference between groups, neither mean nor standard deviation have had to be computed. In fact no direct use has been made of the particular numerical values of the data other than their rank order of size. Non-parametric tests can also be applied to paired data and a paired version of the rank-sum test based upon the same principles can be found in Listing 5.4. In this case the observations are ranked according to the absolute values of the difference between the matched control-treated pairs. The sum of the ranks for positive and/or negative differences, ignoring zero differences, is used as the test statistic. Again if the differences are due only to random variability, it would be expected that the number of positive and negative rank sums should be approximately the same. Details of the paired Wilcoxon rank-sum test can be found in Mendenhall & Sincich (1989).

5.9 A COMPARISON OF PARAMETRIC AND NON-PARAMETRIC TESTS

The relative effectiveness of parametric and non-parametric tests can be seen in Table 5.3 which summarizes the results from the application of both the paired and unpaired forms of the t-test and the Wilcoxon test to the nerve action potential data set. In either case, the paired tests produce higher levels of significance, as indicated by smaller tail probabilities. In the case of the paired t-test, the probability of the observed t-value of 8.27 having occurred by chance is 0.000 04 compared with almost 0.008 for the unpaired test. Similarly, the paired Wilcoxon test produces a value of 0.005 paired compared with 0.025 unpaired. It can also be seen that the parametric tests produce a higher level of significance than the comparable non-parametric tests.

In general, non-parametric tests are less powerful than equivalent parametric tests in the sense that they are less able to detect differences within data. This is not suprising since they use less information about the data (rank order rather than actual value). However, the difference is not great, rarely falling to less than 86% of comparable parametric tests (Snedecor & Cochran, 1967). Bearing in mind the advantages of making no assumption about the normality of the unknown underlying distribution of experimental results, this may be considered to be a small price to pay. A further discussion of the relative merits of parametric and nonparametric methods can be found in Colquhoun (1971).

5.10 LIMITATIONS OF SIMPLE PAIR-WISE TESTS

Statistical tests such as the *t*-test and the others designed to determine the significance of the difference between two groups of observations are widely described in the statistical literature and extensively used in the biological sciences. They do, however, have distinct limitations and there are experimental designs where simple pair-wise tests should be applied carefully or not applied at all. A particularly common case occurs when three or more different treatments are to be compared. Such an experiment might consist of sets of control measurements (C) and then treatment with four different drugs or concentrations of drug ($D_1 \ldots D_4$). In such an experiment, it can be asked whether each treated group is significantly different from the control group (C–D_1, C–D_2, etc.) and also whether treated groups are significantly different from each other (D_1–D_2, D_2–D_3, etc.). With five groups, a total of 10 such pair tests are possible.

There are no difficulties in performing 10 separate *t*-tests, but a serious problem lies in the computation of the significance probability. If a single test typically yields a *t*-value with a significance probability of 0.05 this is equivalent to saying that in 20 repetitions of the test this value is likely to occur only once due to chance variation in the control data when no real difference exists between the groups. This is usually considered an acceptable error rate, and one can therefore have confidence in concluding that a difference exist. However, in the multiple treatment experiment we can (or have the potential to) repeat the test 10 times, and if this were to be done, the probability of finding a falsely significant result in one of the 10 tests becomes 0.5 (10 × 0.05).

This may not be a great concern when *t*-test results produce very small tail probabilities (e.g. <0.01) since, even with repetition of the test, the significance probabilities are well within acceptable limits. However, in such circumstances, the test result is only confirming what is clearly obvious. Unfortunately, it is when results are on the border of significance that statistical tests are most valuable and are most relied upon. In these circumstances, misapplication of pair tests to multiple comparisons are bound to lead to incorrect conclusions.

5.11 ANALYSIS OF VARIANCE

The problems arising from the repetition of pair tests can be avoided by using a single test applied simultaneously to all of the data sets obtained from the experiment. One of the most common is *analysis of variance*. The method is a means of quantitatively partitioning the variance of the observations into variance *within* treatment groups and the variance *between* groups. The ratio of the variance between and within groups is taken as the test statistic which determines whether a significant difference exists between any of the groups.

For a set of *j* treatment groups of various sizes, each containing n_j data points, the within-group variance is obtained from

$$s_w^2 = \frac{\sum_{j=1}^{k} \sum_{i=1}^{n_j} (x_{ij} - m_j)^2}{\sum_{j=1}^{k} (n_j - 1)} \quad [5.21]$$

where m_j is the average for each treatment group and x_{ij} is the *i*th element in group *j*. The between-groups variance is obtained from

$$s_b^2 = \frac{\sum_{j=1}^{k} n_j (m_j - m_g)^2}{k - 1} \quad [5.22]$$

where m_g is the average for the complete data set, irrespective of treatment group. The variance ratio *F* is computed from

$$F = \frac{s_b^2}{s_w^2} \quad [5.23]$$

Given a null hypothesis that there is no difference in mean value resulting from the treatment applied to each group, then s_b^2 and s_w^2 are simply different estimates of the same quantity and therefore F should be close to unity. On the other hand if there *is* a difference due to one (or more) of the treatments s_b^2 will be larger than s_w^2 and F will be larger than unity. The F probability distribution is a bivariate function of the degrees of freedom of the between- and within-group variance. It can be obtained from statistical tables or computed numerically using Paulson's transformation to obtain an equivalent point on the normal distribution (Thistead, 1988).

$$z = \frac{(1 - 2/9n_{f,w})F^{1/3} - (1 - 2/9n_{f,b})}{\sqrt{[(2/9n_{f,w})F^{2/3} + 2/9n_{f,b}]}} \quad [5.24]$$

where $n_{f,b}$ and $n_{f,w}$ are the degrees of freedom of the between- and within-group variances.

The following hypothetical experiment illustrates the use of analysis of variance. To investigate the potency of a number of newly synthesized test compounds as neuromuscular blocking drugs, the effects of the compounds on endplate current amplitude were studied. The experimental protocol consisted of applying the compound to a nerve–muscle preparation then measuring the average EPC evoked in a number of muscle fibres. Four different compounds were studied and in addition a series of experiments were done in the absence of any drug, as a control.

Figure 5.5 shows the results tabulated in an analysis of variance spreadsheet. The results are collected into five distinct groups – Control (no drug), D_1, D_2, D_3, D_4, with the measurements stored in columns B–F. Each group contains the average EPC amplitude from 5–7 muscle cells.

The initial phase of the analysis is to determine whether there is any significant difference between groups at all, using the variance ratio test. Within- and between-group variances are calculated by the formulae in block A8...G15 (see Listing 5.6 for details) resulting in a value of $F = 30.95$. The probability of a value as high as this occurring is obtained by integrating the tail area of the F distribution $F(n_{f,b} = 5, n_{f,w} = 25)$ for $F \geq 30.95$. This evaluates, using Equation 5.24, to a probability of 4.5×10^{-9}. Therefore it can be concluded that there are significant differences between some of the groups.

5.12 MULTIPLE COMPARISON TESTS

The variance ratio test does not provide any information on *which* of the groups are different. For instance inspecting the averages for each group shows that three of the compounds (D_1, D_2 and D_4) have depressed the EPC amplitude while D_3 has produced an increase. It is useful to know which of these is significantly different from the control group and also whether they are significantly different from each other. The attempt to make multiple comparisons between groups brings back the same problems in determining the true significance level as occurred with the simple pair-wise test. However, the additional information provided by the variance analysis can be used to compute a significance level corrected for the multiple comparisons.

Multiple comparison tests are a difficult area in statistical theory which is simply avoided in many elementary statistical textbooks. There is no single all-embracing test that is satisfactory for all occasions. Most differences between tests arise in the means used to determine the critical value of the test statistic used to make the acceptance/rejection decision. The significance of the difference between any pair of groups is dependent upon the total number of pair-wise group comparisons being made which depends on the design and purpose of the experiment. Taking the neuromuscular block experiment discussed earlier as an example, one possible purpose of the experiment might have been simply to determine if the drug responses D_1–D_4 differed from the control (C). In this case only four comparisons are required (C–D_1, C–D_2, C–D_3, C–D_4) and a quite sensitive critical test value could be computed on that basis. However, if it were also to be asked whether there were differences between drug responses (e.g. some were more potent than others) a different set of tests would have to be performed using a higher critical value since a total of 10 comparisons must now be performed.

In general, there is much to be said for taking a simple and slightly conservative approach to the determination of the significance of differences between groups. If 10 comparisons are possible it is reasonable to assume that they may be made. *Tukey's test* is simple to apply and errs towards the side of conservatism, although by no means as much as others. A test statistic, not unlike the *t*-value, is used to assess the difference between the means of any pair of groups (i, j)

$$q = \frac{m_i - m_j}{\sqrt{[(s_w^2/2)(1/n_i + 1/n_j)]}} \quad [5.25]$$

where s_w^2 is the within groups variance from the F test and n_i and n_j are the number of measurements in each group. The probability distribution for q can be obtained from the expected distribution of group mean values for a given number of groups, known as the *Studentized range*, a bivariate function of the total degrees of freedom within the experiment (total number of observations − number of groups) and the number of groups being compared. Tables for the Studentized range can be found in statistical texts (Daniel, 1987; Glantz & Slinker, 1990). In the spreadsheet in Figure 5.5, a table of critical q values for the 5% level of signficance has been stored in a two-dimensional table in a linked spreadsheet which can be accessed using the function @INDEX() which looks up this table.

The q values derived using Equation 5.25 for the 10 pair-wise differences between the five groups in Figure 5.5 are shown in tabular form in block A21 . . . G30. The critical q value for a 5% significance level is $q_{5\%}$ (n_g=5, n_f=25) =4.15. On this basis, drugs D_1, D_2 and D_4 can be seen to have had an effect, being significantly different from the control group, while D_3 has not. It can also be seen that D_2 is significantly more potent than either D_1 or D_4 while D_1 and D_4 are not significantly different.

The appropriate degree of conservatism within the multiple comparison test is a matter of debate and others have suggested tests which achieve a greater sensitivity by, in effect, assuming that all 10 possible comparisons are usually not required. Tukey's test used a single critical value based upon the number of experimental groups. The Student–Neumann–Kuels test, while using the same q-test statistic, derives the critical value from the rank order of the group means. To perform the test, group means are ranked according to their magnitude. When a pair of groups are compared, the critical q-value is determined using the number of group means spanned by the groups being compared, instead of the total number of groups. In the case of the example, if the largest and smallest groups (D_3–D_4) were being compared they would span five group means and the critical q-value of $q(n_g$=5, n_f=25) would be used. On the other hand if D_1 and D_4 were compared they would span only two means, and $q(2,25)$ would be used. This has the effect of making the test more sensitive, by supplying smaller critical q values for groups closer together in the rank sequence. However, a restriction is placed upon which comparisons are permitted. If the difference between a pair of groups proves to be non-significant, all pairwise differences between groups bounded by these groups within the rank order are also deemed non-significant.

Other tests exist which are specific to particular situations such as Duncan's test for the difference between treated groups (e.g. D_1–D_4) and a single control group. Further discussion of concepts underlying multiple comparison test and their relative merits can be found in Steel & Torrie (1980) or Daniel & Coogler (1975).

Analysis of variance is a powerful technique which is worthy of a much more detailed treatment than can be afforded here. The one-way analysis presented above is only its simplest form; a multiple comparison equivalent to the unpaired *t*-test. The technique can be greatly extended to allow the components of variance due to multiple treatment factors within an experiment using *two-way analysis of variance*. It could be used, for instance, to determine whether a particular drug treatment had a different action on different types of animal. Similarly, if multiple treatments (e.g. a series of drug concentrations) can be applied to the same cell during an experiment, the set of results from a series of these cells can be analysed using analysis of variance with *repeated measures*, an equivalent to the paired *t*-test.

A comprehensive introduction to these methods can be found in Glantz & Slinker (1990) and FORTRAN routines for some of the methods in Davies (1971).

Statistical analysis and presentation of results

```
        A         B        C        D        E        F        G        H         I
 1
 2              One-way Analysis of Variance (Borland Quattro)
 3
 4   Within group variance    30.73   d.f=   25.00           0.39894  5.7276
 5   Between group variance  950.89   d.f=    4.00           3.8E-08 16.4027
 6   F (Var.b/Var.w)          30.95                          1.00001  5.1E-09
 7   z (Peizer & Pratt)        5.7276                        0.0004
 8   p(>=F)                    5.1E-09                       1.98615
 9                                                           0.15168
10   Group        1         2        3        4        5   Global  5.2933
11              Control     D1       D2       D3       D4           4.83859
12   In use       1         1        1        1        1           15.1509
13   Mean       106.15    93.45    75.54   107.16    92.61   94.50  0.74238
14   SS         137.98   208.89   170.48    94.48   156.35         30.7899
15              814.86     6.57  2155.78  801.41    24.95           3.99019
16   n            6         6        6        5        7
17   d.f.         5         5        5        4        6
18
19              Multiple Comparison test - q values
20
21   q                    Group    1        2        3        4        5
22                        Mean   106.15   93.45    75.54   107.16   92.61
23   Group     Mean        n       6        6        6        5        7
24   +--------+-------+-------+-------+-------+-------+-------+--------------
25    1       106.15    6                  5.61    13.53    0.42     6.21
26    2        93.45    6                           7.91    5.77     0.39
27    3        75.54    6                                  13.32     7.83
28    4       107.16    5                                            6.34
29    5        92.61    7
30
31   q.critical(5%)               4.15
32
33              Data Entry area
34
35   Group        1         2        3        4        5
36   +- Data -+-------+-------+-------+-------+------------
37             103.35    84.85    78.35   104.81   88.91
38             115.11   100.54    67.13   111.83   94.73
39             105.01    90.06    81.72   107.72   92.76
40             101.20    94.64    80.81   111.31   97.66
41             102.67   101.01    74.73   100.12   98.44
42             109.59    89.63    70.52            92.01
43                                                 83.76
44
```

Figure 5.5 Analysis of variance (one way) spreadsheet containing a set of five groups (Control, D1–D4) of experimental results. Between- and among-group variance and F ratio shown in block A4...F8. Studentized range q-values are shown in the block A21...G29 with the critical q value at the 0.05 significance level for Tukey's test shown in D31.

Analysis of variance is a parametric method and is based, like the t-test, on the assumption that the underlying population of observations is normally distributed. There are also non-parametric equivalents, notably the *Kruskal–Wallis* one-way analysis of variance by ranks and the *Friedman* two-way analysis of variance. They make use of the ranking approach used in the

5.13 ANALYSIS OF TRENDS IN MEASUREMENTS

In the statistical analysis procedure discussed so far there has been an implicit assumption that although a measured waveform parameter, such as peak amplitude, may show some random fluctuation between one record and the next there is no systematic trend. Each record within a sequence collected under the same conditions (control or treated) is equally representative of the phenomena under study. This has allowed us to use single average values to make comparisons between control and treated groups of records. There are, however, many circumstances where a parameter does vary systematically within a series of records. Such changes might occur within a series collected during the application or washout of a drug, or there might be changes in parameters dependent on the frequency of stimulation. When parameters are varying in such a way it is not possible to represent a series of the results with a single average. Rather, the data must be plotted as a time series.

The analysis of time-dependent parameter changes is illustrated by an example from a series of experiments to investigate the effects of high frequency nerve stimulation on transmitter release. Figure 5.6 shows the peak amplitudes of a series of 278 endplate currents, evoked by stimuli produced at a rate of 10 Hz. One of the aims of this experiment was to observe if the EPC amplitude in this tissue type (snake costocutaneous muscle) could be sustained at high rates of stimulation. It can be seen that, while peak current varies quite considerably from one stimulus to the next, there is nevertheless a clear reduction in amplitude between the beginning and end of the train with the last EPC being 37% of the first. Two distinct trends can be seen, a rapid initial decay of around 46% of control amplitude which is completed within the first 10 records, followed by a second phase producing another 16% reduction. In fact this slow phase has not reach completion by the end of the train of stimuli.

A quantitative analysis of such trends in waveform measurements requires more than the simple descriptive statistics such as the mean and standard deviation discusssed so far. It is necessary to develop a hypothesis and some mathematical expression for the shape of the trend observed in the data, a process known as *curve fitting*. In the case of the example in Figure 5.6, it can be seen that there are two components to the decay, suggesting that there may be two quite different underlying processes bringing about the depression in EPC amplitude, or at least two separate rate-limiting steps. It is possible to model the observed decay in amplitude as a process described by the sum of two exponential functions decaying to a steady state

$$y(t) = A_f \exp\left(\frac{-t}{\tau_f}\right) + A_s \exp\left(\frac{-t}{\tau_s}\right) + S$$

[5.26]

Equation 5.26 describes only the general form of the curve fitting the data in Figure 5.6. A set of 'best fit' values for the five parameters (A_f, τ_f, A_s, τ_s, S) must be found by numerical means which produces a curve which matches the trend. In the case of Figure 5.6 the set ($A_f = 45.7$, $\tau_f = 3.4$, $As = 20.4$, $\tau_s = 168$, $S = 27.31$) yields the best fit, and is shown superimposed on the data. The best fit parameters summarize the behaviour of the trend in much the same way as descriptive statistics. Curve fitting and the development of mathematical models is an important analytical technique and will be discussed in detail in Chapter 6.

5.14 ANALYSIS OF RANDOM DISTRIBUTIONS

So far, we have considered the observed random variations in the parameters simply as unwanted noise which we sought to eliminate as much as possible by averaging, filtering or by fitting mathematical models. However, although some of the variation may be due to noise within

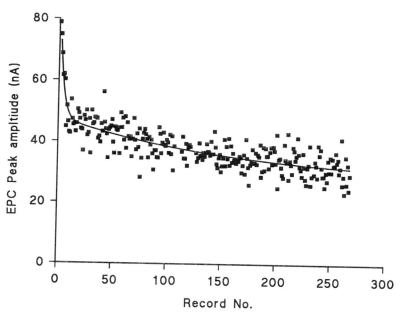

Figure 5.6 Analysis of trends in waveform measurements. Depression of EPC amplitude during a train of 268 EPCs evoked by stimuli at a rate of 10 Hz. EPC peak amplitude is plotted vs record number within train. The shape of the trend is modelled using two exponential functions decaying to a steady state (Equation 5.26) with the best-fit curve shown superimposed. (Data courtesy of Dr T. Searl, University of Strathclyde.)

the measurement instrumention and is of little interest, other components are intrinsic to the system under study and, if appropriately analysed, may yield useful information. For instance, although the discussion of the EPC train data in Figure 5.6 has concentrated upon the overall trends within the data, it is clear that there is also a quite striking degree of random variability of EPC amplitude.

A feature of randomly varying data is that it is difficult to discern clear patterns by direct observation of the data values. In the presentation of such data, it is more profitable to focus upon how often certain values or ranges of values are occurring rather than looking at the values themselves. The frequency of occurrence of data values can be represented with a *frequency histogram*. The range of possible data values is divided into discrete intervals or *bins*. Each data point is assigned to a bin on the basis of its position within the histogram range. For instance, a histogram with n bins ($b_j, j = 1, n$), over a range of values y_l to y_h, could be compiled by obtaining the bin index j for each data point y_i

$$j = \frac{(y_i - y_l)\, n}{y_h - y_l} + 1 \qquad [5.27]$$

and using it to increment the appropriate bin count ($b_j = b_j + 1$). The histogram is then plotted with bin values on the abscissa and the number of data points in each bin on the ordinate.

Frequency histograms are widely used in the analysis of data which have a high inherent random component, such as the amplitude of synaptic currents and the duration of single ion channels (which will be discussed in detail in Chapter 8). It is a particularly useful tool for the analysis of MEPCs which as we have seen are small in amplitude and recorded against a high background noise level. Consequently, the measured signal amplitudes vary a great deal from one record to the next, making it difficult to extract information by direct inspection of individual MEPCs. The amplitudes for a series of 100 MEPCs recorded from snake costocutaneous muscle are shown in the inset in Figure 5.7 (a). Averaging is used to obtain a consistent representative value for MEPC amplitude, but further insight into the underlying physiological processes can be obtained by a closer analysis of the MEPC amplitude distribution.

The MEPC amplitude is considered to be a good measure of the amount of transmitter

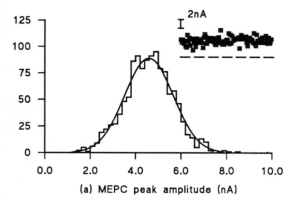

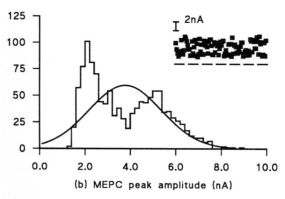

Figure 5.7 Frequency histograms of MEPC amplitudes, compiled from two recording series of 1192 MEPCs from snake neuromuscular junction, under control conditions (a) and after a period of rapid nerve stimulation (b). Amplitudes for a series of 100 MEPCs from each group are shown inset in each figure. (Data courtesy of Dr C. Prior, Strathclyde University.)

precise nature of the distribution of MEPC amplitudes is therefore likely to yield information concerning the distribution of vesicle content.

Figure 5.7 (a) shows the frequency histogram for the MEPC amplitudes from a series of 1192 MEPCs, compiled using Equation 5.27 with 50 bins over the range 0–10 nA. It can be seen that MEPC amplitudes are distributed with a symmetric bell distribution very similar in shape to the normal distribution (see Figure 5.2). The peak of the distribution coincides with the directly computed mean for the series (4.5 nA). This is what would be expected from a normally distributed population of vesicles of a single type and is quite typical of that which is observed at neuromuscular junctions under normal physiological conditions (Searl *et al.*, 1991).

However, there are other conditions where the amplitudes are not homogeneously distributed about a single mean value. Figure 5.4 (b) shows the histogram compiled from another 1192 MEPCs from the same cell, but after a period of rapid nerve stimulation (10 Hz for 2 mins). The distribution of amplitudes now has two quite distinct peaks, suggesting that the vesicles being released are of two different kinds with different transmitter content. These particular results appear to be peculiar to the snake, but similar phenomena have been observed at other neuromuscular junctions (Kreibel & Gross, 1974; Bevan, 1976).

The frequency histograms in Figure 5.7 were computed using features within the SCAN signal analysis program designed for this purpose. However, many spreadsheets also support the production of histograms.

contained within a single transmitter storage vesicle. It can be supposed that variations in MEPC amplitude will have a component due to variation in the amount of transmitter contained in the vesicle in addition to the random background noise superimposed on the signal. Variation from random background noise from sources in the instrumention will normally be symmetrical and described by a normal distribution. However, there is no *a priori* reason why the variation due to vesicle transmitter content should be similarly described. The intraterminal vesicle population may, for instance, be comprised of more than one type of vesicle, containing different amounts of transmitter. Such a population would be likely to produce a distribution with more than one peak. The

5.15 QUANTITATIVE COMPARISON OF FREQUENCY DISTRIBUTIONS

The results in Figure 5.7 are particularly dramatic and there can be little doubt that there has been a real change in the MEPC amplitude distribution. Both the frequency histograms were compiled from a large number of measurements resulting in relatively smooth unambiguous curves. Histograms compiled from smaller numbers of records are, however, subject to large statistical fluctuations, and may show peaks and troughs which can easily be misinterpreted as meaningful features. For these reasons it is valuable to be able to make a quantitative comparison between two frequency distributions and to be able to place a statistical level of significance on any observed differences.

Two statistical tests are commonly available for assessing the significance of the differences between frequency distributions, the *chi-square* test and the *Kolmogorov–Smirnov* test. The chi-square test is slightly simpler to implement and is described here. The quantity χ^2 is a measure of the difference between an experimentally *observed* distribution of measurements and the distribution that would have been *expected* if the measurements were drawn from a population following a particular theoretical distribution.

For instance, when discussing Figure 5.7 (a) it was asserted that the distribution of MEPC amplitudes followed the normal distribution. It is possible to test this assertion using the chi-square test. As discussed in Section 5.3, the normal distribution is described mathematically by its probability density function. In the chi-square test this function (Equation 5.19) is used to determine the expected number of data points falling within each bin by integrating between the limits of the bin, i.e.

$$E_i = N \int_{b_1}^{b_u} \frac{1}{\sqrt{(2\pi\sigma^2)}} \exp\left(\frac{-(x-\mu)^2}{2\sigma^2}\right) dx \quad [5.28]$$

If the bin width $(b_u - b_1)$ is small enough, the integration can be simplified to

$$E_i = \frac{N(b_u - b_1)}{\sqrt{(2\pi\sigma^2)}} \exp\left(\frac{-[(b_u - b_1)/2 - \mu]^2}{2\sigma^2}\right) \quad [5.29]$$

The smooth line in Figure 5.4 (a) shows the expected bin values assuming a normal distribution, computed with Equation 5.29 using the directly calculated mean and standard deviation of the 1192 measurements as μ and σ. χ^2 is computed from the following equation

$$x^2 = \sum_{i=1}^{n} \frac{(B_i - E_i)^2}{E_i} \quad [5.30]$$

where B_i and E_i are the observed and expected contents for the *i*th histogram bin. For the correct application of the test, it is necessary to ensure that the expected value of any bin is not less than one. If necessary, bins must be pooled together to ensure this condition. For data sets such as Figure 5.7, where the data does not completely span the histogram range, bins at the extreme ends of the range, expected to contain less than one event, are simply excluded from the analysis.

The larger the value of χ^2, the greater the difference between the observed and expected distributions. The significance of the difference is determined by the χ^2 probability distribution which can be computed, as for the *t*- and *F*-distributions, by a transformation to an equivalent normal variate (Thisted, 1988).

$$z = \frac{(\chi^2/n_f)^{1/3} - [1 - 2/(9n_f)]}{\sqrt{[2/(9n_f)]}} \quad [5.31]$$

The degrees of freedom n_f for the chi-square test are the number of bins minus the number of parameters defining the theoretical distribution which, in the case of the normal distribution is 2 (μ, σ).

In Figure 5.7 (a) the distribution of MEPC amplitudes is contained in 32 bins. Applying Equations 5.29 and 5.30 to these bins results in a χ^2 of 32.2 with 30 degrees of freedom and a 0.14 significance probability. This suggests that there is no strong evidence for the distribution not being normal. On the other hand, the twin peaked distribution in Figure 7.4 (b) is spread over 42 bins, yielding a χ^2 of 482 ($n_f=40$) which has a probability of 1.3×10^{-20} of occurring by chance.

5.16 STATISTICAL SOFTWARE PACKAGES

While spreadsheet programs provide flexible data manipulation and analysis tools, they are rarely supplied with more than the most

```
MTB > twosample c1 c2

TWOSAMPLE T FOR C1 VS C2
       N      MEAN     STDEV   SE MEAN
C1     8     1.075     0.183     0.065
C2     7     1.957     0.172     0.065

95 PCT CI FOR MU C1 - MU C2: (-1.082, -0.682)

TTEST MU C1 = MU C2 (VS NE): T= -9.62  P=0.0000  DF=  12
```

Figure 5.8 An extract from the log of a session with the Mintab statistics program. An unpaired t-test has been performed on two groups of data stored in columns C1 and C2 of the Minitab spreadsheet.

rudimentary statistical functions built in. Also, as can be seen from the spreadsheet formulae listings at the end of this chapter, the implementation of a complete statistical test can be quite involved. This is particularly true for non-parametric tests where many additional columns, with a somewhat convoluted logic, are required to perform what are conceptually quite simple operations.

Another class of software, the *statistical software package* exists which provides substantially more advanced statistical features. These packages often have their origins in large mainframe- and minicomputer-based packages, and have been ported for use on personal computers. It is quite common to find many versions of the packages running on a wide variety of different types of computer. Notable among these packages are SPSS (Statistical Package for the Social Science), Minitab, BMDP (Biomedical Data Analysis Package) and RS/1.

Given their origin in the minicomputer/mainframe world, most of the well-known statistical analysis programs are command-driven, requiring the user to learn a simple command language to use to program. Their ability to manipulate data in tabular form is also somewhat inferior to the spreadsheet. The means by which the results are computed are not as visible as with the spreadsheet. In their favour it can be said that they provide a complete implementation of a wide range of statistical tests whereas the spreadsheet only provides the basic foundations which allow the user to construct them. They also support better implementations of non-parametric statistics than do spreadsheets. The log of a short session with the Minitab program is shown in Figure 5.8. Examples of the use of such packages can be found in Mendenhall & Sincich (1989).

5.17 SCIENTIFIC GRAPH-PLOTTING PROGRAMS

In addition to statistical analysis, there is often also a requirement to be able to present results in a variety of graphical forms, the most common being line plots, histograms and bar charts. Such plots must be of a high quality, suitable for use as the originals for figures to be published in scientific journals. These plots often require annotation with text, lines and a variety of symbols.

Although many spreadsheet programs have quite sophisticated graph-plotting facilities, they are aimed at producing presentation graphics for business users. They therefore often lack certain stylistic features and the more complex forms of graph which are unique to scientific work, such as logarithmically scaled axes, standard error bars, and conventional scientific symbols such as circles, squares in both open and solid forms. In the last few years, a number of excellent graph-plotting programs specifically designed for scientific work have appeared on the market, in particular, Biosoft Fig.P for the IBM PC, Jandel Sigmaplot and Computer Associates Cricket-Graph (versions of which run on both the PC and Macintosh) and Kalaedagraph for the Macintosh. These programs are distinguished by their extensive range of scientific graph types and flexibility in forms of annotation. (Many of the figures in this book have been produced using the Fig.P program.)

5.18 SUMMARY

This chapter has provided a description of the statistical and data presentation procedures, commonly applied to the results of electrophysiological experiments. Statistical procedures are important tools which, when properly applied, help to prevent the experimenter drawing unjustifiable conclusions from experimental results. Methods for determining the statistical significance of observed differences are particularly important, in this respect.

The spreadsheet program has been demonstrated as a useful tool for both storing numerical results and applying statistical tests. A number of example spreadsheets implementing simple statistical tests have been presented. However, no single application program, at present, provides all the features necessary for the statistical analysis and presentation of experimental data. The spreadsheet, statistical package and scientific graph-plotting program each have their own unique function not well supported by the others. This yet again emphasizes the importance of being able to easily import and export data between packages. Fortunately, almost all of the programs discussed here, have at least the ability to import and export in the standard ASCII format discussed in the previous chapter (4.17). In addition the LOTUS 1-2-3 spreadsheet file format has become somewhat of a standard in itself with many programs being able to interpret it, including Fig.P and Minitab.

Listing 5.1. Unpaired Student's t-test (Borland Quattro, spreadsheet formulae)

```
B2:= 'Student's T-test (unpaired)
A4:= 'Control   B4:= 'Drug   C4:= 'Results   D4:= +A4   E4:= +B4
A5:= '+-------------   B5:= '+-------------   C5:= '+-------------   D5:= '+-------------   E5:= '+----+
C6:= 'Mean   D6:= @AVG(A6..A102)   E6:= @AVG(B6..B102)
C7:= 's.d.   D7:= @SQRT(@VAR(A6..A102)*D9/(D9-1))   E7:= @SQRT(@VAR(B6..B102)*E9/(E9-1))
C8:= 's.e.m.   D8:= +D7/@SQRT(D9)   E8:= +E7/@SQRT(E9)
C9:= '(n)   D9:= @COUNT(A6..A102)   E9:= @COUNT(B6..B102)
C10:= 'Difference (1-2)   E10:= +D6-E6
C11:= 'Pooled s.d.   E11:= @SQRT(((D9-1)*D7*D7+(E9-1)*E7*E7)/(D9+E9-2))
C12:= 'Degrees of freedom   E12:= +D9+E9-2
C13:= 'Student's T   E13:= @ABS(D6-E6)/@SQRT(E11*E11/E9+E11*E11/D9)
C14:= 'z (equivalent)   E14:= (E12-(2/3)+0.1/E12)*@SQRT((1/(E12-(5/6)))*@LN(1+(E13*E13/E12)))
C15:= 'p(>=t)   E15:= +D19
C17:= 0.398942280385
D17:= +E14
C18:= 3.8052E-08   D18:= +D17*D17/2
C19:= 1.00000615302   D19:= +C17*@EXP(-D18)/(D17-C18+C19/(D17+C20+C21/(D17-C22+C23/
                                      (D17+C24-C25/(D17+C26+C27/(D18+C28))))))

C20:= 0.000398064794
C21:= 1.98615381364
C22:= 0.15167911635
C23:= 5.29330324926
C24:= 4.8385912808
C25:= 15.1508972451
C26:= 0.74238092407
C27:= 30.789933034
C28:= 3.99019417011
```

Listing 5.2. Paired Student's t-test (Borland Quattro spreadsheet formulae)

C4:= ' A-B , D4:= 'Results (B-A)

A5:= '+--------------, B5:= '+-------------- C5:= '+-------------- A4:= 'Control B4:= 'Drug D5:= '+-------------- E5:= '+-----+
C6:= +B6-A6 D6:= 'Mean E6:= @SUM(C6..C102)/E9
C7:= +B7-A7 D7:= 's.d. E7:= @SQRT((@SUMPRODUCT(C6..C102,C6..C102)-E9*E6*E6)/(E9-1))
C8:= +B8-A8 D8:= 's.e.m. E8:= +E7/@SQRT(E9)
C9:= +B9-A9 D9:= '(n) E9:= @COUNT(A6..A102)
C10:= +B10-A10 D10:= 'Degs. free. E10:= +E9-1
C11:= +B11-A11 D11:= 'T E11:= @ABS(E6)/E8
C12:= +B12-A12 D12:= 'z (equiv.) E12:= (E10-(2/3)+0.1/E10)*@SQRT((1/(E10-(5/6)))*@LN(1+(E11*E11/E10)))
C13:= +B13-A13 D13:= 'P(>=t) E13:= +E17
C14:= +B14-A14
C15:= +B15-A15 D15:= 0.398942280385 E15:= +E12
C16:= +B16-A16 D16:= 3.8052E-08 E16:= +E15*E15/2
C17:= +B17-A17 D17:= 1.0000061530 E17:= +D15*@EXP(-E16)/(E15-D16+D17/(E15+D18+D19/(E15-D20+D21/(E15+D22-D23/(E15+D24+D25/(E16+D26))))))
C18:= +B18-A18 D18:= 0.00039806794
C19:= +B19-A19 D19:= 1.9861538136
C20:= +B20-A20 D20:= 0.15167911635
C21:= +B21-A21 D21:= 5.2933032492
C22:= +B22-A22 D22:= 4.8385912808
C23:= +B23-A23 D23:= 15.1508972451
C24:= +B24-A24 D24:= 0.7423809240
C25:= +B25-A25 D25:= 30.7899330
C26:= +B26-A26 D26:= 3.99019417011

Listing 5.3. Wilcoxon rank-sum test for unpaired differences .

```
H1:= '{/ Sort;Block}a10..b10~
B2:= 'Wilcoxon Rank-sum test.   H2:= '{/ Sort;Block}{?}~
H3:= '{/ Sort;Key1}A10~A~
A4:= 'Group 1 rank sum  D4:= @SUM(H10..H101)   E4:= 'n=  F4:= '{/ Sort;Go}
A5:= 'Group 2 rank sum  D5:= @SUM(J10..J101)   E5:= 'n=  F5:= @SUM(I10..I101)   H5:= '{quit}

A7:= 'Data   B7:= 'Grp.   C7:= 'Raw  D7:= 'Ties   E7:= 'Tie   F7:= 'Ranks   G7:= 'Group 1   I7:= 'Group 2
C8:= 'rank   E8:= 'sums   F8:= '+ties  G8:= 'n   H8:= 'ranks   I8:= 'n   J8:= 'ranks
A9:= '+--------   B9:= '+-------   D9:= '+-------   E9:= '+-------   F9:= '+-------    G9:= '+----------   H9:= '+---------
I9:= '+----------   J9:= '+---------

C10:= 1   D10:= 1   E10:= @IF(D10=1,C10,E9+C10)   F10:= @IF(D1>D10,F11,E10/D10)   G10:= @IF(B10=1,1,0)   H10:= @IF(B10=1,F10,0)
I10:= @IF(B10=2,1,0)   J10:= @IF(B10=2,F10,0)

C11:= +C10+1   D11:= @IF(A11=A10,D10+1,1)   E11:= @IF(D11=1,C11,E10+C11)   F11:= @IF(D12>D11,F12,E11/D11)   G11:= @IF(B11=1,1,0)
H11:= @IF(B11=1,F11,0)   I11:= @IF(B11=2,1,0)   J11:= @IF(B11=2,F11,0)

C12:= +C11+1   D12:= @IF(A12=A11,D11+1,1)   E12:= @IF(D12=1,C12,E11+C12)   F12:= @IF(D13>D12,F13,E12/D12)   G12:= @IF(B12=1,1,0)
H12:= @IF(B12=1,F12,0)   I12:= @IF(B12=2,1,0)   J12:= @IF(B12=2,F12,0)

.... Repeated until ....

D101:= @IF(A101=A100,D100+1,1)   E101:= @IF(D101=1,C101,E100+C101)   F101:= @IF(D41>D101,F41,E101/D101)   G101:= @IF(B101=1,1,0)
H101:= @IF(B101=1,F101,0)   I101:= @IF(B101=2,1,0)   J101:= @IF(B101=2,F101,0)
```

Notes. Data for groups is entered into block A10..B101 with the data points in column A and 1 or 2 in column B to indicate which group the data point belongs to. Column C contains a raw rank sum. The macro in block H1..H5 sorts the data into rank order. Columns D, E, F are used to correct the group rank order for ties. Column H and J contain the ranks for groups 1 and 2 respectively.

Listing 5.4 Wilcoxon paired rank-sum test.

```
B1:= 'Wilcoxon paired rank-sum test.   H1:= '{/ Sort;Block}a10..d10~
H2:= '{/ Sort;Block}{?}~
A3:= 'No. of pairs (non-zero)   D3:= @SUM(J10..J112)   H3:= '{/ Sort;Key1}d10..A~
A4:= 'T+ (sum positive ranks)   D4:= @SUM(I10..I101)   H4:= '{/ Sort;Go}
A5:= 'T- (sum negative ranks)   D5:= @SUM(H10..H101)-D4   H5:= '{quit}
A7:= 'Group 1   B7:= 'Group 2   C7:= 'Diff.   D7:= 'Abs.   E7:= 'Raw   F7:= 'Ties   G7:= 'Tie   H7:= 'Ranks   I7:= 'Ranks   J7:= 'Non-zero
A8:= 'Control   B8:= 'Drug   C8:= '(2-1)   D8:= 'diff.   E8:= 'Rank   G8:= 'sums   H8:= '+tied   I8:= '(+)   J8:= 'pairs
A9:= '+--------  B9:= '+--------  C9:= '+--------                    E9:= '+--------  D9:= '+--------                G9:= '+--------
                H9:= '+--------                                      I9:= '+--------  P9:= '+--------                J9:= '+--------
```

Formulae for computing pair differences and sorting into rank order

```
C10:= +B10-A10
D10:= @ABS(C10)
E10:= @IF(C10<>0,1,0)
F10:= 1   G10:= @IF(@CELL("type",A10..A10)="v",@IF(F10=1,E10,G9+E10),0)
H10:= @IF(@CELL("type",A10..A10)="v",@IF(F11>F10,H11,G10/F10),0)
I10:= @IF(C10>0,H10,0)   J10:= @IF(C10<>0,1,0)
```

Row C11..J11 is repeated with appropriate row number changes for rows 12..101

```
C11:= +B11-A11
D11:= @ABS(C11)
E11:= @IF(C11<>0,+E10+1,0)
F11:= @IF(@CELL("type",A11..A11)="v",@IF(C11=C10,F10+1,1),0)
G11:= @IF(@CELL("type",A11..A11)="v",@IF(F11=1,E11,G10+E11),0)
H11:= @IF(@CELL("type",A11..A11)="v",@IF(F12>F11,H12,G11/F11),0)
I11:= @IF(C11>0,H11,0)
J11:= @IF(C11<>0,1,0)
```

Notes. Data pairs are entered into block A10..B101. Column C contains difference between pairs, Column D the absolute value of difference. Data in block A10..D101 is ranked into ascending order of absolute value (using column D as the key) with the macro in block H1..H5. Column E contains the raw set of rank orders 1–90 and column F the ranks after correction for ties. Columns H and I contain the ranks for positive and negative differences (zeroes are excluded).

Listing 5.5. One-way analysis of variance with multiple comparisons.

A4..H17 Analysis of variance calculation

```
A4:= 'Within group variance   D4:= @SUM(B14..F14)/(@SUM(B17..F17))  E4:= 'd.f=   F4:= @SUM(B17..F17)
A5:= 'Between group variance  D5:= @SUM(B15..F15)/@SUM(B12..F12,-1)  E5:= 'd.f=   F5:= @SUM(B12..F12)-1
A6:= 'F (Var.b/Var.w)   D6:= +D5/D4
A8:= 'p(>=F)   D8:= +I6
A10:= 'Group  B10:= 1  C10:= 2  D10:= 3  E10:= 4  F10:= 5  G10:= 'Global
B11:= 'Control  C11:= 'D1  D11:= 'D2  E11:= 'D3
A12:= 'In use  B12:= @IF(B17>1,1,0)  C12:= @IF(C17>1,1,0)  D12:= @IF(D17>1,1,0)  E12:= @IF(E17>1,1,0)  F12:= @IF(F17>1,1,0)
A13:= 'Mean  B13:= @IF(B17>1,@AVG(B37..B132),0)  C13:= @IF(C17>1,@AVG(C37..C132),0)  D13:= @IF(D17>1,@AVG(D37..D132),0)
       E13:= @IF(E17>1,@AVG(E37..E132),0)  F13:= @IF(F17>1,@AVG(F37..F132),0)  G13:= @AVG(B37..F125)
A14:= 'SS  B14:= @IF(B17>0,@VAR(B37..B132)*B16,0)  C14:= @IF(C17>0,@VAR(C37..C132)*C16,0)  D14:= @IF(D17>0,@VAR(D37..D132)*D16,0)
       E14:= @IF(E17>0,@VAR(E37..E132)*E16,0)  F14:= @IF(F17>0,@VAR(F37..F132)*F16,0)
B15:= @IF(B12=1,(B13-$G$13)*(B13-$G$13)*B16,0)  C15:= @IF(C12=1,(C13-$G$13)*(C13-$G$13)*C16,0)
D15:= @IF(D12=1,(D13-$G$13)*(D13$G$13)*D16,0)  E15:= @IF(E12=1,(E13-$G$13)*(E13-$G$13)*E16,0)
F15:= @IF(F12=1,(F13-$G$13)*(F13-$G$13)*F16,0)  H15:= 3.9901941701I
A16:= 'n  B16:= @COUNT(B37..B132)  C16:= @COUNT(C37..C132)  D16:= @COUNT(D37..D132)  E16:= @COUNT(E37..E132)  F16:= @COUNT(F37..F132)
A17:= 'd.f.  B17:= @IF(B16>1,B16-1,0)  C17:= @IF(C16>1,C16-1,0)  D17:= @IF(D16>1,D16-1,0)  E17:= @IF(E16>1,E16-1,0)
F17:= @IF(F16>1,F16-1,0)
```

A19..H31 Tukey multiple comparison test

```
B19:= 'Multiple Comparison test - q values
A21:= 'q  C21:= 'Group  D21:= +A25  E21:= +A26  F21:= +A27  G21:= +A28  H21:= +A29
C22:= 'Mean  D22:= +B25  E22:= +B26  F22:= +B27  G22:= +B28  H22:= +B29
A23:= 'Group  B23:= 'Mean  C23:= 'n  D23:= +C25  E23:= +C26  F23:= +C27  G23:= +C28  H23:= +C29
A24:= '+------------  B24:= '+------------  C24:= '+------------  D24:= '+------------  E24:= '+------------
       F24:= '+------------  G24:= '+------------  H24:= '+------------
A25:= +B10  B25:= +B13  C25:= +B16  E25:= @ABS($B25-E$22)/@SQRT(($D$4/2)*(1/$C25+1/E$23))
F25:= @ABS($B25F$22)/@SQRT(($D$4/2)*(1/$C25+1/F$23))  G25:= @ABS($B25-G$22)/@SQRT(($D$4/2)*(1/$C25+1/G$23))
H25:= @ABS($B25-H$22)/@SQRT(($D$4/2)*(1/$C25+1/H$23))
A26:= +C10  B26:= +C13  C26:= +C16  F26:= @ABS($B26-F$22)/@SQRT(($D$4/2)*(1/$C26+1/F$23))
G26:= @ABS($B26G$22)/@SQRT(($D$4/2)*(1/$C26+1/G$23))  H26:= @ABS($B26-H$22)/@SQRT(($D$4/2)*(1/$C26+1/H$23))
```

Listing 5.5 (b)

```
A27:= +D10    B27:= +D13    C27:= +D16    G27:= @ABS($B27-G$22)/@SQRT(($D$4/2)*(1/$C27+1/G$23))
              H27:= @ABS($B27H$22)/@SQRT(($D$4/2)*(1/$C27+1/H$23))
A28:= +E10    B28:= +E13    C28:= +E16    H28:= @ABS($B28-H$22)/@SQRT(($D$4/2)*(1/$C28+1/H$23))
A29:= +F10    B29:= +F13    C29:= +F16

Critical values of Studentized range distribution for 5% sig. level stored in a table in the linked spreadsheet QTABLE.WQ1

A31:= 'q.critical(5%)    D31:= @INDEX([QTABLE]$A$1..$D$125,@SUM(B12..F12)-2,$F$4)

B33:= 'Data Entry area
A35:= 'Group   B35:= 1    C35:= 2    D35:= 3    E35:= 4    F35:= 5
A36:= '+- Data -----    B36:= '+----------    C36:= '+----------    D36:= '+----------    E36:= '+----------    F36:= '+----------

Approximation formulae for calculating F distribution upper tail area probability for F ratio. F is transformed into an equivalent
normal deviate and then AS66 normal area approximation is used to computed tail area.

D7:= ((1-(2/(9*F4)))*D6^(1/3)-(1-2/(9*F5)))/@SQRT((2/(9*F4))*D6^(2/3)+2/(9*F5))
I4:= +D7
I5:= +I4*I4/2
I6:= +H4*@EXP(-I5)/(I4-H5+H6/(I4+H7+H8/(I4-H9+H10/(I4+H11-H12/(I4+H13+H14/(I5+H15))))))

H4:= 0.398942280385
H5:= 3.8052E-08
H6:= 1.00000615302
H7:= 0.00039806794
H8:= 1.98615381364
H9:= 0.151679116635
H10:= 5.29330324926
H11:= 4.8385912808
H12:= 15.1508972451
H13:= 0.742380924027
H14:= 30.789933034
```

CHAPTER SIX

Mathematical modelling and curve fitting

Simple types of signal waveform measurements, such as those discussed in Chapter 4, have the advantage of requiring few *a priori* assumptions about the nature of the signals under study. On the other hand, many electrophysiological signals have complex time courses that cannot be satisfactorily expressed solely in terms of simple amplitudes and durations. A great deal of information is contained in the details of these time courses concerning the rate of opening and closing of the ion channels carrying the currents which generate the signals and their dependence on membrane potential and time.

6.1 MATHEMATICAL MODELS OF ION CURRENTS

To probe further into the kinetics of ion channel behaviour it is helpful to develop theoretical models which describe the observed waveforms and to test such models against the actual experimental data. The classical example of such a modelling approach is the equations developed by Hodgkin and Huxley to describe the properties of the ionic conductances underlying the action potential observed in the squid giant axon (Hodgkin & Huxley, 1952a–d). Hodgkin–Huxley models have subsequently been used with minor variations to describe action potentials in almost all forms of excitable tissue including cardiac muscle (Noble 1975), myelinated nerve (Frankenhaeuser, 1960) and skeletal muscle (Adrian *et al.*, 1970). Similar techniques have been applied to synaptic currents, notably the development of a theoretical basis for the behaviour of the endplate current in skeletal muscle (Magleby & Stevens, 1972) and of the block of endplate ion channels by various drugs (Steinbach, 1968; Adams, 1977; Ruff 1977).

The essence of the modelling approach is to develop a mathematical equation (or set of equations) capable of describing some essential feature, usually the time course, of experimentally recorded signals. For each system under study the general form of a modelling equation has to be devised. Such equations can take forms belonging to many different classes of mathematical function (exponentials, polynomials, etc.). However, much of the theoretical modelling of ionic conductances and channels has drawn heavily on ideas from chemical reaction kinetics. Consequently, much use has been made of functions which are based upon sums or powers of the *exponential* function. Other approaches exist such as the use of polynomial equations or fractals (Leibovitch, 1989) but none at present appear to provide models as acceptable as those provided by modified reaction kinetics.

A simple example of an exponential-based model is that applied to the decay phase of the miniature endplate current from skeletal muscle (Magleby & Stevens, 1972). The time course of these currents can be represented by a function of the form

$$y(t) = A \exp\left(\frac{-t}{\tau}\right) \qquad [6.1]$$

In the above equation the MEPC is characterized by two parameters, the peak amplitude A and the time constant τ of decay. Figure 6.1 shows an MEPC with an exponential function superimposed. During the decay phase of the signal the exponential function almost perfectly describes the time course, with the MEPC only distinguishable by its background noise. Note, however, that the rising phase of the MEPC takes a finite time while the exponential function rises instantaneously. The exponential model is therefore a simplification of the actual MEPC. Rather than attempting to produce a complete model describing every feature of the MEPC, a choice has been made to model only the signal's most distinctive feature.

6.2 CURVE FITTING

The mathematical model in Equation 6.1 is represented in a general form as a mathematical expression with two undefined (or free) parameters (A, τ), the values of which lie outside the scope of the model. Equation 6.1 can therefore represent any number of different MEPCs of different amplitude and duration depending upon the quantitative values of the parameters. To make a model fully represent a particular MEPC, such as the example in Figure 6.1, it is necessary to find the actual values of the parameters which generate an equation which matches the MEPC decay time course. These parameters are known as the *best fit* parameters and are obtained by the process of *curve fitting*.

The best-fit parameters obtained from such modelling and curve-fitting exercises can be used in essentially the same way as the simpler characteristic waveform measurements such as peak amplitude discussed in Chapter 4. However, since they embody aspects of underlying theoretical models, they are much more powerful tools. Similarly, the degree to which the mathematical model provides a good match to the experimental results is itself a powerful means of validating experimental hypotheses. Consequently, curve fitting is a technique of considerable importance in the analysis of electrophysiological data, forming a bridge between hypothetical mathematical models and experimental data.

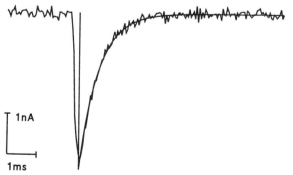

Figure 6.1 An MEPC from rat skeletal muscle with a mathematical model (the exponential function, Equation 6.1 with $A = -3.93$ nA, $\tau = 1.37$ ms) superimposed on its decay phase.

6.3 QUANTIFYING GOODNESS OF FIT

A variety of approaches to finding the best fit curve have been used in the past, even including the purely subjective method of choosing what appears visually to be the best from a range of different curves superimposed over the signal. The dangers inherent in such a subjective approach should be obvious. It is very difficult to eliminate the possibility of conscious or unconscious bias on the part of the curve fitter. In practice, an objective quantitive method is needed for determinining the best fit parameters although, as will become apparent later, this itself is not without pitfalls for the unwary.

The first step in finding the best-fit curve is to devise a measure of *goodness of fit*. The most common measure in current use is the principle of *least squares*. The least squares criterion is based on the straightforward notion that the magnitude of the difference between the data points and the mathematical curve being fitted is a good measure of how well the curve fits to the data. The process can be illustrated using the specific example of the linear (straight line) function. Figure 6.2 (a) shows a plot of a series of 10 data points (x, y) with a straight line function $(f(x) = mx + c)$ superimposed. If the value of the line function is computed for the same x values as the data, the set of *residual differences* $(y - f(x))$ between the the function and the data can be computed as shown in Figure 6.2 (b). In order to obtain a measure of the magnitude of the deviation from the fitted line, the residuals are squared which produces a positive value irrespective of the sign of the residual (Figure 6.2 (c)). The sum of the squared residuals is an overall measure of the extent that the data points deviate from the mathematical equation.

The best-fit function is the one with a set of parameters that minimize the sum of squared residuals. In general terms, the sum of squares can be expressed in the form,

$$S = \sum_{i=1}^{n} (y_i - f(x_i, p_1, \ldots, p_m)_i)^2 \quad [6.2]$$

where y_i is the ith point from a set of n data points and $f(p_1, \ldots, p_m)_i$ is the value of the mathematical function at the ith point with a set of parameters $p_1, \ldots, p_m$.

6.4 FINDING THE LEAST SQUARES

We now need a means of finding which, of all possible sets of function parameters, produces the minimum value of the sum of squared residuals. In the case of the straight line it is possible to find an analytical solution (i.e. one in terms of an algebraic expression) for the parameter values m and c which minimizes S. For the straight line, Equation 6.2 becomes

$$S = \sum_{i=1}^{n} (y_i - mx_i - c)^2 \quad [6.3]$$

This can be shown to be a quadratic equation (see Lee & Lee (1982) for derivation) with its minimum found under conditions where both derivatives of S (with respect to m and c) are zero, i.e.

$$\frac{\partial S}{\partial m} = 2c\Sigma x_i - 2\Sigma x_i y_i = 0, \quad \frac{\partial S}{\partial c} = 2m\Sigma x_i = 0$$

$$[6.4]$$

Rearranging these equations to solve for m and c and also to make them more suitable for computer calculation produces:

$$m = \frac{\Sigma (x_i - x_m)(y_i - y_m)}{\Sigma (x_i - x_m)(x_i - x_m)}, \quad c = y_m - m x_m$$

$$[6.5]$$

where y_m and x_m are the mean values of the x and y points within the data set. Using the above equations the best-fit line to the data in Figure

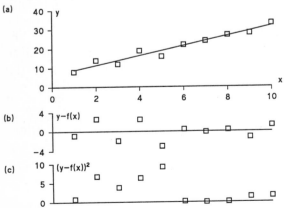

Figure 6.2 Least-squares goodness of fit criterion applied to a straight line function ($y = mx + c$). (a) shows a straight line passing through a set of experimental data points, (b) the residual difference between the data points and the corresponding value of the function at the same x value, and (c) the squared residual differences.

6.2 yields $m = 2.53$ and $c = 6.33$. The above method of curve fitting is used extensively and is also known under the name of *linear regression*. It predates the use of the computer since for small data sets the computations can be performed with pen and paper, and is widely discussed in most elementary statistical texts (e.g. Lee & Lee, 1982).

6.5 LINEARIZING TRANSFORMS

Unfortunately, analytical least squares solutions can only be obtained for a rather restricted class of mathematical functions which are linear functions of the independent variable x. In addition to the straight line, these include the quadratic function and higher order polynomials. Some functions which are not themselves linear can also be transformed into related linear functions. The exponential function, as applied to the MEPC, is of this type. Taking the natural logarithm of both sides of Equation 6.1 results in

$$\begin{aligned}\log_e(y) &= \log_e(A\exp(-t/\tau)) \\ &= \log_e(A) + \log_e(\exp(-t/\tau)) \\ &= \log_e(A) - (-t)/\tau \\ &= c + mt \end{aligned} \quad [6.6]$$

A plot of $\log_e(y)$ vs t results in a straight line with a slope of $-1/\tau$ and a y intercept of $\log_e(A)$. Figure 6.3 shows this logarithmic transform procedure applied to the MEPC from Figure 6.1. The negative-going MEPC has been inverted (since logarithms can only be taken of positive numbers) and converted to natural logarithms (Figure 6.3 (a)). A straight line is then fitted to $\log_e(y)$ vs t using the least squares method which provides best-fit values of m and c. A and τ for the best-fit exponential curve are then obtained from

$$A = \exp(c), \quad \tau = -1/m \quad [6.7]$$

Linearizing transformations have the advantage that they extend the use of simple linear least-squares methods to a wider range of functions. However, the transformations have the side effect of selectively emphasizing any background noise on the record in a non-linear fashion. One requirement for the proper application of linear least squares is that the y data points have a

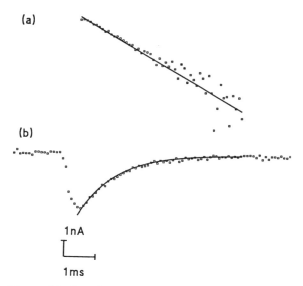

Figure 6.3 Best-fit exponential curve to MEPC decay phase by least-squares fit of straight line to logarithmically transformed data. (a) Log. transformed data with best-fit straight line; (b) original MEPC data record with best fit exponential, $A = -4.12$ nA, $\tau = 1.2$ ms.

constant error variance. This condition holds for the original untransformed MEPC data (Figure 6.3 (b)) which is superimposed on the signal background noise of constant properties. But this is not true for the logarithmically transformed y data, where the error variance increases markedly as the MEPC decays, as can be seen in Figure 6.3 (a).

The non-linear distribution of random errors in the data can bias the best fit line since a large number of widely scattered data points at the end of the record contribute far more to the sum of squared errors than a small number of tightly fitting points at the start. This problem can be mitigated by modifying the least-squares equations to include a weighting factor for each data point which reduces the influence of each point proportional to the variance of the residual error. With weighting, the least squares equations become

$$S = \sum_{i=1}^{n} w_i(y_i - mx_i - c)^2 \quad [6.8]$$

$$m = \frac{\sum w_i(x_i - x_m)(y_i - y_m)}{\sum w_i(x_i - x_m)(x_i - x_m)}, \quad c = y_m - mx_m \quad [6.9]$$

In the case of the logarithmic transform the relative error variance increases in inverse proportion to the square of the signal amplitude and so a suitable weighting factor is

$$w_i = y_i^2 \qquad [6.10]$$

Using simulated data sets where the true best fit is known, it can be shown that the use of weighted least squares generally provides a small but significant improvement of around 5–10% in the accuracy of the fit to logarithmically transformed data, depending on the amount of noise in the original signal. Listing 6.1 provides a routine for finding the amplitude and time constant for the best-fit exponential using the weighted least-squares method outlined above.

Linearizing transformations exist for a variety of other common functions and have been used quite extensively in the study of enzyme kinetics (e.g. Eadie and Lineweaver–Burke plots). However, these often introduce much more intractable non-linearities in the error distributions and must be approached with caution (Colquhoun, 1971; Valko & Vajda, 1989). Such methods have probably been rendered obsolete by the increasingly widespread availability of the iterative curve-fitting methods to be discussed next.

6.6 ITERATIVE NON-LINEAR CURVE FITTING

Linear least squares is limited to linear mathematical functions or ones which can be linearized by transformation and this precludes its application to many functions of interest to the electrophysiologist. In particular it can only be applied to a single exponential function decaying to zero. The equally common example of an exponential decay to a non-zero steady state:

$$y(t) = A_\infty + A_0 \exp(-t/\tau) \qquad [6.11]$$

cannot be linearized, nor can the sum two exponentials

$$y(t) = A_s \exp(-t/\tau_s) + A_f \exp(-t/\tau_f) \qquad [6.12]$$

In the absence of a convenient analytical solution for the set of best-fit parameters, it is necessary to find a numerical solution for the model parameters which minimize the sum of squares. This process is known as *iterative non-linear least squares*. Although quite complex in its actual implemention, the method is simple in essence, and can be summarized as the following procedure:

(a) make an initial guess for model parameters $p_1, \ldots p_n$
(b) compute the sum of squares S from Equation 6.2
(c) change $p_1, \ldots p_n$ by a small amount and compute S again
(d) repeat (b) and (c) using some strategy to ensure that the changes to $p_1, \ldots p_n$ result in a smaller value of S until no further improvement can be obtained.

Methods differ primarily in the *function minimization* strategy, used to find the set of parameters which will reduce S from one iteration to the next. It should be clear, however, that whatever the method used the repeated computation of S, required during the search for the best-fit set of model parameters, is likely to involve more computation than the single analytical solution for linear least squares. Considering the problem from a geometric perspective, finding the minimum of a single parameter function consists of finding the minimum point on the curve $f(x)$ vs x such as in Figure 6.4 (a). For a two parameter function $f(x,y)$ the problem extends itself to finding the minimum point of a surface in three-dimensional space, $f(x,y)$ vs x and y, as in Figure 6.4 (b). In general, functions of n parameters can be represented as surfaces in $n + 1$ dimensional space. Finding the minimum is essentially a matter of traversing this n-dimensional landscape trying to find the deepest valley.

The minimization of functions is a complex subject and a wide variety of methods have been developed over the past 30 years. While reliable methods for finding the minimum of a function of a single parameter are well developed and have been know for centuries (e.g. Gauss–Newton method, method of bisection, etc.), finding the minimum of a multi-parameter function is not so easy and in fact no single universal method exists which is successful for

Figure 6.4 One dimensional function $f(x) = 1.5x^2 - 10x + 20$ with a minimum at $x = 3.33$. Function gradient $df/dx = 3x - 10$ shown at $x = 8$. (b) Two dimensional function surface for sum of squared residuals for the function $(A \exp(-t/\tau) + S)$ where $A = 50$, $\tau = 50$, $S = 20$, plotted for the range of trial parameters $A = 1 \ldots 100$ and $\tau = 1 \ldots 100$. (Produced using the Maple V mathematics program.)

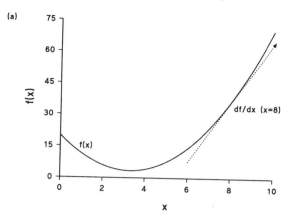

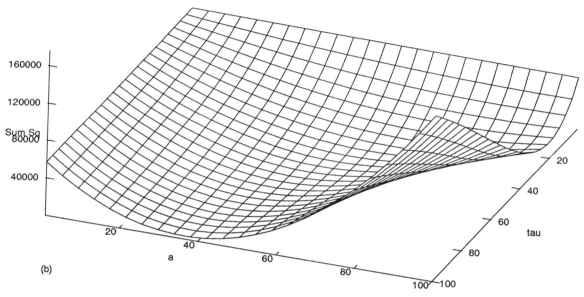

all classes of function. The difficulty lies with the complexity of the topography of the multi-dimensional surface. Even two parameter functions can produce complex landscapes with wide shallow valleys, ridges, and several different minima in different regions of the surface.

It might, at first, be thought that a strategy for finding the minimum of the function would be simply to compute S for a wide variety of parameter values spaced at small regular intervals over the range of values likely to contain the minimim. This, however, proves to be too expensive in terms of computation time for more than a few parameters. To find the minimum of a three parameter function to a 1% accuracy would require S to be computed a million times, taking hours on a personal computer. An algorithm is therefore required which actively seeks the minimum with as few computations of S as possible.

6.7 DIRECT SEARCH FUNCTION MINIMIZATION: THE SIMPLEX METHOD

It should be apparent that the reliability of the minimization algorithm is crucial to the confidence that can be placed on the best-fit parameter values obtained. Function minimization is known to be a tricky business with some methods working well with particular classes of problems and not at all with others. It is therefore essential to verify that a chosen algorithm produces accurate results and ideally with as few iterations as possible.

The simplex method, developed by Nelder & Mead (1965), is perhaps the simplest commonly used minimization algorithm. Descriptions of the method and computer source code for the algorithm are widely found in the literature on numerical methods for function minimization (O'Neill, 1971; Press et al., 1986; Valko & Vajda, 1989). The simplex method is an example of a *direct search* function minimization procedure, meaning that it only makes use of the actual values of the function in its search for the set of parameters which minimize the function and not how rapidly, and in what direction the function is descending towards the minimum, as do the more sophisticated methods to be described later. Instead, direct search methods rely on a clever strategy for exploring the function surface around the current trial parameter set to determine an appropriate direction to try in the next iteration.

In the case of the simplex method applied to an n parameter function, n additional function values each with slightly different parameters are derived from the initial guess. These $n + 1$ parameters form the geometrical shape known as a simplex (hence the name) which in the 2 parameter case is a triangle, as shown in Figure 6.5. The value of the function is computed at each vertex and the vertices with the highest (V_{high}) and lowest (V_{low}) function values identified. A 'downhill' direction on the function surface can be determined by drawing a line from V_{high} through the centre of the line between the remaining vertices V_{low}, V_{next} (the term centroid is used when there are more than two dimensions). The simplex algorithm attempts to find a new point along that line on the function surface with a lower value than any of the existing points.

The first point tested is V_{ref}, the geometric reflection of V_{high} with respect to the centroid. If the function value at V_{ref} proves to be smaller than V_{low}, then a further expansion along the V_{high}–V_{ref} line is attempted with the point V_{exp}. The better of these two trial points is incorporated into the simplex replacing the current worst point (V_{high}).

If the reflection strategy completely fails to produce a reduction in the function value (perhaps because the simplex is positioned close to the bottom of a narrow valley) then the alternative approach of contracting the simplex along the V_{high}–C line producing the point V_{con}. If V_{con} produces a lower function value than the current minimum it is included in the simplex. The contraction strategy has the effect of allowing the simplex to adjust its shape relative to the local topography of the function surface. If V_{con} fails to produce an improvement then the whole simplex is shrunk around the current best value (V_{low}).

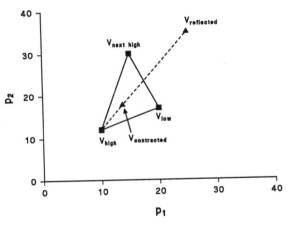

Figure 6.5 The simplex method for searching for the minimum of a two-parameter function. A 'downhill' direction is derived from the sample of the function surface enclosed by the simplex of three trial (p_1, p_2) parameter sets. By reflecting the simplex in this direction and further processes of expansion and contraction the simplex is moved across the function towards the minimum.

The repeated application of the above processes of reflection, expansion contraction, and/or shrinkage moves the simplex slowly across the function surface in the direction of a minimum. The *termination criteria* for determining when a minimum of the function has actually been found is almost an art in itself with a number of different approaches possible (e.g. see Gill et al., 1981).

In general, iteration is stopped when it becomes apparent that continued iteration is producing no further reduction in S. With the simplex method termination is often based upon the fact that the simplex collapses down to a single point when a minimum is encountered.

6.8 FITTING AN EXPONENTIAL CURVE WITH THE SIMPLEX METHOD

A FORTRAN implementation of the simplex algorithm, based on routines presented by O'Neill (1971) and Press *et al.* (1986) can be found in Listing 6.2, where it is being used to fit the three parameter exponential function in Equation 6.11. It is worth noting that there can be quite significant differences between different authors' implementations of the simplex algorithm, particularly in the termination criteria, which can have a profound effect on reliability. Listing 6.2 has a relatively complex procedure for determining when iterations should be terminated. Firstly, the basic simplex loop is terminated when either the difference between the S function at V_{high} and V_{low} becomes negligibly small (i.e. close to the arithmetic precision of the computer), or one of the dimensions of the simplex collapses down to a similarly small value. In either case there is little point in continuing with the simplex iterations. At this point, following O'Neill's procedure, a grid search is performed, computing the sum of square for sets of parameters incremented by small amounts on either side of the putative minimum. If this search reveals that the algorithm has not terminated at a true minimum then iterations are restarted using the new minimum found by the grid search. Not all authors have implemented this grid search (e.g. Press *et al.*), but it proved to be crucial to achieving an acceptable degree of reliability with the algorithm in Listing 6.2.

The performance of a curve-fitting algorithm can be tested by generating simulated data sets from equations with known parameter values which can be compared with best-fit estimates of the parameters obtained using the curve-fitting process. To make such simulated data more realistic, gaussian random noise can be superimposed on to the data. The progress of the simplex algorithm in Listing 6.2, in fitting the three parameter exponential function in Equation 6.11, is illustrated in Figure 6.6. The test data (Figure 6.6 (a)) were compiled using the parameter set ($A_0 = -25$, $\tau = 50$, $A_\infty = 100$) and contained 100 data points over the range 0–99. The algorithm took 193 iterations over a period of 9.5 s (on a 12.5 MHz 286 PC with a numeric data co-processor) to converge to a curve with best-fit parameter values ($A_0 = -26.3$, $\tau = 53.64$, $A_\infty = 101$). Figure 6.6 (c) shows the trial parameter values for each iteration during the search for the minimum. The progress made in minimizing the sum of squares for each set is shown in Figure 6.6 (b). Some examples of trial curves at various points throughout the search are shown superimposed on the data points in Figure 6.6 (a).

Overall a 15-fold improvement is made in the sum of squares during the process of convergence. It can be seen that the path of convergence towards the minimum is quite slow and somewhat erratic with a long period of almost 50 iterations where little improvement is achieved.

It is worth noting that the algorithm's criteria for goodness of fit are significantly more subtle than can be easily distinguished visually. Although it is quite clear from Figure 6.6 (a) that the parameter set at iteration 1 does not produce a good fit, the results from iterations 100 and iteration 193 both look satisfactory. However, at iteration 100 the τ parameter is still less than 50% of the true value. This again emphasizes the dangers inherent in any attempt to find the best fit purely by visual inspection.

The simplex algorithm has performed reasonably well in this particular case, producing a set of best-fit parameters which are all within 7% of the true values. However, it should be emphasized that no minimization algorithm can be guaranteed always to converge to the correct values. Minimization algorithms converge to a *local* minimum, i.e. at a point on the function surface which is lower than all nearby points. However, there may be more than one such minimum on the surface and it cannot be guaranteed that the search algorithm has found the deepest one, or *global* minimum. This should always be borne in mind when interpreting the

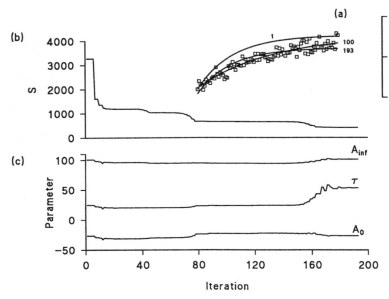

Figure 6.6 Non-linear least-squares curve fitting using the simplex method. An exponential curve has been fitted to the set of data points in (a). The sum of squares (S) as the search for the minimum progresses are shown in (b) and the trial parameter sets (A_0, τ, A_∞) for each iteration in (c). The routine converged after 193 iterations to the values $A_0 = -26.3$, $\tau = 53.6$, $A_\infty = 101$.

best-fit results obtained using an iterative technique. To some extent it is possible to guard against the possibility of converging to a local minimum by repeating the fitting process from a number of different initial parameter sets. If all such trials converge to the same minimum this provides some supporting evidence that it is a global minimum.

gradient and curvature (second derivative) of the function is available it is likely to be helpful in finding that direction. This proves to be the case, and most work in the development of minimization algorithms over the past few decades has therefore concentrated on such *gradient* methods.

6.9 GRADIENT METHODS FOR FUNCTION MINIMIZATION

Direct search methods such as the simplex method are inherently limited in that they ignore useful information contained in the gradient, or slope, of the function for each trial parameter set. As can be seen for the single parameter example in Figure 6.4 (a), the negative gradient indicates both the direction of the minimum and the steepness of decline towards it. For the n parameter case, the gradient consists of a vector of n first partial derivatives, one for each parameter.

The essential problem in finding the minimum of a multi-parameter function is to find a direction in the n-dimensional parameter space which points towards the minimum. Once such a direction has been found it is possible to search along that direction until the minimum is found. Intuitively, if information concerning the

6.10 METHOD OF STEEPEST DESCENT

Adopting vector notation, the sum of squares for a function with m parameters can be expressed as

$$S(\mathbf{p}) = \sum_{i=1}^{n} (y_i - f(x_i, \mathbf{p}))^2 \quad [6.13]$$

where $\mathbf{p}$ is a vector containing the m function parameters $p_1 \ldots p_m$.

$$\mathbf{p} = \begin{bmatrix} p_1 \\ \cdots \\ p_m \end{bmatrix} \quad [6.14]$$

There are m gradients on the function surface at each point $\mathbf{p}$ given by the first partial derivates with respect to each parameter which can be represented by a vector $\mathbf{g}$ containing m elements. For $m = 3$ this is

$$\mathbf{g} = \begin{bmatrix} dS/dp_1 \\ dS/dp_2 \\ dS/dp_3 \end{bmatrix} \quad [6.15]$$

For any particular point **p** in the parameter space, the set of gradients define the 'uphill' direction on the function surface. The downhill direction is therefore given by

$$\mathbf{d}_S = -\mathbf{g} \quad [6.16]$$

In the method of *steepest descent*, the gradients are calculated as well as the function value at each iteration. The set of parameters for the next iteration are calculated by finding the minimum value of S computed along the line $\mathbf{d}_S$. Unfortunately, while this approach certainly arrives at a minimum it does so very slowly, requiring many iterations. Although $\mathbf{d}_S$ always defines a downhill direction it does not necessarily point directly towards the minimum, as can be seen from the gradient shown in Figure 6.4 (a). Consequently, many readjustments are needed to $\mathbf{d}_S$ before the minimum is reached and the steepest descent algorithm is rarely used alone as the minimization strategy.

6.11 NEWTON'S METHOD

The failings of the steepest descent method stem from its implicit use of a linear model as an approximation of the function surface. Clearly, the function surface around a minimum will be curved and cannot be adequately approximated by a straight line or a flat surface, especially close to the minimum. In these circumstances, a much better approximation can be obtained using a quadratic model. This approach is used in *Newton's* method to provide the descent direction

$$\mathbf{d}_n = -\mathbf{H}^{-1}\mathbf{g} \quad [6.17]$$

where **H** is the $m \times m$ matrix of second partial derivatives of $S(\mathbf{p})$, which for $m = 3$ is

$$\mathbf{H} = \begin{bmatrix} dS/dp_1 dp_1 & dS/dp_1 dp_2 & dS/dp_1 dp_3 \\ dS/dp_2 dp_1 & dS/dp_2 dp_2 & dS/dp_2 dp_3 \\ dS/dp_3 dp_1 & dS/dp_3 dp_2 & dS/dp_3 dp_3 \end{bmatrix}$$
$$[6.18]$$

and is known as the *Hessian* matrix. No attempt will be made to derive this expression fully here. However, it can be understood qualitatively that the second derivative information contained in the Hessian matrix is being used to modify the basic direction produced by the gradient. From another point of view, Newton's method is a generalization of the Newton-Raphson method for finding the minimum of single dimensional function, which is often discussed in elementary numerical analysis courses. A detailed discussion of the principles and derivation of Newton's method, and gradient methods in general, can be found in Gill *et al.* (1981).

Unlike $\mathbf{d}_S$, under favourable conditions $\mathbf{d}_n$ points directly at the minimum of the function. If the surface was exactly parabolic the minimum would be located within a single iteration. However, if the surface does not have such a form, $\mathbf{d}_n$ may not even point in a direction of descent. Newton's method is therefore capable of providing very rapid convergence to the minimum, but can be unstable, causing the search to even diverge away from the minimum. In practice, Newton's method proves not to be suitable for the initial stages of the search, some distance from the minimum, where the function may not be well approximated by a quadratic function, but can provide very rapid convergence to the minimum from a position sufficiently close. A second problem exists in the accurate computation of the inverse of the Hessian matrix $\mathbf{H}^{-1}$ due to limitations in the numerical precision of computer arithmetic.

Note that the method of steepest descent and Newton's method have complementary properties, steepest descent always providing a reliable downhill direction but having unusably slow convergence while Newton's method provides rapid convergence but poor stability. Consequently, research into function minimization methods over the last 20 years has focused on algorithms which combine the properties of both methods, starting off with the reliable steepest descent and changing to Newton's method when sufficiently close to the minimum (Dennis, 1977; Jennrich & Raston, 1979; Gill *et al.*, 1981).

6.12 LEVENBERG–MARQUARDT METHOD

The Levenberg–Marquardt (L–M) method, developed by Marquardt (1963) from earlier work by Levenberg (1944), employs a minimum search algorithm which succeeds in combining properties of the method of steepest descent and Newton's method. This is achieved by modifying the Newton formula, adding a constant λ to the diagonal elements of the Hessian matrix. Equation 6.17 becomes

$$\mathbf{d}_{lm} = -\{ \mathbf{H}^{-1} + \lambda \mathbf{I} \} \mathbf{g} \qquad [6.19]$$

When λ is large ($\lambda \gg 1$), $\mathbf{d}_{lm}$ becomes equivalent to a scaled version of the steepest descent formula (Equation 6.16). When λ is small ($\lambda \ll 1$) it returns to Equation 6.17. The minimum search starts off with λ set to a large value working in steepest descent mode. As the search progresses, λ is adjusted at the end of each iteration, being reduced by a factor of 10 for a sucessful reduction in S and increased if a reduction has not occurred. Another feature of the L–M method when applied to least-squares problems is its use of a special approximation to the Hessian matrix. In the case of the least-square minimization problems discussed here, the Hessian matrix has a special structure which allows it to be approximated by a matrix of products of first derivatives, avoiding problems associated with calculating second derivatives. A more detailed discussion of the theory behind the L–M method can be found in Everitt (1987). A good practical introduction, including source code for an L–M algorithm, can also be found in Press *et al*. (1986).

6.13 QUASI-NEWTON METHODS

The L–M method is specifically associated with least-squares minimization, due to its use of the special structure of the Hessian matrix for these problems. The quasi-Newton methods have similar combined steepest descent/Newton properties, but are more generally applicable to the minimization of functions other than least squares. In a quasi-Newton method, attention is focused on the development of an approximation to the Hessian matrix which is built up over a series of search iterations. The initial approximation to H is one which forces the algorithm to have the steepest descent behaviour, and which gradually changes to Newton as the matrix is developed iteration by iteration.

There are numerous practical and theoretical difficulties involved in the development of function minimization routines, particularly associated with maintaining numerical stability within the matrix inversion calculations involved. These considerations make the development of function minimization routines very much a matter for the specialist in numerical analysis. It is much more practical for the non-specialist to obtain such routines from a good numerical subroutine library such as the *NAG Workstation library*. This is an extensive library of numerical analysis routines developed and marketed by the Numerical Algorithms Group (NAG) at Oxford University. The NAG library has been under development for many years and usually contains robust examples of the state of the art in many areas of numerical analysis, including function minimization. It is available for a wide range of computers including the IBM PC family, Sun workstations and mainframes. Using routines from a well-established and widely distributed library also has the advantage of making it simpler for other laboratories to repeat and verify the numerical aspects of one's work.

6.14 A PRACTICAL EXAMPLE USING THE NAG LIBRARY

Listing 6.3 shows how the NAG routine *E04GCF* (a quasi-Newton algorithm) is used to find the best-fit parameters for the three parameter exponential function. E04GCF is a general-purpose minimization routine which must be supplied with an external user-written subroutine containing the code for the function to be minimized. This subroutine which, in this case, is called LSFUN2 supplies values of the function and its gradients when called by E04GCF. In order to maintain the generality of the fitting routine the data points to be fitted are passed to LSFUN2 via the common block /lsfcom/ avoiding the need to pass the data through the fitting routine itself. The function to be fitted is

$$f(x, \mathbf{p}) = p_1 \exp(-x/p_2) + p_3 \quad [6.20]$$

where p_1, p_2 and p_3 correspond to A_0, τ and A_∞ in Equation 6.11. The gradients for this function are obtained as the set of first partial derivates relative to the parameter set $\mathbf{p}$, i.e.

$$\frac{\partial f}{\partial p_1} = \exp(-x/p_2) \quad [6.21]$$

$$\frac{\partial f}{\partial p_2} = \frac{p_1 x}{p_2^2} \exp(-x/p_2)$$

$$\frac{\partial f}{\partial p_3} = 1$$

For most functions commonly applied to the analysis of electrophysiological signals, deriving the equations for the partial derivatives analytically as above is straightforward using the basic rules of differential calculus. In practice the process is somewhat tedious, in terms of the amount of algebra involved for functions with more than 2–3 parameters. However, for those no longer skilled at such tasks, it is possible to use analytical equation solvers such as the *Mathematica* (Wolfram, 1991) or *Maple* (Char et al., 1991) programs. It is also possible to compute a numerical approximation to the derivatives from the forward difference formula

$$\frac{df_i}{dp_i} = \frac{f(x, p_i + dp) - f(x, p_i)}{dp} \quad [6.22]$$

where dp is a small increment (e.g. 0.001). Some minimization routines (e.g. E04FDF from the NAG library) use this approximation to calculate the gradients internally, requiring the user to supply only the function itself. A curve-fitting routine using analytically derived gradients is likely to perform better than one using forward difference approximations (as long as the analytical equations are correct, of course!); see Gill et al. (1981) for more details.

Using the function defined in LSFUN2, Listing 6.2 performs the following steps

(a) scale x and y data to ± 1 range
(b) devise initial estimates for parameters
(c) call E04GCF to find best-fit parameters
(d) calculate standard error of best-fit parameters using E04YCF
(e) correct best-fit results for effects of scaling.

6.15 PARAMETER SCALING AND INITIAL ESTIMATES

When using iterative minimization methods it is important to ensure that all function parameters being fitted are of approximately the same absolute magnitude (Gill et al., 1981). Otherwise the minimization search is likely to neglect parameters which are more than an order of magnitude smaller than the others. This can result in substantial errors in the best-fit results for that parameter. It is therefore good practice to rescale all parameters to lie within approximately the same range. After curve fitting is completed, the true parameter values are derived by dividing by the scaling factors.

Like all iterative fitting procedures, an initial set of parameters is required to start off the iterative process. Bearing in mind that the least-squares function may have several local minima in different regions of the parameter space, it is prudent to provide initial parameter 'guesses' which are reasonably close to what is expected to be the best-fit values, at least within an order of magnitude and of the correct sign. Completely arbitrary guesses which are unrealistic, such as negative time constants for exponentials, can often lead the routine to fail to converge or to converge to results which are meaningless within the physiological context of the mathematical model. Initial parameter guesses are often entered directly by the user but, for many functions, it is also possible to derive reasonably good estimates purely from the data itself. In the case of the exponential function in Listing 6.2, a good set of initial estimates are obtained from

$$p_1 = y_1 - y_n, \quad p_2 = \frac{x_n - x_1}{3}, \quad p_3 = y_n \quad [6.23]$$

where (x_1, y_1) and (x_n, y_n) are the first and last points in the data set.

It is worth noting that the gradient method routines can make extensive demands on available memory. E04GCF requires a 13 kbyte working array in order to fit a three parameter function to a set of 100 data points. Listing 6.3 can be modified to fit other functions by changing the function definition in LSFUN2.

6.16 ESTIMATION OF THE PARAMETER STANDARD ERROR

As for any other measurement, best-fit parameters derived from curve fitting are of limited value without knowledge of how precise an estimate they are. This depends on the shape of the least-squares function surface at the minimum. Best-fit parameters obtained from a minimum at the bottom of steep narrow valleys are well defined, since even small deviations away from the best fit will result in large increases in the sum of squares. Conversely, functions with wide shallow valleys produce poorly defined fits. Given the complexity of multi-dimensional function surfaces it is not unusual to find both well and poorly defined parameters from the same curve fit.

The curvature of the function surface around the minimum provides a measure of shape and steepness and this, in the case of gradient methods, can be obtained from the Hessian matrix (Equation 6.18) calculated using the best-fit parameters. This estimate of the precision of parameter estimation derived from the Hessian is commonly referred to as the parameter standard error. The NAG library provides a routine E04YCF to compute this data as shown in Listing 6.2.

6.17 COMPARISON OF THE SIMPLEX AND GRADIENT METHODS

All of the methods discussed (simplex, Levenberg–Marquardt and quasi-Newton) can be found within the life sciences literature. Many authors have advocated the simplex algorithm (Caceci & Cacheris, 1984; Nicol et al., 1986;

Harmatz & Greenblatt, 1987) as a simple and reliable curve fitting algorithm. Gradient methods have also been discussed by Jennrich & Raston (1979) and Carmenes (1991) and numerical software packages such as NAG or IMSL provide mostly gradient method routines. Recently, scientific graph-plotting packages such as Biosoft Fig.P and Jandel Sigmaplot have begun to incorporate curve-fitting algorithms and these also have been of the gradient variety.

Most recent work on the subject from numerical analysts appears to ignore the simplex algorithm completely, discussing only gradient methods (Brown & Dennis, 1971; Dennis, 1977; Thisted, 1988). Those that do discuss the method at all regard it as the least satisfactory of all methods, relegating its use to situations where for various reasons (such as discontinuities in the function being fitted) the gradient methods prove unusable (Gill et al., 1981). The performance of different algorithms has been compared (Box, 1966; Everitt, 1987) suggesting, for a somewhat restricted range of functions, that the gradient algorithms provide the better performance in terms of computation time.

Function minimization methods are known to be sensitive to the type of function to which they are applied. A method may be quite successful when applied to one type of function and may fail completely when applied to another. In the absence of existing information, the performance of algorithms based on the simplex and gradient methods was systematically compared when fitting the following range of functions, commonly applied to electrophysiological signals.

$$y = \frac{p_1 x}{x + p_2} \quad \text{Hyperbola} \quad [6.24]$$

$$y = p_1 \exp\left(\frac{-x}{p_2}\right) + p_3 \quad \text{Exponential} \quad [6.25]$$

$$y = \frac{1}{1 + \exp\left(\frac{-(x - p_2)}{p_3}\right)} \quad \text{Boltzmann} \quad [6.26]$$

$$y = p_1 \exp\left(\frac{-x}{p_2}\right) + p_3 \exp\left(\frac{-x}{p_4}\right)$$
Sum of two exponentials [6.27]

$$y = p_1 \exp\left(-\left(\frac{x-p_2}{p_3}\right)^2\right) + p_4 \exp\left(-\left(\frac{x-p_5}{p_6}\right)^2\right)$$
Sum of two gaussians [6.28]

In order to test the curve-fitting algorithms, 100 point simulated data records were generated from each of the five functions with known sets of parameters. As before, a random element was added to the data to make the records more realistic of actual experimental results. This allowed the sensitivity of the fitting procedure to random noise to be assessed. For each function tested a series of five records were generated with the same underlying function and overall noise variance but different random noise patterns. Best-fit parameters were then estimated from the records using each of three different types of iterative curve-fitting procedures, the simplex algorithm in Listing 6.2, the E04GCF quasi-Newton algorithm from the NAG library described in Listing 6.2, and an advanced form of the L–M algorithm called SSQMIN (Brown & Dennis, 1971), routinely used in the author's laboratory.

The results from this exercise are shown in Table 6.1 which shows the mean value and standard deviation of the best-fit parameters for the groups of five replicate records for each function studied. The accuracy of each fitting procedure can be determined by comparing the best-fit parameters with the original 'true' value. The efficiency of the algorithms was also tested by counting the number of iterations and time taken to converge to the best fit. All algorithms were implemented in the same language (Microsoft FORTRAN V5.0) and the trials were run on the same computer (IBM PC-compatible with 12.5 MHz 286 with numeric data coprocessor).

All three methods perform well with the hyperbolic function, the simplest of all the five trial functions. They all succeed in converging to within 2% of the true parameter value in less than 2 s. While there was little difference in the computation times between the simplex and the quasi-Newton algorithms, the improved L–M algorithm (SSQMIN) proved to be somewhat more efficient than either of the others.

As the curve-fitting tasks became more demanding with increasing numbers of parameters, the performance advantages of gradient method algorithms became apparent. With the three parameter exponential, E04GCF is at least four times faster and SSQMIN 12 times faster than the simplex method. More importantly, the simplex method begins to fail to produce an acceptable degree of accuracy. In the case of the three parameter exponential function, sets of parameters could be found (see Table. 6.1 (c)) where the simplex method yielded best-fit results in error by as much as 50% while E04GCF and SSQMIN were still within 5%. The simplex method also yielded poor results for the sum of two exponentials with errors as high as 20%. Interestingly, the simplex method did perform with tolerable accuracy when applied to the two gaussian functions, although it was still much slower compared to the others. The fact that the sum of the gaussian functions had six parameters and the sum of exponentials four, indicates that the determinant of the algorithm's performance is more subtle than simply the number of parameters.

In summary, both the quasi-Newton and the Levenberg–Marquardt algorithms proved capable of accurately fitting all five functions to the simulated data over a reasonable range of parameters. SSQMIN was in all cases substantially faster than the E04GCF. The simplex method, however, proved to be unreliable for both of the exponential functions studied.

6.18 EXPONENTIALS ARE 'ILL POSED' LEAST-SQUARES PROBLEMS

The poor performance of the simplex method with functions which are sums of one or more exponentials came initially as a surprise since there was little specific warning in the literature that such problems might be expected (Box, 1966; Press *et al.*, 1986; Everitt, 1987). A possible cause lies in the nature of the topography of the

Table 6.1 Comparative performance of three iterative least-squares curve-fitting methods: simplex, NAG quasi-Newton E04GCF, Brown & Dennis' improved Levenberg-Marquardt (SSQMIN)) applied to six functions commonly used in electrophysiological analysis.

	p_1	p_2	p_3	p_4	p_5	p_6	Iterations	Time
(a)	Hyperbola							
True	100.00	60.00						
Simplex	100.0 ± 1.4	60.1 ± 1.8					78	1.7
E04GCF	100.0 ± 1.4	60.1 ± 1.8						1.9
SSQMIN	100.0 ± 1.4	60.1 ± 1.8					6	0.4
(b)	Exponential							
True	−50.0	50.0	100.00					
Simplex	−50.0 ± 0.8	50.4 ± 2.3	100.1 ± 0.9				183	9.5
E04GCF	−50.1 ± 0.8	50.4 ± 2.2	100.1 ± 0.9					3.3
SSQMIN	−50.1 ± 0.8	50.4 ± 2.2	100.1 ± 0.9				5	0.7
True	−10.00	50.0	100.0					
Simplex	−9.2 ± 0.3	24.4 ± 0.3	97.9 ± 0.1				58	3.3
E04GCF	−10.3 ± 1.1	54.7 ± 16	100.3 ± 1.3					4.3
SSQMIN	−10.3 ± 1.1	54.4 ± 16	100.3 ± 1.3				4	0.5
(c)	Sum of two exponentials							
True	80.0	5.0	20.0	50.0				
Simplex	77.4 ± 2.1	4.4 ± 0.4	25.3 ± 3.2	38.4 ± 7.1			187	16
E04GCF	79.5 ± 1.0	5.0 ± 0.1	20.4 ± 0.8	48.8 ± 1.8				3.9
SSQMIN	79.5 ± 1.0	5.0 ± 0.1	20.3 ± 0.8	48.9 ± 1.7			5	1.3
(d)	Sum of two gaussians							
True	50.0	40.0	20.00	100.00	70.0	20.0		
Simplex	54.5 ± 1.3	42.5 ± 0.5	21.8 ± 0.4	94.2 ± 0.9	71.2 ± 0.3	19.2 ± 0.2	434	40
E04GCF	50.0 ± 0.9	40.0 ± 0.4	20.0 ± 0.4	99.8 ± 0.9	70.0 ± 0.2	20.0 ± 0.2		7.1
SSQMIN	49.9 ± 1.0	40.0 ± 0.4	20.0 ± 0.4	99.9 ± 0.9	70.0 ± 0.2	20.0 ± 0.2	5	1.9
(e)	Boltzmann							
True	100.00	70.00	20.00					
Simplex	99.9 ± 3.1	70.0 ± 1.5	19.9 ± 0.4				205	9.7
E04GCF	101.2 ± 1.8	70.6 ± 0.8	20.2 ± 0.3					2.9
SSQMIN	101.2 ± 1.8	70.6 ± 0.8	20.2 ± 0.3				5	0.6

least-squares function close to the minimum. It consists of a type of asymmetric shallow valley (e.g. see Figure 6.2(b)) which is known to be difficult to traverse (Box, 1966). Problems with such properties are known in mathematical terms as 'ill posed'. They are difficult to solve accurately using numerical methods with finite precision and are very sensitive to small changes in values in the data set being fitted. Many writers advise caution when applying iterative least squares to exponentials and others have even suggested not to do it at all (Acton, 1970; Van Mastright, 1977).

It appears that the direct search algorithm used in the simplex method proves to be particularly poor at traversing this complex surface close to the minimum. This might be inferred from the plot of trial parameters during the simplex search in Figure 6.6. It performs well enough in the beginning, reducing the sum of square to within a factor of three of the final value within 20 iterations, and bringing at least the A_0 and A_∞ parameters to within 10% of their goal. However, it then takes another 173 iterations finally to reach the minimum. It can also be seen that for many iterations parameters remain virtually constant being changed only by small amounts, close to the limits of the computer's numerical precision, before suddenly jumping to different values. In case the performance problems were due to some limitation or error in coding of the simplex algorithm used, a variety of other implementations were investigated, including the NAG E04CCF routine. However, these routines performed no better and sometimes worse.

Algorithms using gradient information, on the other hand, appear to work much more effectively in traversing this kind of topography. As discussed earlier, these methods switch to Newton steps when they get close to the minimum. If the function being fitted resembles a quadratic function, which is likely to be the case of a sum of squares, then a quasi-Newton or Levenberg–Marquardt algorithm can use knowledge of the function gradients and curvature to make a direct jump close to the minimum. In comparison, the simplex method has no alternative but to contract itself and move in tiny steps towards the minimum. This is not only inefficient but, if the simplex size becomes so small that it encroaches on the arithmetic precision of the computer, the process may terminate before the minimum is reached.

If anything, the above discussion should have emphasized the difficult nature of both writing and assessing the performance of curve-fitting routines. Approaches such as the simplex method may be easy to understand and implement, but they can have severe limitations. These results reinforce the need to make use of well-tried library implementations of the more complex gradient methods. It is also worth emphasizing that these results are specific to least-squares miminization problems. Interestingly, the simplex method appears to work much more reliably when applied to maximum likelihood problems which require a similar iterative function minimization, as will be discussed in Chapter 8.

6.19 ERROR PROFILE FOR THE EXPONENTIAL FUNCTION

Even the most robust curve-fitting methods have unavoidable limitations in their range of application. Care must therefore be taken to ensure that the set of data points to be fitted contain sufficient information to accurately quantify the function being fitted. For any given function, it is necessary to know the range of parameter values for which the method can be shown to work accurately and reliably. Without this information it is impossible to assess what degree of reliability can be placed in best-fit values derived from actual experimental data. Given the complexities of the iterative curve-fitting process it is difficult to produce an estimate of accuracy solely on a theoretical basis. Some functions are demonstrably harder to fit than others for reasons related to the shape of their sum of squares function surface. These difficulties can be circumvented by validating the curve fitting process using simulated data generated with known parameters over the range expected to be found experimentally.

Such considerations are of particular importance when considering fits to exponential functions which are known to be difficult. For example, the time course of an exponential relaxation is 98.2% completed within a period of time equal to four time constants. Therefore, such signals should be acquired and stored in digital records of equal or slightly longer duration. This does not always prove possible, due to an unexpectedly long time constant, or part of the curve being obscured by an artifact. Given such limitations, it is useful to know what proportion of the complete exponential curve is required for accurate estimation of the time constant. For instance, a curve fit to a data set lasting 10 ms is unlikely to yield an accurate estimate of an exponential process with a time constant of 1 s. Similarly, it is useful to know the effect of random background noise on the estimation of the time constant.

As for the comparison between fitting algorithms, a series of simulated data records containing exponential functions were computed using Equation 6.11 with parameters $A_0 = -50$, $A_\infty = 100$, and τ within the range 10–150. A small amount of gaussian random noise (rms amplitude 1 unit) was added to each record. For each value of τ studied, four records were generated with different random noise patterns. Equation 6.11 was then fitted to the data records using the NAG E04GCF quasi-Newton fitting routine, producing a set of best fit parameters for each record. The error in the best-fit results (derived from the standard deviation of the best-fit parameters) was expressed as a percentage of the true parameter value.

Figure 6.7 (a) shows the fitting errors for τ and A_0. When the record contained at least two time constants of the exponential time course, the fitting procedure succeeded in reliably determining both parameters to within 5%. However, with less than two time constants the error in the estimation, of τ in particular, started to increase markedly to 9% at $\tau = 0.6$. It is also worth noting that the error in τ is always greater than that for A_0. Figure 6.7 (a) is only one of a family of curves required to validate the fitting process fully.

6.20 ERROR PROFILE FOR THE DOUBLE EXPONENTIAL

Another common parameter estimation problem is the determination of amplitude and time constants of the sum of a pair of exponentials. Examples of double exponential curves in electrophysiological signals are numerous and appear wherever multi-stage kinetic processes operate. The decay of endplate currents in the presence of ion channel blocking drugs (Ruff, 1977) and single ion channel open or closed time distributions (Colquhoun & Hawkes, 1983) are some examples. Curve fitting is used here to separate out the two components and estimate their time constants and amplitudes. It is useful to know what degree of separation between components is necessary for their time constants to be accurately measured.

Figure 6.7 (b) shows the results from a

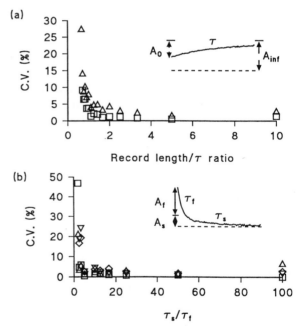

Figure 6.7 (a) Error in estimation of amplitude (A, ■) and time constant (τ, ▲) for an exponential function (Equation 6.11) fitted to 100 point simulated data sets using the NAG E04GCF quasi-Newton least squares iterative fitting routine. True parameters, $A_0 = -50$, $A_\infty = 100$, $\tau = 10\ldots 150$. Error expressed as a percentage of the true parameter value, plotted vs the number of time constants contained in the data record (length 100 units). An example of one of the simulated data sets is shown inset. (b) Error in estimation of parameters of a sum of two exponentials (A_f ■, τ_f ▲, A_s ▼, t_s ♦) (Equation 6.12) fitted to 100 point simulated data sets. True parameters $A_f = 80$, $\tau_f = 0.5 \ldots 25$, $A_s = 20$, $\tau_s = 20$, an example shown inset. Error plotted vs τ_s/τ_f ratio.

series of simulated double exponential curves generated using Equation 6.12 with the τ_s/τ_f ratio varying between 2 and 100, and A_f/A_s fixed at 5. For τ_s/τ_f ratios between 5 and 50 the curve-fitting procedure proves capable of correctly estimating all parameters with an error of 5% or less. A small increase in error can be seen to occur for $\tau_s/\tau_f = 100$. This is not surprising since this corresponds to a rapidly decaying exponential component with $\tau_f = 0.5$, which is represented in only the first three points of the data set. Since an exponential decay requires at least three points to define it this represents the upper limit for measurable τ_s/τ_f ratios within a 100 point data set. Interestingly, it seems possible to approach this limit without an excessive increase in error.

In contrast to the results for large τ_s/τ_f ratios, for values less than five the error rapidly escalates to almost 50% at $\tau_s/\tau_f = 2$. These results indicate that curve-fitting methods cannot be expected accurately to resolve exponential components whose time constants differ by less than a factor of five. As mentioned above, this is a property of the exponential function itself and the topography of its least-squares function surface. It is unlikely that other curve-fitting methods would produce better results and some, like the simplex method, produce worse results.

Simulation studies, such as the two just discussed, are a valuable aid in interpreting experimental results derived using curve-fitting procedures. It could even be argued that curve-fitting procedures should not be used unless the range of meaningful parameter values have been determined. Such prior knowledge can often be valuable, in that it may be possible to alter experimental conditions to ensure that experimental parameters fall within an interpretable range. Also, it is often found that the exercise of creating and analysing the simulated data gives the experimenter an insight into the nature of the mathematical functions being used, of benefit when interpreting the real results.

6.21 CHOOSING THE BEST MODEL TO FIT THE DATA

Although curve-fitting procedures can determine the best-fit parameters for a given mathematical model of the observed data, they do not directly determine whether that model is a good one, in the sense that it describes the data well, or whether a better model might exist. Models are often chosen based on the intuition of the researcher or derived from a theory of the physiological mechanisms underlying the phenomena under study. One of the most common situations where choices have to be made between models occurs when analysing signals which exhibit multi-exponential time courses. Usually, there is no *a priori* knowledge of the number of exponential components to be expected. In fact, determining the number of exponentials objectively from the experimental data provides important information concerning the number of distinct observable kinetic states of the system under study.

Debates have arisen in the past as to the appropriate number of exponential components required to describe experimentally observed signals. For instance, drugs which block the ion channel at the neuromuscular junction endplate typically alter the shape of the, normally exponential, decay phase of the endplate current. Some studies have reported that the decay phase is represented by the sum of two exponentials (Ruff, 1977) whereas other have reported that three exponentials are necessary (Beam, 1976). Such differing results may, of course, be due to real differences in the experimental procedures, drugs, or the cells used by the particular experimenters. However, it may also be due to differences in the curve-fitting procedures used and the criteria the authors used to discriminate between competing models.

Clearly, quantitative criteria are required to compare the relative merits of different models. A typical example of a model selection problem can be seen in Figure 6.8 which shows (a) an exponential, (b) sum of two exponentials and (c) sum of three exponentials fitted to an experimental data set. In order to illustrate the effectiveness of the model selection procedures the data set has been generated using a two exponential function with known parameters. The curve fits were carried out using the Levenberg–Marquardt algorithm within the Biosoft Fig.P software package. The best-fit results are shown in Table 6.2.

Table 6.2 Model discrimination. Best fit results from 1, 2 and 3 exponential models to the simulated two exponential curve shown in Figure 6.8, with chi-square and residual runs test results, parameter standard errors and correlations derived from Hessian matrix. Curve fits performed with the Biosoft Fig.P program using the Levenberg–Marquardt method.

	A_1	τ_1	A_2	τ_2	A_3	τ_3	χ^2 (n_f)	Runs (n_+, n_-)
Simulation								
$A_1 \exp(-x/\tau_1) + A_2 \exp(-x/\tau_2)$								
True value	80	5	20	50				
Curve fits								
$A_1 \exp(-x/\tau_1)$								
Best fit	83	11.2					2969 (98)	24 (47,53)
s.e.	0.55	0.11					$p \approx 0$	$p = 5.5 \times 10^{-8}$
$A_1 \exp(-x/\tau_1) + A_2 \exp(-x/\tau_2)$								
Bestfit	80.6	4.97	19.6	51.0			103.46 (96)	57 (45,55)
s.e.	0.91	0.11	0.72	2.11			$p = 0.29$	$p = 0.078$
Correlation								
A_1	1	0.12	−0.58	0.55				
τ_1		1	−0.77	0.64				
A_2			1	−0.91				
τ_2				1				
$A_1 \exp(-x/\tau_1) + A_2 \exp(-x/\tau_2) + A_3 \exp(-x/\tau_3)$								
Best fit	69.6	4.84	11.1	5.82	19.5	51.2	109 (94)	57 (45,55)
s.e.	2640	16.9	2638	128	2.0	4.3	$p = 0.14$	$p = 0.078$
Correlation								
A_1	1	−1.0	0.99	0.99	−0.79	0.71		
τ_1		1	−0.99	0.99	−0.77	0.68		
A_2			1	−0.99	0.79	−0.71		
τ_2				1	−0.81	0.73		
A_3					1	−0.97		
τ_3						1		

6.22 CHI-SQUARE AND VARIANCE RATIO TESTS

A useful indicator of the quality of the fit between a model and experimental data is the magnitude of the residual sum squares which has been used to guide the fitting process. In general, it is to be expected that better fitting models will produce smaller residuals. The chi-square statistic described in Section 5.15, is closely related to the sum of squares and can be used to determine to a given level of significance, whether a model is consistent with the experimental data. A chi-square analysis indicates that the single exponential (a), yielding $\chi^2 = 2969$ with 98 degrees of freedom (number of x,y pairs of data points − number of model parameters) is inadequate to fit the experimental data since $p(\chi^2 \geq 2969, n_f = 98) \approx 0$. On the other hand both the two and three exponential functions are adequate fits to the data with χ^2 values of 103 ($p(\chi^2 \geq 103, n_f = 96) = 0.29$) and 109 ($p(\chi^2 \geq 109, n_f = 96) = 0.14$) respectively.

While the chi-square test has allowed us to reject the single exponential model it does not, in itself, provide a means of distinguishing between the other two models which have very similar χ^2 values. However it can be argued on the principle of parsimony, that if two models provide fits of similar quality the one with the fewer numbers of parameters should be chosen which promotes the choice of the two exponential model (four parameters) over the three exponential (six parameters).

The exponential models discussed here can be described as a set of *nested* models, in that the one and two exponential functions are representable as sub-sets of the three exponential functions. For such models a test similar to the variance ratio (Section 5.11) can be applied to determine whether the residual sums of squares yielded by two models are significantly different. For models f and g from a nested set, where f is a subset of g, the statistic

$$F = \frac{S_f - S_g}{S_g} \frac{n - k_g}{k_f} \quad [6.29]$$

where S_f, k_f and S_g, k_g are the residual sum of squares and number of parameters for each model and n is the number points in the data set. The significance probability is determined from the F distribution with k_f and $n - k_g$ degrees of freedom. Applying Equation 6.29 to the two and three exponential models, yields an F value close to zero indicating that the residual sums of squares are not significantly different.

These tests can be used to implement a model selection strategy for a set of nested models such as the sums of exponentials. Starting with the model with the least number of parameters, a series of models are fitted to the data until the sum of squares ceases to improve. The chi-square test is used to reject those models which to not fit the data. Of the remaining valid models, the variance ratio test is used to select the model with the least number of parameters which does not have a significantly higher residual sum of squares than the last model in the sequence fitted. On this basis the single exponential model is rejected by the chi-square test because it does

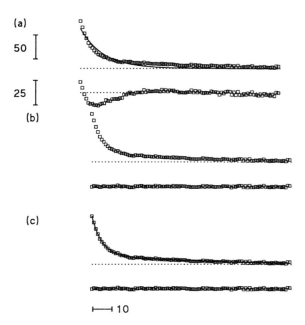

Figure 6.8 Residual difference plots illustrating fitting errors for single (a), double (b) and triple (c) exponential functions fitted to a 100 point simulated data record containing a double exponential function. Best fit lines for each function are shown superimposed upon the data points. Residual differences between data points and best fit are shown below each figure (scaled up ×2 to make difference more apparent).

not fit the data, and the three exponential is rejected because its residual variance is not significantly different from the two exponential, leaving the two exponential as the model of choice.

A different approach must be used when non-nested models are being compared. A set of models are described as *non-nested* when they have sufficiently different mathematical forms that no model can be simply expressed as a subset of another. For instance, the exponential and the hyperbolic (Equation 6.24) function make up a non-nested set. The chi-square test can still be used, but may suggest that both functions provide adequate fits. In such circumstances, the *log error ratio (LER)* (Horn, 1987; Akaike, 1974) can be used to discriminate between models. The LER is computed from the residual sum of squares for each model f and g

$$LER = \log_e \left(\frac{S_f}{S_g} \right) \quad [6.30]$$

A positive LER value would suggest that model g is a better fit than f and vice versa. Unfortunately it is not possible to place an exact significance probablity on the LER value (Horn, 1987). However the *asymptotic information criterion (AIC)* (Akaike, 1974)

$$AIC = \frac{2(k_g - k_f)}{n} \quad [6.31]$$

is often used as a critical value with the two models being deemed not significantly different if $|LER| < AIC$. Further details of these approaches to mode discrimination can be found in Horn (1987), Rao (1973), Schwarz (1978) and Leamer (1983).

6.23 DISTRIBUTION OF RESIDUALS

An alternative approach to determining whether a model fits the experimental data set is to observe the distribution of the residual differences between the fitted function and the data points. The residuals for the three exponential models are shown in Figure 6.8. Residual differences, arising purely from background noise, should be randomly distributed with no correlation between neighbouring points. Consequently, long sequences or runs of adjacent residuals with the same sign are unlikely to occur. Conversely, poorly fitting models show non-random residual distributions, consistently positive in one region and negative in another. For instance, the residuals plot in Figure 6.8 (a) shows clearly that the single exponential function does not fit the data well, with the residuals clustered into three distinct phases.

The degree of clustering within the residual is used by the *runs* test as a means of assessing goodness of fit. A run is defined as a series of residuals of the same sign. If the model provides a good fit to the data, the residuals should be distributed randomly, resulting in a large number of short runs, no more than a few data points in length. On the other hand, poor fits should result in a small number of long runs.

The number of discrete runs (positive or negative) observed within the set of residuals is used as the test statistic u. The probability of observing u (or fewer) runs is determined as a function of the totals of positive and negative residuals $p(\leq u, n_+, n_-)$. These can be determined from tables, but for large data sets $(n_+, n_- > 10)$ a normal approximation exists (Rawlings, 1988)

$$z = \frac{u - \dfrac{2n_+ n_-}{n_+ + n_-} - \dfrac{1}{2}}{\sqrt{\left(\dfrac{2n_+ n_- (2n_+ n_- - n_+ - n_-)}{(n_+ + n_-)^2 (n_+ + n_- - 1)} \right)}} \quad [6.32]$$

where z is the standard normal deviate and the significance probability is obtained as $p(\leq z)$. The results of the runs test applied to the three exponential fits can be seen in Table 6.2. Only 24 runs were observed in the residuals for the single exponential for a total of 47 positives and 53 negatives. The probability of this occurring by chance, computed using Equation 6.32 and the normal area approximation from Section 5.3, is $p(z \leq 5.31) = 5.48 \times 10^{-8}$, clearly indicating that the residuals are not distributed uniformly. On the other hand, both the two and three exponential fits yielded 57 runs with a non-significant probability of $p(z \leq 1.42) = 0.078$.

It should be noted that the runs test is based upon the assumption that residuals are independent of each other, i.e. that the probability of a residual taking a positive or negative value is not related to the polarity of adjacent values. This may not always be the case, especially when the data points are obtained from a signal record which has been low pass filtered (see Figure 3.3, for instance). In such circumstances, the runs test should be used with caution, perhaps reducing the number of data points in the set being fitted, and only rejecting the hypothesis that the residuals are homogeneous if a high level of significance is obtained (e.g. < 0.0001).

6.24 STANDARD ERROR OF BEST-FIT PARAMETERS

Both the runs test and the chi-square test can determine whether or not a model seems to fit the data. For instance, both tests indicate that the single exponential function does not fit the data set well. However, they do not help to discriminate between models which fit the data equally well, such as the two and three exponential functions. In such circumstances the standard error of the best-fit parameters themselves can provide additional indications as to whether a model has too many parameters.

As was discussed in Section 6.16, when using a curve-fitting method such as the Gauss–Newton or Levenberg–Marquardt methods, the parameter standard errors can be computed from the Hessian matrix. These values provide a measure of how tightly the parameters are specified by the fitting process. A large standard error indicates that a parameter is poorly specified and could be varied to a large degree without greatly degrading the quality of the fit. In general, a model which has more parameters than necessary to fit the data tends to produce poorly specified parameters.

This can be seen from the parameter standard errors for the the three models shown in Table 6.2. The three exponential model has much larger standard errors than the two exponential, in some cases orders of magnitude larger than the parameter values themselves. For instance, the amplitude A_2 is 9.88 ± 2219 which is consistent with $A_2 = 0$, i.e. the exponential component not actually existing. Such poor localization of a parameter strongly suggests that a third exponential is not necessary to fit the data.

The Hessian matrix also provides covariance information indicating the degree to which parameters are correlated with each other. A high degree of correlation between parameters also indicates that there are too many parameters in the model. The correlation matrix for the two and three exponential models is shown in Table 6.2. A value of one indicates perfect correlation while zero indicates no correlation. A high degree of correlation between parameters also indicates that there are too many parameters in the model. In the case of the two exponential model, parameter correlations lie in the range 0.12–0.91 compared with a range of 0.71–0.99 for the three exponential model with six parameters actually in the range 0.99–1.

6.25 SUMMARY

Mathematical modelling and iterative curve fitting provide a powerful tool for the development and quantitative testing of theories concerning the mechanisms underlying electrophysiological signals. With the increased availability of both sufficiently powerful personal computers and appropriate software, such as the NAG library or applications programs such as Fig.P or Sigmaplot, the technique is now accessible to most laboratories.

However, curve fitting is not without its pitfalls. It cannot be guaranteed that an iterative fitting routine will necessarily converge upon a meaningful set of best-fit parameters, or converge at all. Modern gradient-based methods such as Levenberg–Marquardt or Gauss–Newton appear to be more robust and have a wider range of applicability than direct search methods such as the simplex method. Nevertheless, it is essential to test any method against realistic data sets generated from simulations with known parameter values, in order to build up an awareness of the limits of the effective range of the curve-fitting process. Similarly, discriminating which of several potential models provides the best fit to the data must also be done with care.

It should be noted that the issue of the relative merits of the simplex versus gradient curve-fitting methods in general is not as clear cut as it may seem from this discussion of least-squares curve fitting. In particular, the simplex method performs much better when applied to other fitting methods such as the maximum likelihood techniques for fitting probability density functions to distributions to be discussed in Chapter 8.

Mathematical modelling and curve fitting

```fortran
c       Listing 6.1. Exponential function fitting routine.
c       Finds the best fitting exponential curve to a set of data points
c       using logarithmic transform and linear least squares.
        subroutine exponential_fit( x, y, np, invert, amplitude, tau, weight )
        real x(np),y(np)                        ! X and Y data arrays (IN)
        integer*2 np                            ! No. of pairs of data points (IN)
        logical invert                          ! If invert = .TRUE. invert signal before fit.
        real*4 amplitude,tau                    ! Best fit amplitude and time constant (OUT)
        real*4 weight(np)                       ! Working array for weights

        if( invert ) then                       ! If we are dealing with
                do i = 1,np                     ! negative-going signal
                        y(i) = -y(i)            ! invert it to allow log.
                end do
        end if

        i= 1
        do while( (y(i).gt.0.) .and. (i.le.np) )   ! Convert to log(e)
                weight(i) = y(i)*y(i)              ! and calculate weighting
                y(i) = log(y(i))                   ! factor
                i = i + 1
        end do                  ! Stop when first negative value encountered
        nfit = i - 1

        call least_squares( x, y, weight, nfit, slope, y_intercept )
        amplitude = exp( y_intercept )
        if( invert ) amplitude = -amplitude
        tau = -1./slope
        return
        end

        subroutine least_squares(x,y,weight,nfit,slope,y_intercept)
c       Find best-fit straight line ( y = slope*x + y_intercept)
        real*4 x(nfit),y(nfit)                  ! X and Y data point
        real*4 weight(nfit)                     ! Error variance weighting factor
        integer*2 nfit                          ! No. of data point pairs (IN)
        real*4 slope,y_intercept                ! best fit parameters

        xavg = 0.                               ! Find mean values
        yavg = 0.                               ! of X and Y data
        do i = 1,nfit
                xavg = xavg + x(i)
                yavg = yavg + y(i)
        end do
        xavg = xavg / float(nfit)
        yavg = yavg / float(nfit)

        xdiff = 0.
        ydiff = 0.
        do i = 1,nfit
                xdiff = weight(i)*(x(i) - xavg)*(x(i) - xavg) + xdiff
                ydiff = weight(i)*(y(i) - yavg)*(x(i) - xavg) + ydiff
        end do
        slope = ydiff / xdiff
        y_intercept = yavg - slope*xavg
        return
        end
```

```
c       Listing 6.2. Simplex iterative curve-fitting routine.
c       An iterative least squares routine for fitting a 3 parameter exponential
c       function using Nelder & Mead's simplex method.
        subroutine fit_exponential( par, npar, x, y, np )
        real*4 x(np),y(np)                          ! (In) Data point arrays
        real*4 par(npar)                            ! (In/Out) function pars.
        integer np                                  ! (In) No. of data points
        integer npar                                ! (In) No. of parameters

        xmax = -1E30                                ! Normalize x and y data
        ymax = -1E30                                ! to lie in range +/-1
        do i = 1,np
            if( abs(x(i)) .gt. xmax ) xmax = abs(x(i))
            if( abs(y(i)) .gt. ymax ) ymax = abs(y(i))
        end do
        do i = 1,np
            x(i) = x(i)/xmax
            y(i) = y(i)/ymax
        end do
        par(1) = (y(1) - y(np))                     ! Define initial values
        par(2) = (x(np) - x(1))/3.                  ! of function parameters
        par(3) = y(np)

        call simplex( par, npar, iteration, fmin, x, y, np )  ! Do simplex curve fit

        par(1) = par(1)*ymax                        ! Correct best fit
        par(2) = par(2)*xmax                        ! for data normalization
        par(3) = par(3)*ymax
        return
        end

        subroutine simplex(par,npar,iteration,fmin,x,y,np)
c       Simplex algorithm for finding the minimum of the function sum_of_squares
        real*4 par(npar)                            ! (In)=Initial, (Out)=best fit pars.
        real*4 x(np),y(np)                          ! (In) Data arrays
        real*4 fmin                                 ! (Out) Minimum sum of squares
        integer iteration                           ! (Out) No. of iterations to convergence
        real*4 ssq(10),s(10,10),centroid(10)        !
        parameter(max_iteration = 2000 )
        logical converged

        ep = epsilon(ep)*2.                         ! Numerical precision
        nvertices = npar+1
        ilow = 1
        do ipar = 1,npar
            s(ilow,ipar) = par(ipar)
        end do
        iteration = 0
        converged = .false.
c ... continued
```

```fortran
c Listing 6.2 (b)
          do ivertex = 1,nvertices                  ! Compute initial simplex
             if( ivertex .ne. ilow ) then           ! with npar+1 vertices
                do ipar = 1,npar                    ! from initial parameter
                   s(ivertex,ipar) = s(ilow,ipar) * ! guesses supplied
     &                (float(ivertex)/float(nvertices) +0.5)
                   par(ipar) = s(ivertex,ipar)
                end do
             endif
             ssq(ivertex) = sum_of_squares(par,x,y,np)
          end do

c     Iterative loop to minimize sum_of_squares
          do while( (.not. converged) .and. (iteration.lt.max_iteration) )
             iteration = iteration + 1
             ilow = 1
             ihigh = 2
             do ivertex = 1,nvertices                       ! Find
                if( ssq(ivertex) .lt. ssq(ilow) ) ilow = ivertex  ! highest
                if( ssq(ivertex) .gt. ssq(ihigh) ) ihigh = ivertex ! and lowest
             end do                                         ! vertices
             do ipar = 1,npar
                centroid(ipar) = 0.                         ! Find centroid
                do ivertex = 1,nvertices
                   if( ivertex .ne. ihigh ) then
                      centroid(ipar) = centroid(ipar) +
     &                   s(ivertex,ipar)/float(npar)
                   endif
                end do
             end do
             do ipar = 1,npar                               ! Find reflection of
                par(ipar) = 2.*centroid(ipar) - s(ihigh,ipar) ! highest vertex
             end do                                         ! in centroid.
             f = sum_of_squares(par,x,y,np)
             if( f .le. ssq(ilow) ) then                    ! If it is better than
                do ipar = 1,npar                            ! current lowest,
                   s(ihigh,ipar) = par(ipar)                ! keep it.
                end do
                ssq(ihigh) = f
                ilow = ihigh
                do ipar = 1,npar                            ! Try a further
                   par(ipar) = 2.*par(ipar) - centroid(ipar) ! expansion along
                end do                                      ! the Vhigh-centroid line
                f = sum_of_squares(par,x,y,np)
                if( f .le. ssq(ilow) ) then                 ! If expansion is
                   do ipar = 1,npar                         ! even lower
                      s(ihigh,ipar) = par(ipar)             ! keep it.
                   end do
                   ssq(ihigh) = f
                endif
c ... continued
```

```
c Listing 6.2 (c)
      else
          n = 0                                         ! Expansion strategy has
          do ivertex = 1,nvertices                      ! failed.
              if( f .lt. ssq(ivertex) ) n = n + 1
          end do
          if( n .gt. 1 ) then                           ! Reflection is, at least,
              do ipar = 1,npar                          ! better than the current
                  s(ihigh,ipar) = par(ipar)             ! two highest vertices,
              end do                                    ! so substitute it for the
              ssq(ihigh) = f                            ! highest
          elseif( n .eq. 1 ) then
              do ipar = 1,npar
                  s(ihigh,ipar) = par(ipar)
                  ssq(ihigh) = f
              end do
              do ipar = 1,npar
                  par(ipar) = (par(ipar) + centroid(ipar))/2.
              end do
              f = sum_of_squares(par,x,y,np)
              if( f .lt. ssq(ihigh) ) then
                  do ipar = 1,npar
                      s(ihigh,ipar) = par(ipar)
                      ssq(ihigh) = f
                  end do
              endif
          else
              do ipar = 1,npar                          ! Contract the size
                  par(ipar) = (centroid(ipar)+s(ihigh,ipar))/2.
              end do
              f = sum_of_squares(par,x,y,np)
              if( f .lt. ssq(ihigh) ) then              ! If this result is better
                  do ipar = 1,npar                      ! than the current worst
                      s(ihigh,ipar) = par(ipar)         ! parameter set,
                  end do                                ! incorporate it into the
                  ssq(ihigh) = f                        ! simplex
              else
                  do ivertex = 1,nvertices              ! No improvement
                      if( ivertex .ne. ilow ) then      ! by contraction
                          do ipar = 1,npar              ! so shrink
                              s(ivertex,ipar) =         ! the whole
     &                          (s(ivertex,ipar) + s(ilow,ipar))/2.   ! simplex around
                              par(ipar) = s(ivertex,ipar)             ! best vertex
                          end do
                          ssq(ivertex) =
     &                        sum_of_squares(par,x,y,np)
                      endif
                  end do
              endif
          endif
      endif
c ... continued
```

c Listing 6.2 (d)

```
        ilow = 1
        ihigh = 2                                          ! TERMINATION CRITERIA
        do ivertex = 1,nvertices
           if( ssq(ivertex) .lt. ssq(ilow) ) ilow = ivertex   ! Find
           if( ssq(ivertex) .gt. ssq(ihigh) ) ihigh = ivertex ! highest
        end do                                             ! and lowest
                                                           ! vertices
        converged = .false.
        if( (ssq(ihigh)-ssq(ilow)) .le. ep ) then          ! Terminate if sum of
           converged = .true.                              ! squares between
        end if                                             ! highest and lowest
                                                           ! vertex is at limit
        do ipar = 1,npar                                   ! Terminate if any
           if( abs(s(ihigh,ipar)-s(ilow,ipar)).le. 2.*ep ) then ! of the simplex
              converged = .true.                           ! dimensions has
           end if                                          ! collapsed
        end do

        if( converged ) then
           ssq_min = 1E30                                  ! If the simplex
           do istep = -1,1,2                               ! algorithm has
              do ipar = 1,npar                             ! converged, check
                 par(ipar) = s(ilow,ipar) + 0.001*float(istep) ! that this is a
                 ssq_min = sum_of_squares(par,x,y,np)      ! real minimum.
                 if( ssq_min .lt. ssq(ilow) ) then         ! If not, re-start
                    do i = 1,npar                          ! the algorithm
                       s(ilow,i) = par(i)
                    end do
                    ssq(ilow)=ssq_min
                    converged = .false.
                 endif
              end do
           end do
        endif
     end do
     do ipar = 1,npar
        par(ipar) = s(ilow,ipar)
     end do
     fmin = ssq(ilow)
     return
     end


     real*4 function sum_of_squares(par,x,y,np)
     real*4 par(1),x(np),y(np)
                                                           ! Residual sum of
     sx = 0.                                               ! squares for
     do i = 1,np                                           ! exponential
        r = y(i) - (par(1)*exp(-x(i)/par(2)) + par(3))     ! function
        sx = sx + r*r
     end do
     sum_of_squares = sx
     return
     end
```

```
c     Listing 6.3. Quasi-Newton iterative curve fitting using NAG library
c     to fit a 3-parameter exponential curve ( y = p(1)*exp(-x/p(2)) + p(3) )

      subroutine fit_exponential( xin, yin, np, p, se )
      parameter(npmax=200,npar=3,liw=1,lw=(8+2*npar+2*npmax)*npar+3*npmax)
      real*8 xin(np), yin(np), p(npar), se(npar)
      real*8 w(lw),sumsq,xmax,ymax,x(npmax),y(npmax),cvar(npar)
      integer iw(liw)
      common /fitcom/ x,y
      xmax = -1E30
      ymax = -1E30
      do i = 1,np
          x(i) = xin(i)                              ! Copy x,y data arrays to be fitted
          y(i) = yin(i)                              ! into LSFUN2 common block.
          xmax = max( abs(x(i)), xmax )              ! Find max. absolute value
          ymax = max( abs(y(i)), ymax )              ! of x,y data
      end do
      do i = 1,np                                    ! Scale x,y data points
          x(i) = x(i)/xmax                           ! to lie in the range
          y(i) = y(i)/ymax                           ! -1 to 1.
      end do
      p(1) = (y(1)-y(np))                            ! Calculate initial parameter
      p(2) = (x(np) - x(1))/3.                       ! needed to start off curve fit.
      p(3) = y(np)
      ifail = 1
      call E04GCF( np, npar, p, sumsq, iw, liw, w, lw, ifail )   ! Call fitting routine
      p(1) = p(1)*ymax                               ! Correct best-fit parameters
      p(2) = p(2)*xmax                               ! for effects of data scaling
      p(3) = p(3)*ymax
      ns = 7*npar + 2*np + npar*np + (npar*(npar+1))/2 +
     & 1 + max(1,(npar*(npar-1))/2)                  ! Calculate variance of
                                                     ! best-fit parameters
      nv = ns + npar                                 ! from Jacobian matrix (w(ns))
      ifail = 1
      call E04YCF(0,np,npar,sumsq,w(ns),w(nv),npar,cvar,w,ifail) ! created by E04GCF
      se(1) = dsqrt(cvar(1))*ymax                    ! Convert to standard deviation
      se(2) = dsqrt(cvar(2))*xmax                    ! and re-scale
      se(3) = dsqrt(cvar(3))*ymax
      return
      end

      subroutine lsfun2( np, npar, p, f, gradient, lg )
c     Called by E04GCF to calculate array of residuals f(np) and gradients
c     in gradient(np,npar), using set of parameters supplied in p(npar) and x,y data
c     supplied via common block /fitcom/
      integer np,npar,lg
      real*8 p(npar),f(np),gradient(lg,npar)
      parameter(npmax=200)
      real*8 x(npmax),y(npmax)
      common /fitcom/ x,y
      do i = 1,np
          f(i) = (p(1)*exp(-x(i)/p(2)) + p(3)) - y(i)              ! f(p)
          gradient(i,1) = exp(-x(i)/p(2))                          ! df/dp1
          gradient(i,2) = p(1) * (x(i)/(p(2)*p(2))) * exp(-x(i)/p(2)) ! df/dp2
          gradient(i,3) = 1.                                       ! df/dp3
      end do
      return
      end
```

CHAPTER SEVEN

Analysis of voltage-activated currents

Having established the general basis of electrophysiological signal analysis, in this and the two remaining chapters, we now turn to the discussion of the analysis techniques which are specific to a number of particularly important electrophysiological methods. This chapter will cover the procedures involved in the recording and analysis of the signals produced using the voltage clamp technique, in particular the analysis of the voltage-activated currents which underlie the generation and propagation of the action potention in excitable cells. The development of the voltage clamp allowed these currents to be studied for the first time, and greatly advanced our understanding of the electrical activity of cells.

7.1 THE VOLTAGE CLAMP

Under normal conditions the cell membrane potential is free to vary in response to changes in the trans-membrane current flow. If the membrane conductance mechanisms are themselves potential- and time-dependent, an initial change in conductance causes a change in membrane potential which then feeds back, further changing the conductance. The effects of such feedback can be dramatic producing the wide variety of cell action potential waveforms seen in nerve, muscle and other cells. In such unconstrained conditions it is difficult to derive the behaviour of the membrane currents unambiguously from potential measurements alone. A comprehensive introduction to ion currents and their role in cell excitability can be found in Hille (1984).

The development of the voltage clamp technique by K.S. Cole (Cole, 1949; Cole & Moore, 1960) provided an elegant solution to this problem. The voltage clamp is essentially a feedback control system for fixing the cell membrane potential at a level set by the experimenter, in spite of changes in membrane current, thus removing the complicating current/

membrane potential interactions. The voltage clamp allows membrane currents to be studied directly under simpler, more controlled conditions. This technique allowed Hodgkin and Huxley to perform their classical quantitative analysis of the role of the Na$^+$ and K$^+$ conductances in the generation of the nerve action potential (Hodgkin & Huxley, 1952a–d). The voltage-clamp is now one of the basic tools for studying ionic conductance mechanisms in excitable cells.

Voltage clamping can be achieved with a wide range of techniques and types of instrumentation (Cole & Moore, 1964; Nonner, 1969; Adrian et al., 1970; Katz & Schwarz, 1974; Hille & Campbell, 1976; Smith et al., 1980; Finkel & Redman, 1984), but all are essentially varieties of feedback control systems which measure the cell membrane potential and inject an appropriate amount of current into the cell to maintain the potential constant, in spite of changes in membrane conductance. This chapter is only concerned with the analysis of the current and voltage signals derived from the voltage clamp and consequently a detailed treatment of its operation will be avoided. A discussion of the the principles and practice of voltage clamping can be found in Standen et al. (1987).

7.2 MEMBRANE EQUIVALENT CIRCUIT MODEL

In order to make the principles of analysis understandable it is helpful to outline a few essential elements of the theoretical basis for cell membrane ionic currents. Again, the reader is directed to Hille (1984) for a more substantial treatment of this subject. The electrical potential of a cell is regulated by a heterogeneous population of ion channels within the cell membrane which allow a variety of ions to pass in and out of the cell. Ion channels are more or less selective for particular types of ions. Some are cation selective passing a wide range of positive ions but not negative (e.g. the endplate channel), others are highly selective for specific ions (e.g. Ca^{2+} channels, K$^+$ channels, etc.). The electrical behaviour of a cell can be represented by means of an electrical *equivalent circuit* model, consisting of a set of voltage sources and conductances in parallel (one for each type of channel existing in the membrane), with a capacitance element representing cell membrane capacity (Hodgkin & Huxley, 1952d).

Figure 7.1 shows the equivalent circuit model for a cell with three ionic conductances, selective for Na$^+$, K$^+$ and Cl$^-$. The species of ion passed by a channel, combined with the relative concentration of these ions on either side of the cell membrane, determines the direction of the current flow through the channel. At some potentials the net current flow will be inward, at others it will be outward, and a potential can be found where the net flow is zero – the *reversal potential*. Under voltage clamp conditions, the current flowing through one conductance arm of the circuit is

$$I_i = G_i(V_m - V_i) \quad [7.1]$$

where i is the ion (Na$^+$, K$^+$, Cl$^-$), V_m is the cell membrane potential. V_i is the potential of the voltage source for the ion which can be seen from Equation 7.1 to be the reversal potential for I_i. In the case of a channel which is selective for a single ion, V_i is related to the intra- and extracellular concentrations of that ion by the Nernst equation

$$V_i = \frac{RT}{F} \log_e \frac{[I]_o}{[I]_i} \quad [7.2]$$

The conductance term G_i is not generally a constant value. It may be a function of membrane potential V_m and may also have time-dependent properties when V_m is changed. A key element of voltage clamp experimentation is the observation of how G_i behaves in response to voltage stimuli.

It is worth noting that the equivalent circuit model described here predates the development of the patch voltage clamp, and the ionic conductances in the model represent the voltage- and time-dependent properties of *populations* of ion channels, observable with whole cell voltage clamp techniques. This approach still proves to be useful, even though it is now possible to observe the current flowing through single

Analysis of voltage-activated currents 135

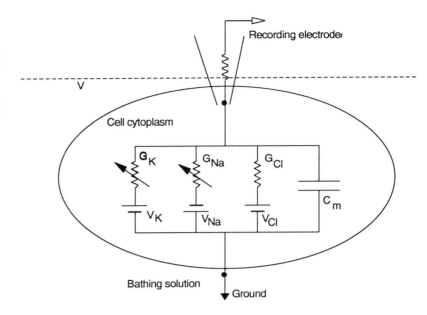

Figure 7.1 Electrical equivalent circuit model of a simple spherical cell permeable to Na$^+$, K$^+$ and C$^-$ ions. Each ion is represented by a conductance (G_K, G_{Na}, G_{Cl}) and an electrochemical voltage source (V_K, V_{Na}, V_{Cl}). G_K and G_{Na} are active voltage-sensitive conductances while G_{Cl} is constant.

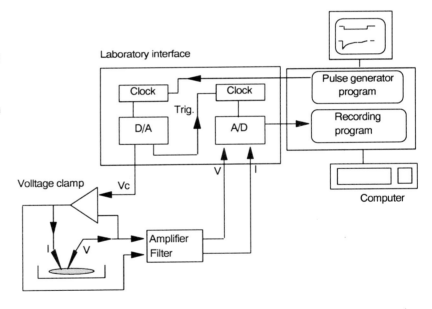

Figure 7.2 A computer-based voltage clamp command and recording system. Software provides families of command voltage pulses at timed intervals and simultaneously records resulting cell current and voltage, stores the data on disc, and displays waveforms on screen.

channels directly. This chapter is restricted to the analysis of such whole cell currents with techniques for single-channel analysis treated separately in Chapter 8.

By stepping the membrane potential to different holding potentials and observing the direction of the resulting current flow, it is possible to determine the reversal potential. Repeating the experiment in bathing solutions of different ionic compositions, reveals how V_i depends on the concentration of different ions. In summary, the aims of voltage clamp experimentation is to characterize the range of ionic conductances within a cell membrane and to determine their selectivity for different ions and other properties. This is done by applying a variety of voltage stimuli to the cell and observing the current responses.

7.3 DIGITAL RECORDING OF VOLTAGE CLAMP SIGNALS

Typical implementations of computer-based voltage clamp recording systems have certain characteristic features. The voltage clamp circuitry provides two analog output signals; cell membrane potential and current. On the input side, the voltage clamp requires a *command voltage* which allows the user to define the potential at which the cell is to be clamped and to apply a variety of voltage stimulus waveforms. A typical voltage clamp experiment consists of recording the current responses to command voltage waveforms which may differ in amplitude shape and duration, in order to infer the underlying potential- and time-dependence of the ionic conductances being studied. Waveforms may be single or multiple rectangular steps, linear ramps or sine waves. In the past these waveforms were produced by complex and relatively inflexible hardwired timing and voltage generation circuits which often had to be built in-house.

The generation of command voltage waveforms was an obvious application of the computer and now many laboratories use computer-based systems. In fact, it is possible to combine command voltage generation with simultaneous digital recording of membrane current and potential on a single computer. Figure 7.2 shows a block diagram of such a computer-based voltage clamp control and recording system. Command voltage waveforms are generated as arrays of numbers within computer memory which are converted to analog voltage levels via a D/A converter at strictly timed intervals and output to the voltage clamp. Similary, two analog channels are digitized using the A/D converter to collect the resulting membrane current and potential signals.

7.4 SIMULTANEOUS VOLTAGE GENERATION AND RECORDING

The simultaneous production of an analog voltage output waveform and digitization of analog signals is technically a difficult task, especially at the higher sampling rates (10–100 kHz) used in some voltage clamp experimental work. It can be done in a variety of ways with varying degrees of flexibility. One approach is to use the programmed data transfer technique described in Chapter 2, creating a program loop to perform the following operations

(a) get command voltage value from memory and output to digital to analog converter (DAC)
(b) wait for A/D conversion to complete
(c) read analog to digital converter (ADC) and store sample in memory
(d) repeat (a)–(c) until the required number of samples have been collected.

Although simple, this approach has the disadvantage of closely linking the timing of D/A outputs and the A/D samples. It is awkward to produce a command voltage sweep and a recording sweep of different duration and/or sampling rate. It also has all the disadvantages associated with the programmed data transfer method discussed in Section 2.7. The computer must be completely dedicated to this task alone, excluding all other operations such as monitoring the keyboard or time of day clock. It also has to be programmed in assembler language to achieve sampling rates much greater than 1–2 kHz. Nevertheless, given these limitations a great deal can be done using this approach. It was successfully used by Kegel *et al.* (1985) in the early versions of the pClamp software and was the best that could be done at the time within the limitations of the Labmaster interface.

Ideally, the timing of the A/D recording and D/A output sweeps should be quite separate and independent of each other. This requires that laboratory interface ADC and DAC hardware work independently and that two separately programmable clocks are available, one for A/D sample timing and the other D/A output timing. Similarly, two separate direct memory access or interrupt channels are also required to handle the A/D-to-host computer and host-D/A data transfer. Not all laboratory interfaces can meet these specifications. The design of the widely used Data Translation DT2801A, for instance, precludes writing to its DAC while

A/D sampling is in progress. Similarly, the DT2821 only has a single clock for both A/D and D/A timing. However, the Labmaster DMA, Data Translation DT2831, National Instruments LAB-PC and Cambridge Electronic Design 1401 all have sufficient clocks and the necessary data transfer channels.

Direct memory access (DMA) data transfer facilities are crucial for the development of a flexible high performance DAC/ADC system. DMA enhances performance by devolving the transfer of A/D samples into memory on to the DMA controller (see Sections 2.12–2.16) supporting high-speed sampling with negligible load on the computer. The CPU time freed by the DMA can then be devoted to other tasks, in this case handling the D/A output. Most interfaces can only make use of a single DMA channel (the DT2831 is an exception) so D/A output must be handled using interrupts. The relatively limited performance of interrupt driven data transfer (<10 kHz) does not usually prove to be a problem since it is rarely necessary to update the DAC at rates much greater than 1 kHz, even in circumstances where A/D sampling is being performed at 100 kHz.

A simple program using the above techniques to implement a combined DAC/ADC sweep is provided in Listing 7.1. The program produces rectangular command voltage pulses at regular intervals and records current and voltage channels from the voltage clamp. Two DAC channels are used, one producing the voltage clamp command voltage waveform the other producing a brief 5 V ADC synchronization pulse step. Two A/D channels are used, recording current and voltage. The synchronization pulse from the second D/A channel is fed into the ADC external trigger input of the laboratory interface. The ADC is used in external trigger mode so that the A/D sampling sweep does not start until this pulse is received. Depending on where the synchronization pulse occurs relative to the command voltage pulse, the ADC can be made to begin at any point during the D/A waveform.

The program generates pulses according to the following protocol:

(a) wait until time for next pulse
(b) create command voltage and trigger pulse waveform
(c) set up an A/D recording sweep using DMA, using the synchronization pulse DAC channel as the external trigger
(d) start an interrupt-driven DAC output
(e) when the sweep is complete, save ADC data to file
(f) repeat (a)–(e) until all sweeps have been done

A/D conversion is performed by the routine *adcdma* from Listing 2.3, using the DMA data transfer method. This routine takes care of the programming of the A/D sampling rate, number of channels and the setting up of the DMA controller as discussed in Sections 2.11–2.13. Listing 2.3 is specific for the National Instruments LAB-PC interface but a routine to perform the same functions can usually be found within the software libraries supplied with most interfaces.

The D/A conversion sweeps are performed by the routine *dac–sweep* which uses the system clock of the host PC to perform the timing of DAC outputs. The IBM PC family of computers makes use of one channel of an Intel 8253 programmable clock to support the MS-DOS time of day functions. In normal operation, this clock produces pulses at a rate of 18.2 Hz, invoking the hardware interrupt IRQ0 which, in turn, calls an interrupt service routine to update the MS-DOS time and date. It is possible to intercept IRQ0 and use this timed interrupt to perform other operations such as updating D/A output channels. One can also change the clock rate, increasing it to the 1 kHz needed for D/A output timing.

The routine *dac_sweep* does this by inserting the interrupt service routine *dac_isr* (Listing 7.1 (b)) into the interrupt vector table address for IRQ0 and reprogramming the PC system clock to produce a higher rate. When an IRQ0 interrupt occurs this routine writes a pair of D/A output values to the DAC output ports. The MS-DOS time of day functions are maintained by calling the original DOS service routine at appropriate intervals to preserve the original 18.2 Hz rate required for that function. After the D/A output sweep is completed, the routine *disable_dac* replaces the address of the original

DOS service routine into the IRQ0 vector table. As always, if any use is made of interrupts it is important for a program to restore the system to its original condition before exiting, to avoid system crashes. Further details concerning the programming of the PC system clock can be found in Eggebrecht (1990).

The program in Listing 7.1, although functional, was deliberately simplified to clarify the principles involved. In practice it would be enhanced to display signals on screen while they were being recorded and to provide more complex command voltage waveforms. The routine can also be modified to work with a wide range of laboratory interface boards, the only condition being that the DACs can be directly programmed while A/D sampling is in progress. However, if this cannot be done (e.g. DT2801A, DT2821), it is also possible to obtain simple D/A output cards at low cost such as the Amplicon PC-24 or Data Translation DT2816.

The above approach to combined DAC/ADC recording works on the principle that all the complex operations are performed by the host computer. It is therefore most appropriate for use with relatively simple laboratory interface designs. An alternative approach is to use an intelligent interface with an on-board processor such as the Cambridge Electronic Design 1401 or 1401plus. These interfaces can be programmed to execute sequences of independently timed A/D and D/A operations in a similar fashion to that described above but completely within their own internal memory, data being transferred back to the host only when the complete DAC/ADC sweep is finished. The CED 1401plus can support simultaneous independent A/D and D/A operations at rates of around 100 kHz.

7.5 COMMAND VOLTAGE PROTOCOLS – SINGLE STEP

The simplest and most commonly used command voltage protocol consists of step pulses shifting the cell membrane potential from a fixed holding potential to a series of test potentials in order to study the voltage dependence of an ionic current.

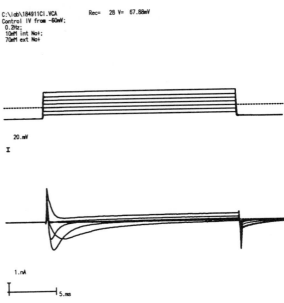

Figure 7.3 A typical set of voltage clamp data records. Examples from a family of voltage and current records showing Na^+ currents from rat dorsal root ganglion neurones, evoked by 20 ms steps to voltages (-55 mV ... $+55$ mV) from a holding potential of -60 mV. (Records courtesy of Dr A. Elliot, University of Dundee.)

Figure 7.3 shows some examples from a family of voltage and current records, obtained using the author's voltage clamp analysis program VCAN. The current under study in this case is the Na^+ current from rat dorsal root ganglion neurones, voltage clamped using the whole cell patch clamp technique. The currents have been evoked by 20 ms voltage steps from a holding potential of -60 mV to a range of levels between -55 mV and $+60$ mV. As is normal in such experiments, other ionic currents (K^+ and Ca^+) present in the cell have been eliminated by the addition of blocking agents. The Na^+ current shown is an example of a current which is both highly voltage sensitive and shows complex time-dependent properties. The current, initially absent at the -60 mV holding potential, is activated by depolarizing voltage steps. The Na^+ current signal is negative going, indicating that it is an inward current, flowing from the bathing solution into the cell. It rises rapidly to a peak within a few milliseconds then decays exponentially. Rate of rise, peak amplitude and rate of decay of the waveform all vary with the magnitude of the voltage step. The records also display the sharp spikes of capacity current lasting approximately 0.5 ms associated with the abrupt change in membrane potential at the beginning and end of the voltage pulse.

7.6 CURRENT–VOLTAGE CURVES

One consequence of the voltage sensitivity of the Na$^+$ current is that the peak current evoked by a voltage step varies with potential in a highly non-linear way. This can be shown by plotting peak current versus the voltage level to produce a *current–voltage (I–V)* curve. The peak current within each record can be obtained using one of the signal waveform measurement techniques described in Chapter 4. The signal may be displayed on screen and a readout cursor used to pick out the peak value manually (Section 4.10) or an automatic peak finding method used (Section 4.13). If an automatic method is used, the analysis area should be carefully set to exclude the capacity current. The peak *I–V* curve for the family of Na$^+$ current records in Figure 7.3 is shown in Figure 7.4 (a). It is now clear that the Na$^+$ current does not significantly activate until the membrane potential is stepped to voltages more positive than −25 mV. In fact up till then only a small outward current is evoked by the voltage pulse. However, for steps from −20 mV onwards a significant inward current appears which increases in size and reaches a peak at +15 mV; with steps to more positive potentials, it diminishes and reverses sign at potential positive of 53 mV.

Complex as this curve might seem, it can readily be interpreted using the equivalent circuit model. The reversal potential V_{Na} can be estimated from the *I–V* curve by interpolating between the pair of voltage levels at which the peak current is seen to change sign, resulting in a value of 56.3 mV. This is close to the value of 50 mV expected for an Na$^+$ current from the Nernst equation (Equation 7.2), given the known Na$^+$ concentrations inside (10 mM) and outside (70 mM) the cell. This value of V_{Na} can then be used in Equation 7.1 to compute the

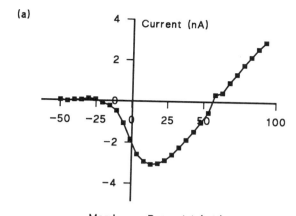

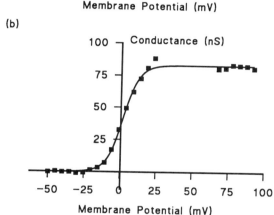

Figure 7.4(a) Peak current–voltage curve derived from the family of records in Figure 7.3 by an automated peak-finding method. (b) Conductance–voltage curve computed from the data points in (a), using Equation 7.1 with an estimate of 56.3 mV for V_{Na}, the Na$^+$ current reversal potential. Points close to V_{Na} have been eliminated from the plot due to the voltage step being too small for accurate calculation of the conductance. The smooth line drawn through the data points in the best-fit Boltzmann function (Equation 7.3) with parameters G_{max} = 82.7 nS, V_{half} = 2 mV, V_s = 5.6 mV.

conductance–voltage (G–V) curve, shown in Figure 7.4 (b). In comparison with the *I–V* curve, the *G–V* curve provides a more direct representation of the underlying voltage sensitivity of the ionic channels since it has eliminated the effects of the different current driving forces ($V - V_{Na}$) at different potentials. The sigmoid *G–V* curve seen here is typical of many voltage-activated current systems. As first shown by Hodgkin & Huxley (1952a–c), curves of this shape can be fitted quite well by the Boltzmann function

$$G = \frac{G_{max}}{1 + \exp\left(\frac{-(V - V_{half})}{V_s}\right)} \quad [7.3]$$

where G_{max} represents the maximum attainable conductance, V_{half} the potential at which it is half activated, and V_s a slope factor representing the steepness of the curve. The line through the data points in Figure 7.4 (b) has been obtained by fitting Equation 7.3 to the data points using the non-linear curve-fitting procedures described in Chapter 6 (Levenberg–Marquardt method, Fig.P program). This resulted in a set of best-fit parameter values, $G_{max} = 82.7\,nS$, $V_{half} = 2\,mV$, $V_s = 5.6\,mV$.

7.7 VOLTAGE RAMP PROTOCOLS

Current–voltage curves have been discussed so far in the context of the peak Na^+ current, however, they are equally applicable to currents such as the delayed or inward rectifier K^+ currents which, although voltage sensitive, do not inactivate. In such cases it is also possible to produce an I–V curve based upon the average current measured once the current has achieved its steady state. Steady state I–V curves can also be generated very rapidly by evoking the current with a *voltage ramp* rather than a series of voltage steps. A steady state I–V curve, produced directly by a voltage ramp between −100 mV and −100 mV over a period of 1.5 s, is shown in the signal record in Figure 7.5. The ramp command pulse is generated by updating the D/A output in a series of small incremental steps at frequent intervals (4 ms in this case) to produce a staircase waveform which is then passed through a low pass filter to achieve a smooth ramp.

The rapid generation of the I–V curve using voltage ramps can be of particular value in experiments to determine the ion selectivity of agonist-activated channels. Computing the I–V curve from a family of rectangular voltage steps can easily take 20–30 s. However, many types of channel show a rapid desensitization to the effects of the agonist, resulting in the current decaying away to zero within a few seconds after its application. For instance, endplate

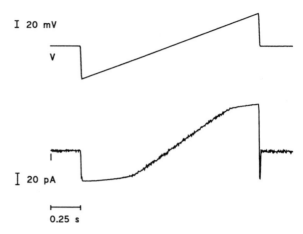

Figure 7.5 Steady-state I–V curve generated directly by applying a ramp command waveform to the voltage clamp, from −100 mV to 100 mV at a rate of 133 mV/s.

channels show this behaviour and $5HT_3$ channels (Lambert et al., 1989) even more so. Generating the I–V curve using a voltage ramp such as in Figure 7.5 takes less than a second and can therefore be accomplished before desensitization sets in. Examples of the use of the voltage ramp to generate I–V curves can be found in Bormann et al. (1987) or Yellen (1984). It should be noted, however, that although generating an I–V curve while a currrent is desensitizing is quite satisfactory for determining the reversal potential, the overall shape of the curve may be distorted.

7.8 ANALYSIS OF CURRENT TIME COURSE

The time course of the Na^+ current is also strongly voltage dependent with current activation and decay becoming faster the more depolarizing the voltage step. Quantitative analysis of these time courses and their voltage sensitivity can yield information concerning the kinetic behaviour of the ionic conductance system. The first stage in analysis is to find a mathematical function which accurately describes the current time course and to fit that function to the data, using the techniques discussed in Chapter 6. Current time courses for a wide range of types of ionic conductance, have

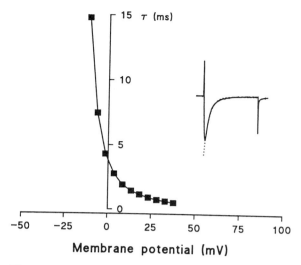

Figure 7.6 Voltage dependence of the time constant of Na$^+$ current decay from the records in Figure 7.3, obtained by fitting an exponential function (Equation 7.4) to the decay phase of the current as shown inset.

been shown to be readily described mathematically by simple combinations (sums, products or powers) of exponential functions (Hodgkin & Huxley, 1952a–c).

The currents in Figure 7.3 are complex, but it is possible to treat the activating and decaying phases of the currents separately. The decay phase of the Na$^+$ current is well fitted by a three parameter exponential relaxation from an initial value A_0 at the onset of the voltage step (t_0) to a steady state A_∞ with a time constant τ

$$I_{Na} = A_0 \exp(\frac{-(t - t_0)}{\tau}) + A_\infty \quad [7.4]$$

The three parameters (A_0, τ, A_∞) can be obtained for a current record by fitting Equation 7.4 to the sample points in the digitized record between the current peak and the end of the voltage pulse, using iterative curve fitting. Figure 7.6 shows the voltage dependence of τ obtained from a series of curve fits (Levenberg–Marquardt method, SSQMIN) applied to the Na$^+$ current records from Figure 7.3. This technique is often used to quantify changes in the time constant of decay of Na$^+$ currents, or the number of exponentials required to fit, induced by a variety of drugs, particularly local anaesthetics (Chiu, 1977; Schneider & Dubois, 1986).

7.9 THE HODGKIN–HUXLEY EQUATIONS

Hodgkin and Huxley in their seminal study developed an important mathematical model of the voltage-activated Na$^+$ and K$^+$ currents. On the basis of records similar to those in Figure 7.3, they proposed that the Na$^+$ current could be described as the product of two separate parameters each with its own voltage and time dependence; an activation parameter m regulating the rate of rise of the conductance and an inactivation parameter h regulating the rate of subsequent decay of conductance. The Na$^+$ current I_{Na} was therefore given by the equation

$$I_{Na} = G_{Na.max} m^3 h (V - V_{Na}) \quad [7.5]$$

Time-dependent properties are conferred on the model by having the activation and inactivation parameters each governed by first-order rate equations

$$\frac{dm}{dt} = \alpha_m(1 - m) - \beta_m m, \quad \frac{dh}{dt} = \alpha_h(1 - h) - \beta_h h$$

$$[7.6]$$

The rate constants α_m, β_m, α_h and β_h are all functions of membrane potential, chosen so as to make m increase with depolarization and h decrease. If the membrane potential is held constant, m and h tend to the steady-state values

$$m_\infty = \frac{\alpha_m}{\alpha_m + \beta_m}, \quad h_\infty = \frac{\alpha_h}{\alpha_h + \beta_h}$$

$$[7.7]$$

An abrupt step of the membrane potential from one level to another, causes m and h to vary exponentially towards the steady state appropriate to the new voltage level

$$m = m_\infty - (m_\infty - m_0) \exp(\frac{-(t - t_0)}{\tau_m})$$

$$[7.8]$$

and

$$h = h_\infty - (h_\infty - h_0) \exp(\frac{-(t - t_0)}{\tau_h})$$

$$[7.9]$$

where m_0 and h_0 are the steady-state values at the initial membrane holding potential. The time constants of the exponential rise τ_m and decay τ_h are related to the rate constants by

$$\tau_m = \frac{1}{\alpha_m + \beta_m}, \quad \tau_h = \frac{1}{\alpha_h + \beta_h} \quad [7.10]$$

The Hodgkin–Huxley (H–H) model is a semi-empirical model in that the voltage dependence of the rate constants α_m, β_m, α_h, β_h must be extracted from experimental data. The methods for doing this are thoroughly and clearly discussed in the original papers (Hodgkin & Huxley, 1952a–c). However, the process has been made somewhat easier with the use of iterative curve fitting. Equations 7.5, 7.9 and 7.10 can be combined and simplified for the case of voltage steps from a sufficiently hyperpolarized holding potential to allow the assumption that $h_0 = 1$ and $m_0 = 0$. The Na$^+$ current time course can then be represented by

$$I_{Na} = I'\,[1 - \exp(\frac{-(t-t_0)}{\tau_m})]^3\,[h_\infty - (h_\infty - 1)\exp(\frac{-(t-t_0)}{\tau_h})] \quad [7.11]$$

This equation has four free parameters, the activation and inactivation time constants τ_m, τ_h, the maximum current I' that would have been achieved in the absence of inactivation, and the steady-state inactivation h_∞. By fitting Equation 7.11 to the current time course from a family of voltage steps, it is possible to determine the voltage dependence of τ_h, τ_m and h_∞. The voltage dependence of m_∞ can be determined from

$$m_\infty = \frac{G'}{\text{Max}(G')} \quad \text{and} \quad G' = \frac{I'}{(V - V_{Na})} \quad [7.12]$$

where $\text{Max}(G')$ is the maximum obtainable value of G'.

7.10 TWO-STEP COMMAND VOLTAGE PROTOCOLS

Single-step command voltage protocols provide a significant amount of information concerning the voltage dependence of the H–H rate constants. However, there are distinct gaps in the voltage range that can be studied in this way, particularly for the steady-state inactivation parameter h_∞. Clearly, inactivation of the Na$^+$ current can only be directly measured after the current has first activated which requires a depolarizing step to at least -20 mV. However, h_∞ is already a small value by this potential. A complete analysis such as performed by Hodgkin & Huxley, over a wide range of potentials, requires the use of two-step command voltage protocols. Such protocols exploit the fact that conductance systems obeying H–H kinetics are dependent not only on the voltage during the test step but on the voltage level before the step. This is due to differences in the initial steady state levels of m_0 and h_0.

Figure 7.7 is an example of a two-step experiment to estimate the voltage dependence of h_∞ over a much wider range than is possible with a single step. The voltage protocol consists of an initial *pre-conditioning* step V_{pre} of duration 1 s followed by a 20 ms test step V_t during which the evoked Na$^+$ current is measured. A family of records was obtained with different values of V_{pre} from -150 mV to 0 mV while V_t was held constant at 0 mV. V_{pre} strongly influences the magnitude of the peak Na$^+$ current I_{pk} during the test pulse (as can be seen from the records inset in Figure 7.7) by modulating the initial level of steady state inactivation. The test step V_t can activate only whatever fraction of the conductance remains available. Since h has much slower kinetics than m, I_{pk} is a measure of h_∞ during the pre-pulse. The voltage dependence of h can therefore be shown by plotting I_{pk} versus V_{pre} as in Figure 7.7 and h_∞ can be calculated from

$$h_\infty = \frac{I_{pk}}{\text{Max}(I_{pk})} \quad [7.13]$$

In a similar fashion, pre-pulses of varying duration can be used to estimate values for τ_h over a wide range of potentials (see Hodgkin & Huxley, 1952d for details).

In practice, one- or two-step command voltage protocols prove sufficiently powerful tools for extracting the kinetic properties of most voltage-activated currents. However, triple step

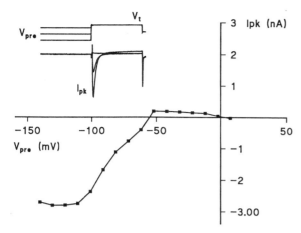

Figure 7.7 The voltage dependence of the Na$^+$ current H–H parameter h_∞ estimated using a two pulse command voltage protocol. Three currents from the family are shown inset. A pre-pulse (V_{pre}) is used to set the level of steady-state Na$^+$ current inactivation h_∞. The peak Na$^+$ current achieved by the subsequent test pulse (V_t) is then used an an indirect measure of h_∞ (Records courtesy of Dr A. Elliot, University of Dundee.)

protocols have occasionally been used in the analysis of the action of local anaesthetic agents on the Na$^+$ channel.

The H–H analysis described above has proved to be a satisfactory description, not only of the Na$^+$ conductance but also of a wide range of voltage-activated ionic conductances. Hodgkin & Huxley showed that the voltage-activated K$^+$ conductance could be described by an essentially similar set of equations and other workers have subsequently extended the approach to a wide variety of excitable cells, including cardiac muscle (Noble, 1966), skeletal muscle (Adrian et al., 1970) and myelinated nerve (Dodge & Frankehaeuser, 1959; Chiu et al., 1979). The effects of the many agents which can alter the voltage and time dependence of the Na$^+$ channel are often interpreted within the context of H–H models; in particular local anaesthetics (Hille, 1977).

If the H–H equations were simply a convenient description of the properties of the Na$^+$ and K$^+$ conductances under voltage clamp conditions they would be of relatively limited importance. However, as Hodgkin & Huxley demonstrated (1952d), when used as the basis of a mathematical simulation of the squid axon under natural unclamped conditions, they produce very realistic action potential waveforms barely distinguishable from real recordings. This suggests that the equations accurately describe the voltage and time dependence of these classes of ion channel.

A key feature of the H–H model is the use of multiple sets of first-order differential equations to express the time dependence of the system, with the voltage dependence separated out and embodied in non-linear rate constants (α_m, β_m, etc.). This is yet another extension of the methodologies of chemical reaction kinetics discussed in Chapter 6. It leads naturally to the idea that each parameter (m,h) in the system represents the actions of a distinct gate to the flow of ions through the channel, with the Na$^+$ channel representable by four gates in series, three m and one h. These gates can be considered to be actual macromolecular structures within the channel which change configuration in response to trans-membrane potential.

However, it should be borne in mind that these aspects of the model are hypothetical and are not experimental observations. The fact that one model fits the experimental results well does not mean that it is the only possible model. The H–H model has the advantage of being a relatively simple model, based on concepts which are familiar to the electrophysiologist. Other models based upon quite different second- or third-order differential equations might be possible, but so far a substantially better model has not been developed and some of the key concepts of the H–H model have been experimentally verified (e.g. separate m and h gates, Bezanilla & Armstrong, 1977b). Nevertheless, questions still remain concerning the exact interpretation of Na$^+$ channel gating (Aldrich & Stevens, 1987).

7.11 DIGITAL SUBTRACTION OF LEAK CURRENTS

Detailed kinetic analyses of ionic currents such as the Na$^+$ current described above are crucially dependent upon the assumption that the current signal being measured arises solely from the ionic conductance system under study; that it has been possible to block, or otherwise remove, all other conductances present in the cell membrane. Typically, steps would be taken to eliminate unwanted current components by applying selective blocking agents such as Tetrodotoxin for the Na⁺ channel, or replacement with impermeant ions (TEA or Cs⁺ for K⁺ channels). However, not all unwanted components of the signal can be completely eliminated this way and, in practice, some contamination by unwanted currents is unavoidable. A voltage clamp current signal $I_t(t)$ almost always consists of the sum of at least three distinct components.

$$I_t(t) = I_c(t) + I_i(t) + I_{lk} \qquad [7.14]$$

$I_c(t)$ is the *capacity current* supplied by the voltage clamp to charge and discharge the cell membrane capacity. Capacity currents only occur when the membrane potential is changing and manifest themselves as large transient spikes with rapid exponential decays restricted to the leading and trailing edges of step changes in membrane potential. $I_i(t)$ is the primary *ionic current* under study which, it is hoped, constitutes the main component of the current signal. It may have both transient and steady components in response to a voltage step. I_{lk} is the *leak current*, consisting of the remaining membrane ionic currents which it has not been possible to eliminate, plus current leaking through portions of the cell membrane damaged during the process of insertion or attachment of the electrodes required for controlling the membrane potential. I_{lk} is not time dependent like the other two components.

In some circumstances I_c is short lasting and I_{lk} small allowing both to be neglected, as was the case in the Na⁺ currents examples presented earlier. But in other tissues they can significantly overlap the ionic current under study. Figure 7.8 shows a Ca²⁺ current from a cardiac muscle cell where significant leak current is also being passed. The current in response to a voltage step from −40 mV to +5 mV (Figure 7.8 (a)) is inward at the peak of the Ca²⁺ current but outward at the end of the pulse, indicating that there must be at least one other ionic current present with a different reversal potential.

In order to analyse the Ca²⁺ current in the same fashion as the Na⁺ current, it is first

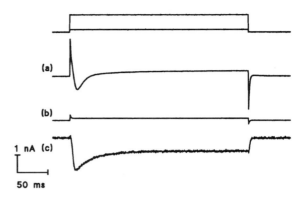

Figure 7.8 Digital subtraction of leak and capacity currents from a record of the Ca²⁺ current from cardiac muscle cells. (a) Raw 1000 point current record in response to a voltage step from −40 mV to +5 mV. (b) Background record produced by −40 mV to −35 mV step which does not activate Ca²⁺ current. (c) Background subtracted record showing true Ca²⁺ current after a scaled version of (b) (× 45/5) has been subtracted.

necessary to separate it from the other components of the whole cell current. In certain favourable circumstances this can be done using the process of *digital leak current subtraction*. As we have seen, many of the voltage-activated currents only occur in response to relatively large depolarizing voltages. For instance, the Na⁺ current is inactive at the cell resting potential (−60 mV to −90 mV) and activated only by depolarizations of at least 30 mV.

Leak and capacity currents on the other hand are evoked by voltage steps of any polarity or magnitude. A current signal evoked by a hyperpolarizing voltage step (or a depolarizing step too small to activate a voltage-activated current) will therefore contain only leak and capacity current components. We can therefore produce a subtraction current record $I_s(t)$ containing only these two components.

$$I_s(t) = I_c(t) + I_{lk} \qquad [7.15]$$

We cannot directly subtract $I_s(t)$ from $I_t(t)$ since the two records have been evoked by different voltage steps and therefore experience different current driving forces. However, leak and capacity currents are usually linearly related to the size of the voltage step which makes it possible to calculate the contribution of these components to the current evoked by the test voltage step from the ratio of the voltage step amplitudes. The leak and capacity currents can therefore be subtracted leaving the ionic current using the formula

$$I_i(t) = I_t(t) - I_s(t) \frac{V_t}{V_s} \quad [7.16]$$

where $I_t(t)$ is the test current signal evoked by the depolarizing voltage step V_t and $I_s(t)$ is the subtraction current signal evoked by the hyperpolarizing step V_s. The Ca^{2+} current in Figure 7.8 has been separated out in this way. A subtraction record (b) was collected using a small 5 mV depolarizing step, too small to evoke any Ca^{2+} current. Scaling this record up by a factor of nine ($V_t = 45$ mV/$V_s = 5$ mV) and digitally subtracting it from the test record (a) results in the record (c) containing the Ca^{2+} current only.

7.12 LEAK SUBTRACTION PROTOCOLS

The digital leak subtraction method was initially developed by Armstrong & Bezanilla, to permit the analysis of Na^+ channel gating currents in the squid giant axon (Armstrong & Bezanilla, 1974). Gating currents are small, very fast, voltage-sensitive components of the capacity transient current I_c believed to be due to the movement of charged structures within the Na^+ channel molecular structure associated with channel gating. These currents only become apparent when almost all of the ionic currents are blocked and the remaining linear leak and capacity currents removed.

The initial approach described in their original paper, was fairly simple and avoided the need to scale the subtraction current record. Subtraction and test records were evoked by voltage steps of the same magnitude but opposite polarities and the leak subtraction was achieved simply by adding the two records together, within a waveform averager (the commonly used digitizing device at the time). This protocol is effective but has the drawback that it requires large magnitude voltage steps which may cause membrane breakdown during the hyperpolarizing step. For instance, Na^+ and Ca^{2+} ionic currents have reversal potentials of at least +40 mV and to study currents in this region depolarizing voltages steps of magnitudes of 150 mV and more are applied. This is not excessively harmful since the step is applied from the existing holding potential of -90 mV, resulting in an absolute membrane potential of only +40 mV. However, a hyperpolarizing step of the same magnitude results in an absolute potential of -240 mV.

To avoid such problems Bezanilla & Armstrong subsequently refined the method, developing what is known as the *P/4 method* and applied it more widely than just to gating currents (Bezanilla & Armstrong, 1977b). In the P/4 method, four subtraction records are produced for each test record using a hyperpolarizing voltage step one quarter the size of the depolarizing test pulse. Leak subtraction is performed by adding up all five records, i.e.

$$I_i = I_t + \sum_{i=1}^{4} I_s \quad [7.17]$$

There is, of course, nothing significant about the use of four subtraction records. The method is easily generalized to the P/N method where N subtraction records are produced using step 1/Nth the size of the test voltage step.

It is worth noting that these early leak subtraction methods had to work within the constraints of the available computer hardware. In particular, while addition operations could be performed rapidly and efficiently with the waveform averager, general scaling or multiplication operations could not. Modern computers no longer have these limitations. While the P/N methods are still in common use today, they are only one example of many different protocols that might be used.

For instance, the subtraction voltage step need not be a fixed fraction of the test step as long as the V_t/V_s voltage ratio can be determined at some stage. A general subtraction formula using arbitrary numbers and sizes of test and leak records is given by

$$I_i = \frac{1}{m} \sum_{i=1}^{m} I_t + \frac{V_t}{nV_s} \sum_{j=1}^{n} I_s \quad [7.18]$$

where m is the number of I_t records averaged, n the number of I_s records and V_t/V_s the ratio of test and subtraction record voltage steps. A leak current subtraction subroutine implementing the above algorithm is shown in Listing 7.2.

There is also scope to vary when and how many subtraction records are collected during the experiment. In the P/4 method the size of subtraction record voltage step must be closely matched to the size of the test voltage step. Therefore it is normal to use a voltage program consisting of four subtraction and one test record produced in close sequence for each voltage level studied. More flexibility is available using the general method in Equation 7.17 since the subtraction records need not be closely associated with the test records. For example, a single series of subtraction records of fixed size might be collected at the end of the experiment and scaled appropriately for subtraction from each test record.

Digital leak current subtraction has proved an effective method for separating voltage-activated ionic currents and gating currents from background currents. However, it has a significant limitation in that the leak current must be linear, i.e. with a constant conductance. Non-linear currents do not scale proportionally to the size of the voltage step and consequently it is not possible to estimate the leak current during the test pulse from the leak current during the subtraction pulse. Such circumstances can occur in normal cells when pharmacological blocking agents have not been used. For instance, in many cell types a prominent element of the passive membrane conductance is the inward rectifier; a K^+ conductance which, unlike the H–H type voltage-activated Na^+ and K^+ conductances, is activated strongly by hyperpolarizing voltage steps and almost absent for depolarizing steps. Clearly, if a subtraction record contains the inward rectifier current and the test record does not, scaling and subtracting that current from the test record will produce erroneous results. Digital leak current subtraction should therefore be used as a complement, rather than an alternative, to the more conventional pharmacological methods.

Care must also be taken to ensure that the leak subtraction process does not add excessive amounts of background noise to the signal. Although the subtraction records may contain a small signal they contain background noise with the same magnitude as the test record. The process of scaling up the leak record before subtraction also magnifies this noise which is imposed on the difference record. To minimize this added noise more than one leak record must be collected per test record and these records averaged to reduce the background noise before scaling and subtraction from the test record. The magnitude of the background noise, σ_i is given by

$$\sigma_i = \sqrt{\frac{\sigma_t^2}{m} + \frac{\sigma_s^2 V_t}{n V_s}} \qquad [7.19]$$

where σ_t and σ_s are the test and background record rms noise levels. From this formula, it can be seen that the P/4 method ($m = 1, n = 4, V_t/V_s = 4$) increases the signal background by a factor of $\sqrt{2}$ after leak subtraction. In order to avoid such an increase in background noise the number of leak records averaged for a given V_t/V_s ratio should be increased. If it is required actually to reduce background noise, more than one of the test records must be averaged.

7.13 VOLTAGE CLAMP SOFTWARE PACKAGES

A good software package for recording and analysing voltage clamp signals should support many of the operations discussed in this chapter in addition to general facilities for recording and displaying the analog signal waveforms discussed in Chapter 4. In particular, it should support the recording of both the current and voltage signals from the voltage clamp while simultaneously generating the voltage clamp command voltage. The voltage pulse generator should be capable of automatically generating a timed series of one step, two step or ramp pulses of varying sizes and durations, allowing the automatic production of families of record for I–V curves, etc. Digital leak current subtraction should be supported with the production of subtraction pulses integrated into the command voltage program. On the analysis side, it should be possible to measure families of records automatically, generate *I–V*, curves and fit single and double exponential curves to signal waveforms.

At present there are a small number of voltage clamp software packages available either as commercial products or distributed free by individual researchers. Most of these run on the IBM PC family although some are now beginning to appear for the Apple Macintosh. The most widely used among the IBM PC packages are the Axon Instruments pCLAMP package, Cambridge Electronic Design's Patch and Voltage Clamp software, the author's VCAN program. These programs have developed over a number of years and have evolved through numerous versions.

7.14 AXON INSTRUMENTS PCLAMP

As mentioned earlier, the pCLAMP suite of programs, developed by Kegel *et al.* (1985), were among the first IBM PC programs produced specifically for electrophysiological work. It is now developed and commercially marketed by Axon Instruments with the latest version at the time of writing being V5.5.1. It is probably the most widely used software package of its class, particularly in the United States. pCLAMP is generally supplied by Axon Instruments as part of a package including the Scientific Solutions Labmaster or Labmaster DMA laboratory interface and an input/output box providing input and output via BNC sockets.

pCLAMP is written mostly in Microsoft QuickBasic with a number of assembler modules which handle A/D and D/A conversion. The source code is supplied as standard. pCLAMP is marketed as a general-purpose electrophysiological analysis program with features for single-channel analysis as well as for whole cell voltage clamp experiments. Figure 7.9 shows a screen display from the CLAMPEX program while it is being used to record a family of voltage-activated currents. The program supports many of the operations discussed in this chapter, including simultaneous command voltage generation and recording.

Within the pCLAMP system the basic digitized signal record is known as an *episode*. An episode consists of up to 2048 interleaved samples from up to four A/D channels at a maximum aggregate rate of 125 kHz (using a 16 MHz 80386 PC). Each episode is evoked by a command voltage waveform produced by CLAMPEX, according to a protocol laid down in a parameters file. Command voltage pulse types include single-and two-step rectangular pulses, ramps

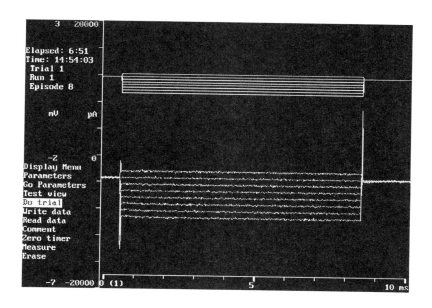

Figure 7.9 Screen display from the CLAMPEX program from Axon Instruments pCLAMP package.

and series of conditioning pulses. A family of episodes can be produced automatically with different voltage pulses grouped together into a *run* of episodes. Complete runs can be repeated a number of times, and the corresponding episodes within each run averaged to produce a *trial* which is stored on file. A trial can contain up to 32 768 samples which is adequate to collect a family of 16, 2 channel × 1024 sample, episodes.

CLAMPEX also implements the P/N leak subtraction protocol, automatically producing a series of appropriately scaled leak subtraction pulses before the test pulse and then summing the complete set together. It should be noted that, since the leak subtraction process is performed 'on-line', the leak records themselves are not preserved on file.

The digitized signal records from the trials performed by CLAMPEX are stored in a data file in binary form. These files can be analysed using the CLAMPAN program which provides a variety of waveform measurements including the generation of *I–V* plots, integration, trend subtraction and digital filtering. Alternatively the CLAMPFIT program provides multi-exponential curve fitting using the Levenberg–Marquardt algorithm. Both CLAMPAN and CLAMPFIT produce ASCII format output files which can be exported to other more general programs such as spreadsheets as discussed in Section 4.17. pCLAMP can plot digitized signal records and *I–V* curve results on HPGL-compatible digital plotters (e.g. Hewlett Packard HP7475, HP7550) or laser printers (HP Laserjet III).

7.15 CAMBRIDGE ELECTRONIC DESIGN

In the United Kingdom, Cambridge Electronic Design Ltd has played a similar role in the development of computer data acquisition within the electrophysiological laboratory as Axon Instruments has done in the United States. Their main product, the CED 1401 laboratory interface, enjoys a similar degree of popularity here to the Labmaster in the USA. CED is a computer systems company providing integrated solutions to laboratory data acquisition problems over a wide area, ranging from mechanical engineering to clinical neurophysiology, not just electrophysiology.

The CED Voltage and Patch Clamp suite is a set of programs for the recording and analysis of single-channel and whole-cell voltage clamp signals. The software is written in Microsoft Pascal with the source available on request. It supports simultaneous recording and voltage pulse generation, in quite a different way from pCLAMP. The CED 1401 is an intelligent interface with an on-board CPU capable, as discussed in Section 2.19, of running complex programs independent of the host computer. The basic working unit is the *frame*, which defines a one- or two-step rectangular or ramp command voltage pulse. Recording of up to 2048 A/D samples from 1–4 channels can be programmed to occur during *sampling windows* defined within the frame. A frame can be made to repeat a number of times with an increment defined for pulse amplitude (or duration) which is added between successive frames allowing families of pulses to be generated. Frames can be collected together into *protocols* allowing a series of different kinds of pulse stimuli or recording durations to be programmed automatically.

Results can be exported as ASCII text files or plotted graphically on an HPGL-compatible digital plotter or laser printer. Data files are analysed using a voltage clamp analysis module which provides features similar to CLAMPAN and CLAMPFIT in pCLAMP; cursor measurement, *I–V* curve generation, exponential curve fitting, averaging. A general 'off-line' leak subtraction protocol, similar to that described by Equation 7.18, is also supported by this module. Unlike the 'on-line' pCLAMP approach, leak frames are preserved in the data file and any series of frames can be used for subtraction.

7.16 STRATHCLYDE SOFTWARE: VCAN

pCLAMP and the CED software are both commercial software packages. They are quite costly packages, as the small market for this type of specialist software dictates, and restricted to the particular laboratory interfaces produced by their suppliers. However, a great deal of software has been developed by electrophysiologists themselves, primarily for use within their own laboratories. It is often possible to obtain this software either free or for a nominal charge. Such software may not always be produced to the highest professional standards, nor can one expect the kind of technical support obtainable from a commercial operation. It may nevertheless be extremely useful, especially if its features closely match the kind of experiments you wish to perform. There is a sufficient diversity between experimental methods and protocols to ensure that no single program can currently provide an ideal solution to all requirements.

The VCAN (Voltage Clamp ANalysis) program is part of the Strathclyde Electrophysiology Software suite of programs (Dempster, 1987; 1989), developed by the author at the University of Strathclyde over the past decade. The software is distributed free of charge to users within the academic community and other non-profit-making organizations (a version of the software is also marketed by Dagan Corp. (Minneapolis, USA)). VCAN is designed to support a wide range of laboratory interfaces which currently include, the Data Translation DT2801, DT2801A, DT2821, DT2812, Scientific Solutions Labmaster and Labmaster DMA, CED 1401, and National Instruments LAB-PC. Unlike the others, which are suites of separate recording and analysis programs, VCAN is a single program integrating both functions. It is written in Microsoft FORTRAN and 8086 assembler language with the source code distributed. VCAN was used to generate the voltage clamp analysis results used to illustrate the procedures earlier in this chapter. Like the other two programs, VCAN supports simultaneous recording and voltage pulse generation.

Pulse generation and recording are not as tightly linked in VCAN as they are in the other programs. The basic working unit in VCAN is the *record* which contains up to 2048 samples from two A/D channels. One- or two-step rectangular or ramp voltage pulses are supported with an incrementing feature similar to that of the CED software. Two separate voltage programs can be defined which interleave with each other (primary and alternate) allowing a variety of leak subtraction protocols to be implemented.

The program supports the generation of I–V curves, multi-exponential and Hodgkin–Huxley curve fitting (Equation 7.11) using an improved Levenberg–Marquardt method (SSQMIN, see Section 6.17). Leak subtraction is performed off-line using the generalized leak subtraction equation (Equation 7.18). Each record in the data file can be classified as a TEST record or a LEAK record (or REJECTED if the record is flawed). LEAK records are averaged and scaled as discussed in Section 7.12 and subtracted from the TEST records.

Each of these programs takes a slightly different approach to the recording and analysis of electrophysiological signals which determine their strengths and weaknesses. The CED package provides the most flexible voltage pulse generation but is more limited on the analysis side. VCAN has a simpler, and consequently less flexible, voltage generator but is easier to use. Within the range of tasks that VCAN can perform, it is more efficient than the other programs but only has two A/D channels. pCLAMP is unique in supporting split sampling rates, so that the initial half of a record can be at a fast rate to resolve the rapid rising phase of a signal, followed by a slow rate to include a slow decay phase. However, pCLAMP's on-line leak subtraction protocol is limited, as is the total number of records that can be acquired per trial.

7.17 OTHER SOFTWARE PACKAGES

Voltage clamp packages are also available for computers other than IBM PC, in particular the Atari ST and the Apple Macintosh. Two packages exist for the Atari ST, one originally developed by Frank Sigworth and now marketed by Instrutech and another produced by Hera Electronik GMBH which is an integral part of their EPC-9 voltage clamp amplifier. The Atari ST is known mostly as a computer games machine, but is also used extensively as a controller for music synthesizers. However, in some countries, it has achieved a wider popularity being used extensively in laboratories in Germany. It is similar in many ways to the Apple Macintosh, based upon the Motorola 68000 microprocessor and employing a graphical user interface (GUI), but at a much lower cost. However, except in the niches markets mentioned earlier, it never achieved the same degree of acceptance as the Apple products. Consequently, the ST family has not developed at the same rate as either the IBM PC or the Macintosh, with the latest versions of both the PC and the Mac greatly outperforming any comparable ST. This, and the lack of a standard expansion bus, limits the prospects for the ST.

The Apple Macintosh, on the other hand, has established itself as a valid alternative to the IBM PC. However, although it is in widespread use within the laboratory, data acquisition and particularly specialized electrophysiological applications have been slow to appear. This is probably due to a number of factors. It is more difficult to produce high performance laboratory interface cards for the Macintosh due to the lack of built-in features such as DMA channels of the IBM PC. The complexity of the Macintosh GUI also makes it more difficult to program. Nevertheless there are signs that progress is being made in this area with Axon Instruments announcing the *Axodata* program for the Macintosh using the Instrutech ITC-16 laboratory interface. Given the large number of Macintosh users, it is likely that more of such programs will appear.

7.18 SUMMARY

It is now the norm to use a computer as the central controller for a voltage clamp experiment, both generating voltage pulse waveforms and recording the resulting current responses. This approach permits a wide range of stimulus waveforms to be easily applied to the cell. In addition, techniques such as digital leak current subtraction provide the capability to enhance the original records greatly by stripping away unwanted background currents.

Laboratory interfaces suitable for this kind of work must be able to handle A/D and D/A operations simultaneously, such as the Labmaster or the CED 1401 for the IBM PC. A number of specialist software packages are available for this area, particularly for the IBM PC and more recently for the Apple Macintosh. These packages have improved greatly over the past few years, both in the range of features supported and ease of use. Given the widespread use of the voltage clamp technique it is likely that this rate of development will continue.

```
c     Listing 7.1 (a). Simultaneous A/D recording and D/A waveform generation routine.
c     Records from A/D channels 1 & 0 (current & voltage) using DMA and
c     PC system clock interrupt for D/A waveform output.
      subroutine adcdac
      parameter( np_adc=1024, np_dac=512, n_channels=2)
      integer*2 iadc(np_adc*2)              ! A/D data buffer
      integer*2 idac(np_dac*2)              ! D/A waveform buffer
      parameter( volts_per_bit = 5./2048., izero=2048 )  ! 12 bit, +/-5V DAC
      real*4 first_amplitude  / 0.1 /       ! Initial command voltage pulse (V)
      real*4 step             / 0.1 /       ! Increment in size between pulses (V)
      real*4 pulse_duration   / 0.5 /       ! Duration of pulse. (s)
      integer*2 npulses       / 10 /        ! No. of pulses in sequence
      real*4 pulse_interval   / 2. /        ! Time between pulses (s)
      integer*2 interval      / 2000 /      ! Sampling interval (µs)
      integer*4 iperiod4,idiv4              ! D/A clock settings
      integer*2 istatus(256)                ! Record status buffer
      character status
      equivalence( status, istatus )

      nsectors_per_record = np_adc*2/256 + 1
      open(unit=1,file='adc.dat',access='direct',form='binary',recl=512)
c     File format: H SDDDDDDDD SDDDDDDDD ... (H=header sector, S=record status,
c                    Rec.1      Rec.2          D=data sector)
      amplitude = first_amplitude
      time_for_pulse = 0.
      time = -1.
      do ipulse = 1,npulses
         dt_dac = (pulse_duration*3. + 0.15)/float(np_dac)   ! D/A update interval
         call check_dt_system( dt_dac, iperiod4, idiv4 )
         ipulse_amplitude = int(amplitude/volts_per_bit)     ! Create command voltage
         isync_amplitude = int(5./volts_per_bit)             ! and sync. pulse waveforms
         i_sync = int(0.15/dt_dac)                           ! Sync. pulse at 0.15secs
         istart = i_sync + np_dac/10                         ! Start of command pulse
         iend = min(istart + int(pulse_duration/dt_dac),np_dac-1)
         do i = 1,np_dac
            idac(2*i-1) = 2048
            if( (i.ge.istart) .and. (i.le.iend) ) idac(2*i-1) = ipulse_amplitude + 2048
            idac(2*i) = 2048
            if(i.eq.i_sync) idac(2*i)=min(isync_amplitude+izero,4095)
         end do
         do while( time .lt. time_for_pulse )                ! Wait until it
            call gettim( ih, im, is, ics )                   ! is time for
            time = 3600.*float(ih) + 60.*float(im) + float(is) + float(ics)/100.
         end do                                              ! next pulse
         time_for_pulse = time + pulse_interval
         call dac_sweep(idac,np_dac,idiv4,iperiod4)          ! Start DAC output
         call adc_dma(2,0,interval,iadc,np_adc,0,1,iadc0)    ! Start ADC sweep
         call dac_stop()                                     ! Turn DACs off.
         isector = (ipulse-1)*nsectors_per_record + 2
         status = 'ACCEPTED'
         write(unit=1,rec=isector) istatus
         write(unit=1,rec=isector+1) (iadc(j),j=iadc0,iadc0+np_adc-1)
         amplitude = amplitude + step                        ! Next pulse
      end do
      close(unit=1)
      return
      end
```

```
c Listing 7.1 (a) continued

      subroutine check_dt_system( dt, iperiod4, idiv4 )
      integer*4 idiv4,iperiod4,idiv4a
      idiv4a = 1
10    continue
         idiv4 = idiv4a
         iperiod4 = 16#10000 / idiv4
         dt1 = 1. / ( 18.2*float(idiv4) )
         idiv4a = idiv4a * 2
      if( (dt1 .gt. dt) .and. (idiv4.lt.64) ) goto 10
      dt = dt1
      return
      end
```

```
;     Listing 7.1 (b). Assembler interrupt routine for DAC sweep.
      title    dac.asm
      include macros.inc             ; Macro definitions, see Listing 2.2
  dac_code    segment para public 'code'
              assume cs:dac_code
  pic_eoi_reg equ 20h                ; End-of-interrupt register
  irq0 equ 8h                        ; PC system clock interupt
  dac0 equ 264h                      ; DAC output registers
  dac1 equ 266h
  dac_buffer_address  label dword    ; Starting address of DAC buffer
  dac_buffer_offset   dw ?
  dac_buffer_seg      dw ?
  dac_points_to_do    dw 99          ; No. of DAC updates still to be done
  divide_count        dw ?           ; No. of clock pulses before next time-of-day
call
  initial_count       dw ?           ; No. of clock pulses / time-of-day call
  vector     struc
  ipr        dw ?
  csr        dw ?
  vector     ends
  dos_clock           vector <>      ; Storage location for old clock interrupt
                                     ; service routine address.
  dac_isr proc far                   ; IRQ0 service routine to
     push ax                         ; update DAC0 and DAC1 when PC system
     push dx                         ; clock interrupt occurs.
     push ds
     push si
     lds si,cs:dac_buffer_address    ; Load ds:si with DAC buffer pointer
     cmp cs:dac_points_to_do,0
     cld
     jle skip_output
        lodsw                        ; ax = DAC0 value from buffer
        mov dx,dac0
        out dx,al                    ; Send lo byte to DAC 0
        inc dx
        mov al,ah                    ; Send hi byte to DAC 0 (triggers D/A output)
        out dx,al                    ; Send lo byte to DAC 0
        lodsw                        ; ax = DAC1 value from buffer
        mov dx,dac1
; ... continued
```

```
        Listing 7.1 (b) continued
            out dx,al                        ; Send lo byte to DAC 1
            inc dx
            mov al,ah                        ; Send hi byte to DAC 1
            out dx,al
            mov cs:dac_buffer_offset,si ; Save new buffer pointer
        dec cs:dac_points_to_do
    skip_output:
        dec cs:divide_count              ; DOS 18.2Hz time-of-day clock
        jnz skip_clock                   ; If <divide_count>=0, call
        mov ax,cs:initial_count          ; DOS time-of-day service routine
        mov cs:divide_count,ax
        pushf                            ; (PUSHF put CPU flags back on stack
        call cs:dos_clock                ;  because DOS time-of-day routine will to an IRET
        jmp dac_isr_exit                 ;  rather than an RET return)
    skip_clock:
        mov al,20h                       ; Acknowledge Interrupt
        out pic_eoi_reg,al               ; to 8259 controller
    dac_isr_exit:
        pop si
        pop ds
        pop dx
        pop ax
        iret
    dac_isr endp


    ; Listing 7.1 (c). Routine to install DAC interrupt service routine into IRQ0 vector
    ; and to initiate the D/A output sweep.

    ;       FORTRAN: call dac_sweep(ibuf,nupdates,idiv,iperiod)
    ;       ibuf    = DAC buffer (DAC0,DAC1,DAC0,DAC1....)
    ;       idiv    = DOS clock multiplier (update rate = <idiv>*18.2Hz)
    ;       iperiod = 8253 counter 0 time period

    clock_mode    equ 43h                 ; PC 8253 system clock mode register
    clock_counter0 equ 40h                ; PC 8253 system clock counter register

    dac_sweep    proc far
        public dac_sweep
        nargs = 4
        save_regs
        unpack 1,nargs                    ; ds:si = address of ibuf
        mov ax,ds                         ; stored in memory for use by dac_isr
        mov cs:dac_buffer_seg,ax
        mov cs:dac_buffer_offset,si
        unpack 2,nargs                    ; cs:dac_points_to_do =
        mov ax,ds:[si]                    ; No. of DAC updates still to
        mov cs:dac_points_to_do,ax        ; be done.

        unpack 3,nargs                    ; Get clock divider <idiv>
        mov ax,ds:[si]                    ; for DOS clock which is called
        mov cs:divide_count,ax            ; every <idiv> clock interrupts
        mov cs:initial_count,ax           ; to maintain an 18.2Hz rate
    ; ... continued ...
```

```
Listing 7.1 (b) continued
    mov ah,35h                      ; DOS call to get address of
    mov al,irq0                     ; DOS clock routine from IRQ0 vector
    int 21h                         ; (Returned in ES:BX)
    mov cs:dos_clock.ipr,bx         ; Save address of DOS clock routine
    mov cs:dos_clock.csr,es

    mov ah,25h                      ; Replace IRQ0 vector with call
    mov al,irq0                     ; to D/A int. service routine
    mov dx, seg dac_isr             ; Put seg:offset of routine
    mov ds,dx                       ; into DS:DX
    mov dx, offset dac_isr
    int 21h

    cli                             ; Disable hardware interrupts
    mov al, 00110110b               ; Load counter 0 of 8253 timer
    out clock_mode,al               ; (set mode = 3, square wave rate )
    unpack 4,nargs                  ; Get new count interval from <iperiod>
    mov ax, ds:[si]
    out clock_counter0,al           ; Load into counter
    xchg al,ah
    out clock_counter0,al
    sti                             ; Enable interrupts again

    restore_regs
    ret nargs*4

dac_sweep  endp
; Listing 7.1 (c) ... continued
; Routine to disable interrupt-driven DAC sweep and restore IRQ0 to normal.

dac_stop proc far
    public dac_stop
;       Disable operation of dac_isr by detaching interrupt service
;       routine and restoring the orginal DOS time-of-day service routine
;       (N.B. THIS ROUTINE MUST BE CALLED BEFORE THE PROGRAM EXITS
;       TO PREVENT SYSTEM CRASHES)
    save_regs
    cli                             ; Disable hardware interrupts
    mov al, 00110110b               ; Load new count value into
    out clock_mode,al               ; counter 0 of P.C.s 8253
    xor ax,ax                       ; ax = 0 = 1000h count value
    out clock_counter0,al           ; 1.19MHz clock. (Clock is in
    xchg al,ah                      ; mode 3, square wave rate generator)
    out clock_counter0,al
    sti                             ; Enable interrupts again
    mov ah,25h                      ; Replace IRQ0 vector with
    mov al,irq0                     ; DOS clock routine
    mov dx,cs:dos_clock.ipr         ; Get address of routine from
    mov ds,cs:dos_clock.csr         ; from <dos_clock> and put in DS:DX
    int 21h
    restore_regs
    ret
 dac_stop endp
 dac_code    ends
 end
```

```
c Listing 7.2. Leak current subtraction routine.
c Calculates average of series of two channel (current and voltage)
c voltage-clamp records, removing leakage/capacity current by
c averaging, scaling, and subtracting a series of leak current records.
c
        subroutine subtract_leak( icurrent, ivolts, np,
     &       itest_start,itest_end,ileak_start,ileak_end )
        integer*2 icurrent(np)              ! (Ret) leak-subtracted current array
        integer*2 ivolts(np)                ! (Ret) voltage data array
        integer*2 np                        ! (In) No. of sample points per channel
        integer*2 itest_start,itest_end     ! (In) Range of test record numbers
        integer*2 ileak_start,leak_end      ! (In) Range of leak record numbers
        integer*4 isum_current(2048)        ! Current summation array
        integer*4 isum_volts(2048)          ! Voltage summation array
        integer*2 ileak_current(2048)       ! Average leak current array
        integer*4 izero_volts,izero_current ! voltage & current channel zero levels.
        integer*2 ibuf(4096)                ! File I/O buffer
        character*8 status                  ! Record status
        equivalence( status, ibuf )         ! 'ACCEPTED' or 'REJECTED'

        nsectors_per_record = (np*2)/256 + 1
        open(unit=1,file='adc.dat',access='direct',form='binary',recl=512)
c       File format: H SDDDDDDDD SDDDDDDDD ... (H=header sector, S=record status,
c                      Rec.1     Rec.2                          D=data sector)

c       Calculate average of leak records

        do i = 1,np
           isum_current(i) = 0              ! Clear summation arrays
           isum_volts(i) = 0
        end do
        navg = 0
        do irecord = ileak_start,ileak_end
        isector = (irecord-1)*nsectors_per_record + 2
           read(unit=1,rec=isector) (ibuf(i),i=1,256)        ! Read record
           if( status .eq. 'ACCEPTED' ) then                 ! Add
              read(unit=1,rec=isector+1) (ibuf(i),i=1,np*2)  ! from file
              do i = 1,np                                    ! to
                 isum_current(i) = isum_current(i) + ibuf(2*i-1) ! summation
                 isum_volts(i) = isum_volts(i) + ibuf(2*i)       ! buffers
              end do
              navg = navg + 1
           endif
        end do
        if( navg .gt. 0 ) then
           do i = 1,np
              ileak_current(i) = isum_current(i) / navg      ! Divide to
              ivolts(i) = isum_volts(i) / navg               ! get average.
           end do
        endif
c
c       Calculate leak record voltage step size as difference
c       between voltage zero level ( average of points 1-10) and
c       voltage level in middle of record (average of np/2 - np/2+9
c ... continued ...
```

```
c Listing 7.2 ... continued
c       Calculate average of test records

        izero_current = iaverage_level( ileak_current, 1, 10 )
        izero_volts = iaverage_level( ivolts, 1, 10 )
        ileak_voltage_step = iaverage_level(ivolts,np/2,10) - izero_volts
        do i = 1,np
            isum_current(i) = 0                 ! Clear summation
            isum_volts(i) = 0                   ! arrays.
            ileak_current(i) = ileak_current(i) - izero_current
        end do
        navg = 0
        do irecord = itest_start,itest_end
            isector = (irecord-1)*nsectors_per_record + 2
            read(unit=1,rec=isector) (ibuf(i),i=1,256)        ! Read record
            if( status .eq. 'ACCEPTED' ) then
                read(unit=1,rec=isector+1) (ibuf(i),i=1,np*2)  ! from file
                do i = 1,np
                    isum_current(i) = isum_current(i) + ibuf(2*i-1)
                    isum_volts(i) = isum_volts(i) + ibuf(2*i)
                end do
                navg = navg + 1
            endif
        end do
        if( navg .gt. 0 ) then
            do i = 1,np
                icurrent(i) = isum_current(i) / navg
                ivolts(i) = isum_volts(i) / navg
            end do
        endif
        izero_volts = iaverage_level( ivolts, 1, 10 )              ! Test voltage
        itest_voltage_step = iaverage_level(ivolts,np/2,10 ) - izero_volts  ! step size
c
c       Scale leak current up by ratio of the test/leak voltage steps
c       and subtract from the test current (Note use of REAL arithmetic)
        if( ileak_voltage_step .ne. 0 ) then                       !
            scale = float(itest_voltage_step)/float(ileak_voltage_step)
        else
            scale = 0.
        endif
        do i = 1,np
        icurrent(i) = icurrent(i) - int(scale*float(ileak_current(i)))
        end do
        close(unit=1)
        return
        end

        integer*2 function iaverage_level( ibuf, istart, navg )
        integer*2 ibuf(1),istart,navg       ! Find average value of a series of
        integer*4 isum4                     ! A/D samples
        isum4 = 0
        do i = istart,istart+navg-1
            isum4 = isum4 + ibuf(i)
        end do
        iaverage_level = isum4 / max(navg,1)
        return
        end
```

CHAPTER EIGHT

Analysis of single-channel currents

The development of the patch voltage clamp method by Erwin Neher and others in the late 1970s (Neher & Sakmann, 1976; Hamill et al., 1981) created a revolution in electrophysiological methods. They discovered that under appropriate circumstances the tip of a fire-polished glass micropipette forms a tight seal when pressed against a cell, effectively isolating a patch of cell membrane a few square micrometres in area underneath the pipette tip. Measurements of the current flowing through this membrane, using a very low noise measuring circuit, revealed the presence of tiny randomly occurring rectangular current pulses. These proved to be currents flowing through single ion channels in the cell membrane. Although Hodgkin & Huxley had introduced the concept of the ion channel as a means of mediating the flow of current across the cell membrane as early as 1952 (Hodgkin & Huxley, 1952d), and it had received general acceptance, Neher's work was the first to demonstrate their existence directly. The technique developed rapidly and is now applied to a very wide range of cell and tissue types. A comprehensive introduction to patch clamp principles and practice can be found in Sakmann & Neher (1983).

The widespread use of the patch clamp technique and, as we will see, the key role that the computer plays in the analysis of single-channel currents, merit detailed consideration. This chapter will discuss the basic principles and practice of these analysis methods, with particular attention being paid to the actual coding of computer routines. These can be used as the foundation of a single-channel analysis package.

The importance of the patch clamp technique lies in its ability to provide detailed *microscopic* information on the behaviour of the ion channels which mediate trans-membrane ionic currents. Hitherto it had only been possible to observe the *macroscopic* properties of the combined currents from the large populations of ion channels in the whole cell.

Several powerful experimental techniques are

made possible using the patch clamp. The simplest mode of operation is the initial *cell-attached* configuration, obtained when the pipette is first sealed against the cell. Currents can be recorded from the ion channels in the isolated patch under relatively natural conditions with the integrity of the cell maintained. The cell-attached mode is of limited use since it is difficult to apply agonists to the membrane surface without a system for perfusing the patch pipette. However, if the membrane–pipette seal is strong enough, the patch can be ripped completely off the cell exposing the inner membrane surface to the bathing solution. In this *inside-out* configuration it is possible to apply substances to the inner membrane surface of the patch and is especially useful for the study of the effect of intracellular second messengers on ion channel function. Alternatively, it is possible to apply a pulse of suction to the pipette and suck away the inside of the patch providing a low resistance access pathway into the cell itself. This *whole cell* configuration allows the patch clamp to be used for conventional whole cell voltage clamping of the type described in Chapter 7. Finally, if the pipette is pulled away from the cell while in a whole cell configuration, it is not uncommon to find that a flap of membrane seals over on to the pipette forming a patch of membrane with its outer surface exposed to the bathing solution. This *outside-out* configuration is extensively used for studying ligand-gated channels where various concentrations of agonists and antagonists can be applied and removed.

Ion channels collectively form the basis of the cell membrane's ionic conductance mechanism. In that respect, many of the issues discussed in Chapter 7 also apply to single channels. In particular, each specific ion selective conductance (Na^+, K^+, Ca^{2+}, Cl^-, etc.) is mediated by a distinct population of ion channels with that selectivity. Similarly, the membrane equivalent circuit model is still used as a representation of the electrical behaviour of the membrane. However, the magnitude of the conductance elements in the model are much smaller, in the order of picosiemens (10^{-12} S) rather than nanosiemens (10^{-9} S).

The time course of single ion channel currents is also quite different from the macroscopic currents through multi-channel populations. Macroscopic currents generally display smooth deterministic time courses, often exponential in shape (e.g. see Figure 7.3), whereas single-channel currents consist of randomly occurring rectangular pulses of current of relatively constant amplitude but varying markedly in duration. A typical single channel current with an amplitude of around -3 pA (picoamperes) can be seen in Figure 8.1. Consequently, a quite different range of techniques are required for the analysis of single ion channels compared with macroscopic currents.

The first stage in the analysis of single-channel currents is usually an analysis of the amplitude of the currents, with the generation of *I–V* curves and the determination of the ionic selectivity of the channel. One of the most striking features that was first observed of single channels was the constancy of the current amplitude. *I–V* curves generated from mean current amplitudes measured over a range of cell holding potentials are most often linear. Channels can therefore be characterized by their conductance which is a constant for a particular class of channels, and the ionic selectivity determined from the current reversal potential and/or ion substitution experiments.

In contrast to amplitude, the duration and frequency of opening of single channels show marked variablity often with voltage and time dependences. It is, in fact, these properties which underlie the non-linear voltage and time-dependent properties of macroscopic currents. However, due to the random variability in channel open and closed times the essential nature of the system is not apparent from any individual channel opening taken in isolation. It is only revealed within the statistical distribution of the open and closed times, observed for a large number of openings. Macroscopic currents are, in a sense, a measure of the distribution of the open and closed times of the population of channels in the cell. Single-channel analysis acquires similar information by studying the distribution of repeated opening and closures of one channel.

Single-channel analysis has the advantage of being able to measure both the channel open and closed time distributions whereas macroscopic

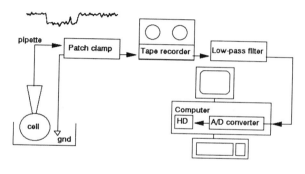

Figure 8.1 A computer system for recording agonist-activated single-channel currents. Signals from the patch clamp amplifier are stored on an instrumentation tape recorder, played back via an eight pole Bessel low-pass filter, digitized and stored on hard disc. A typical single-channel current is shown above.

periods to acquire the required number of channel openings. For instance, approximately 101 s would be required to record 1000 openings of a channel with a mean open time of 1 ms and closed time of 100 ms.

Figure 8.1 shows a typical system for recording agonist-activated single-channel currents. Current passing through the channels in the patch (a cell-attached patch is illustrated) is sensed by the patch clamp device (a low noise current–voltage converter) amplified and output as an analog voltage signal. During the experiment this signal is recorded on an instrumentation tape recorder, which is afterwards played back via a low pass filter, digitized by the A/D converter, and stored on the computer hard disc for analysis.

8.1 RECORDING SINGLE-CHANNEL CURRENTS

The procedures for the digital recording of single-channel currents depend on the nature of the channels being studied. Ligand-gated channels such as the endplate channel are activated by the application of an agonist and will continue to fluctuate between open and closed states for long periods. In order to analyse such channels a continuous unbroken digital record containing 1000 or more openings is required. It is not unusual to find that channel open and closed times can differ by several orders of magnitude, particularly if low concentrations of the agonist have been used to avoid desensitization. Consequently, it may be necessary to record for long periods recording can only reveal their combined effect. This permits more detailed kinetic models of channel behaviour to be constructed (e.g. Colquhoun & Sakmann, 1985; Sine & Steinbach, 1986). However, in doing so it is necessary to record and measure the amplitude and duration accurately of many thousands of channel openings. This makes computer analysis almost a requirement for the successful analysis of single-channel currents.

8.2 ANALOG SIGNAL STORAGE AND CONDITIONING

The long durations of the signals combined with the requirement for a wide recording bandwidth, to resolve very brief openings or closures, requires the high storage capacity of tape, unlike many other areas of electrophysiology where signals are often recorded directly on to the computer system. A variety of tape recording formats are in current use. The oldest technology is FM (Frequency Modulation) $\frac{1}{4}$ or $\frac{1}{2}$ inch width reel to reel magnetic tape, a good example being the Racal Store 4 (Racal, Southampton, UK).

FM tape recorders are expensive since they require precision tape transport systems for high quality, and have been largely superseded by a variety of digital systems. Digital pulse code modulation (PCM) systems which adapted standard domestic VCRs as the tape medium proved a popular and inexpensive alternative to FM during the 1980s. These were based on modified versions of a digital encoder (Sony PCM 701) developed by Sony orginally for high quality audio work (Bezanilla, 1985; Lamb, 1985). Encoders are available from Medical Systems (Greenvale, NY, USA) (based upon the Sony design) and from Instrutech (Mineola, NJ, USA). VCR-based systems have themselves been made obsolete by Sony's development of Digital Audio Tape (DAT) recorders, a much more compact system based upon a rotating tape head system similar to VCRs but designed specifically for audio (Pohlmann, 1989). Modified DAT recorders are now also available for laboratory use (Biologic DTR1200 two channel and DTR1600 eight channel (Biologic, Echirolles, France)). These recorders provide at least two wideband (DC-20 kHz) recording channels with additional voice and digital trigger channels on 1–2 h length tapes.

The tape recorder is played back into the computer via a Bessel type low pass filter. Low pass filtering is more important to single channel currents than for most other electrophysiological signals. The signals produced by the patch clamp current–voltage converter can often have a signal–noise ratio little better than two. In order to distinguish the channel openings from the background noise it is necessary to apply low pass filtration (see Section 3.4–3.9). However, rectangular pulse signals contain large amounts of high frequency components and can often be significantly distorted by their removal. In particular, short duration events may be reduced in amplitude to such an extent as to be lost from the signal. The optimal filter cut-off frequency is therefore a compromise between background noise reduction and the avoidance of distortion, dependent upon both the background noise level and the distribution of event durations. Usually, cut-off frequencies in the 1–10 kHz range are used. Colquhoun & Sigworth (1983) have discussed criteria for choosing an optimal cut-off frequency.

It is advisable to use a high quality multi-pole device as the low pass filter, with a wide range of switch selectable cut-off frequencies. The filter should also be of the Bessel type in order to avoid the introduction of unnecessary phase distortion into the signal. Typical filters are the Frequency Devices LPF902 (Haverhill, USA), Barr & Stroud (Glasgow, UK) and the Kemo VBF8 (Beckenham, UK).

8.3 DIGITIZATION OF SINGLE-CHANNEL CURRENTS

The continuous digital recording of single-channel currents for durations of several minutes results in records containing several Megabytes of data. Due to the limited RAM memory capacity of most computers in current use, the digitization of records of this size usually requires the use of continuous sampling-to-disc recording procedures. As discussed in Section 2.15, very large continuous digital records can be collected in this way and stored on disc while recording is in progress, using only a relatively small temporary buffer (16 kbytes) in RAM. The size of the record is limited by hard disc capacity rather than RAM capacity. Again there are distinct advantages to using a laboratory interface which support the DMA data transfer method to minimize the load on the host computer's CPU.

The A/D sampling rate used for the recording is matched to the low pass filter cut-off frequency. Generally, a rate of 5–10 times higher than the cut-off frequency (see Section 3.4) is used, in order to ease the process of detecting channel open/close transitions and to improve the quality of visual display. For instance, with 2 kHz filtering and a sampling rate of 20 kHz the 101 s continuous record referred to earlier would be 4.04 Mbytes in length. Continuous sampling to disc at rates of at least 30 kHz can be achieved using DMA, compared with around 10 kHz using interrupt-driven data transfer. In fact, with DMA data transfer in use, the hard disc data transfer rate provides a more significant limitation to sampling-to-disc performance. Nevertheless, with a 25 ms access time hard disc, sustained transfer rates of at least 30 kHz can be easily achieved using standard MS-DOS file access procedures as long as care is taken to ensure that the data is stored continuously on disc. Rates as high as 70 kHz can be achieved, by using faster discs, and by circumventing the DOS file system and writing directly to disc sectors.

Continuous sampling-to-disc is in most circumstances the preferred technique to use for recording ligand-gated single-channel currents since it ensures that a complete record of the signal is acquired. However, it can be extremely inefficient to record channels whose open times are several orders of magnitude shorter than their closed times. For instance, a 40 Mbyte data file would be required to record 1000 openings of a channel with a mean open time of 1 ms and a mean closed time of 1000 ms. An alternative approach is to record only the short channel openings discarding the long closed intervals (Sigworth, 1983). A spontaneous signal detection and capture routine such as that in Section 2.16 can be used for this purpose. While this approach is quite satisfactory where channel amplitude and open time are of the primary interest, it cannot be guaranteed that a completely faithful record of all channel openings has been acquired. Like all detection methods, it is not possible from later inspection of the collected records to determine how many events have been missed by the detector.

8.4 VOLTAGE-ACTIVATED SINGLE-CHANNEL CURRENTS

The recording procedures for voltage-activated channels are quite different from that required for ligand-gated channels. In general it is possible to record ligand-gated channels in approximately steady-state conditions where large numbers of channel openings can be collected. However, the common Na^+, K^+ or Ca^{2+} voltage-activated channels rarely produce more than short bursts of openings in response to voltage steps. Such channels must be studied by the repeated application of a voltage step at intervals designed to allow the channels to recover. Since the channel openings naturally occur as discrete bursts there is no need for continuous sampling-to-disc. Instead, the same voltage pulse generation and recording techniques for whole cell currents, as discussed in Chapter 7, are used. The voltage steps required to evoke the channel openings also produce substantial capacity and leakage current artefacts which must be removed from the current record using the leak current subtraction techniques discussed in Sections 7.11 and 7.12. Once the bursts of single-channel currents have been cleaned up in this way, single-channel analysis procedures can then be applied.

8.5 ANALYSIS OF CHANNEL CURRENT AMPLITUDES

As mentioned earlier, the analysis of the amplitude of single-channel currents allows channels to be classified on the basis of conductance and ionic selectivity. Current amplitude measurements are also an essential precursor to the measurements of channel open and closed times since it is necessary to have an accurate measure of the current levels associated with these states, in order to detect the channel open/close transitions.

At its simplest a single-channel current recording is a series of rectangular fluctuations between zero current and a constant open channel current level. Unfortunately, most recordings are rarely as simple as that. There are usually several channels active within the patch. Even a single channel may have more than one open state with different conductances. The measurement of current amplitudes provides information on the number of different types of channel, or sub-conductance states of a single type, located within the patch.

8.6 CURRENT AMPLITUDE HISTOGRAMS

The simplest way of obtaining a measure of current amplitude within a recording is to compute the current amplitude histogram. The total measurable range of the current signal is divided into a set of equal-sized bins. Each A/D sample I_i in the signal record is allocated a bin j using the equation

$$j = \frac{(I_i - I_{lo}) n_b}{I_{hi} - I_{lo}} + 1, \quad Y_j = Y_j + 1$$

[8.1]

where I_{hi} and I_{lo} are the limits of the measurable current range and n_b is the number of bins in the histogram. The histogram is accumulated by applying Equation 8.1 to every sample in the record, and incrementing the histogram bins Y_j accordingly.

Figure 8.2 (a) shows the current amplitude histogram computed from a single-channel record (simulated) containing one channel type with an amplitude of 2.5 pA. The histogram exhibits two peaks, one centred around the zero current level and another at 2.5 pA. In general, an amplitude histogram will consists of a series of peaks corresponding to each distinct amplitude level within the recording, due to either channel sub-conductance levels or multiple channel openings.

The area underneath each peak reflects the amount of time that the channel has spent in that state. The width of the peak is a measure of the variability of the current fluctuations within each conductance state. A significant part of this is due to noise in the recording instrumentation which is dependent on the quality of the gigaseal. However, it can also be due to small poorly resolved variations in the channel conductance (Sigworth, 1985).

8.7 OPEN CHANNEL PROBABILITY

The current amplitude histogram can be used to derive estimates for the lower limit for the number of channels in the patch, the number of distinct channel conductance states, and the *open channel probability*, p_{op}. This is a measure of the proportion of time that the channel spends in the open state, and is useful as a measure of the overall channel activity. It can be computed from the amplitude histogram using the equation

$$P_{op} = \frac{1}{n_t n_c I_u} \sum_{j=1}^{n_b} Y_j (I_j - I_z) \quad [8.2]$$

where n_b is the number of bins in the histogram, n_c the number of channels in the patch, Y_j the number of samples within bin j, n_t the total number of samples in the histogram, and I_z and I_u the zero and unitary current levels. Note that Equation 8.2 is crucially dependent on the number of ion channels contained within the patch which is rarely only one and may often be as high as 100–200 (Dilger & Brett, 1990). In practice, it is difficult to obtain a valid estimate for n_c. At low open probabilities, counting the number of peaks only provides a lower limit for the true number of channels since the probability that all channels are open at the same time may be so low as to make it an extremely unlikely event within the recording period. This makes

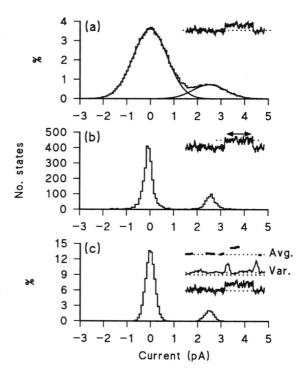

Figure 8.2 Single-channel current amplitude histograms. (a) Amplitude histogram from a simulated single-channel current record with a 2 ms mean open time, 10 ms mean closed time and unit current of 2.5 pA. The area of open and closed state peaks have been estimated by fitting a pair of gaussian curves (shown superimposed) using Equation 8.3. (b) Channel average amplitude histogram from the same record as (a). Transitions detected using 50% threshold method. (c) Running average amplitude histogram computed using Patlak's method with a 16 sample averaging window and variance acceptance threshold set to the variance of a long closed interval.

comparison of p_{op} values between different patches difficult. However, applying Equation 8.2 to the data in Figure 8.2 (a) with an assumption that there is only one channel in the patch provides a figure of 0.16 for p_{op}.

8.8 FITTING GAUSSIAN CURVES TO AMPLITUDE HISTOGRAMS

Under good recording conditions the current amplitude distribution can usually be described by the sum of a series of gaussian functions, one for each peak as in the equation

$$Y_j = \sum_{i=1}^{n} \frac{A_i w}{\sqrt{(2\pi\sigma_i^2)}} \exp\left(\frac{-(I_j - I_i)^2}{2\sigma_i^2}\right) \quad [8.3]$$

where A_i, I_i and σ_i are the fractional area, mean current and standard deviation of the ith gaussian component, w is the bin width. A_i, I_i and σ_i can be estimated by fitting Equation 8.3 to the amplitude histogram using the iterative least squares curve fitting procedures discussed in Chapter 6. This approach proves particularly useful in cases where the peaks overlap, preventing estimation of area and means by simpler approaches. This is illustrated in Figure 8.2 (a) where the areas of the two overlapping peaks have been obtained by fitting Equation 8.3 with $n = 2$ to the amplitude histogram. The best-fit results indicate that 83.5% of the samples within the record are contained in the peak close to zero current ($I_1 = -0.01$ pA) while 16.5% is contained within a peak centred at 2.5 pA. Note that this is consistent with the value of 0.16 for p_{op} computed from Equation 8.2. Both peaks have similar standard deviations (0.723 and 0.704 pA). The two individual gaussian curves are shown superimposed on Figure 8.2 (a).

Gaussian curve fitting can successfully separate the components of two peaks even when there is a considerable degree of overlap. However, care must be taken to ensure that the results are meaningful within the context of the experiment. It is not unusual to find that the best fit produces a component with a very large standard deviation, which cannot be associated with any observed channel opening within the original record. It is often necessary to choose the initial parameter estimates carefully and in some circumstances also to constrain the values of certain parameters. These problems often arise when a small peak overlaps with a substantially larger one. Often, the symmetrical nature of the gaussian curve can be exploited to perform fitting as a two-stage process. A good initial estimate of the larger peak can be obtained from a single gaussian fit to a region of the large peak which does not overlap with the small one. These values can then be used as the initial estimates for that peak in a two gaussian fit which includes the small component. This is often sufficient to ensure that the curve fit converges to a meaningful solution. Programs such as Biosoft Fig.P which allow curve fitting with constrained and unconstrained parameters can be very useful in this application.

The gaussian function usually provides a good fit to the amplitude histogram. However, in circumstances where channels have a large number of brief events, and the signal is heavily low pass filtered, the amplitude histogram can be distorted to such an extent as to be no longer fittable with a gaussian function. Yellen (1984) describes the use of beta functions as an alternative in such circumstances.

8.9 AVERAGE AMPLITUDE HISTOGRAM

The amplitude histogram has the merit of simplicity and provides a good estimate of the proportion of time a channel spends in each of its possible states. However, the width of the peaks within the distribution often makes it difficult to resolve closely spaced current levels even using gaussian curve fits. To enhance the resolution it is necessary to apply some kind of averaging process to obtain a better estimate of channel current levels. If the location of the transitions between channel states is known, it is possible to compute the average current level for each individual open or closed interval. Since the variance of a set of averages is less than that of the original data set, the histogram of the channel current averages will consist of a narrower set of peaks than the amplitude histogram.

Figure 8.2 (b) shows the histogram of channel average amplitudes compiled, using the channel average routine in Listing 8.2, from the same data record used to produce Figure 8.2 (a). The beginning and end of each open and closed interval within the record were located using a half-amplitude threshold crossing method (see Section 8.13). The average current during the interval was calculated from the average of the sample points contained within each interval, according to the formula

$$\text{Avg} = \frac{1}{i_e - i_s + 1 - 2n} \sum_{i=i_s+n}^{i_e-n} I_j \quad [8.4]$$

where i_s and i_e are the transition points into and out of the channel state. A number of sample points n at each end of the interval were excluded to ensure that no transitional points biased the average. A histogram of average currents was accumulated using Equation 8.1. The average amplitude histogram is a plot of number of channel openings rather than A/D samples.

Since open or closed intervals vary in duration, each average is compiled from different numbers of samples with those from long openings much more precisely defined than for short. Consequently, the peaks within an average amplitude histogram are not adequately described by gaussian functions, having narrow centres due to the long-lasting openings and with wide tails due to the short events, as is evident from a comparison between Figures 8.2 (a) and (b).

Although the average amplitude histogram provides a more precise estimate of channel current levels, its application depends on having located the positions of the channel openings within the record. This creates a difficulty since most of the commonly used automatic transition detection methods depend upon an estimate of channel unitary current level already having been obtained. Consequently, in circumstances where the average amplitude histogram is of value it may be necessary for the user to carry out this process semi-manually by displaying the signal records on the screen and indicating the start and end of the interval to be averaged with a cursor.

8.10 PATLAK'S RUNNING AVERAGE AMPLITUDE

To get round these limitations of the average current amplitude histogram, Patlak (1988) developed an averaging technique which retains many of the advantages of channel current averaging but does not require *a priori* knowledge of the location of the channel openings. The technique achieves this by computing a selective running average of the current signal excluding points close to and during transitions between states. As discussed in Section 4.20, a running average is a simple form of digital filtering where each point in the orginal record

I_i is transformed by averaging a window of the previous n sample points using the formula

$$\text{Avg}(I_i) = \frac{1}{n} \sum_{j=i-n+1}^{i} I_j \quad [8.5]$$

This procedure has the effect of reducing the background noise variance by a factor of n. Applying Equation 8.5 alone to the current signal proves unsatisfactory since it also smooths the transitions between states distorting the histogram. It is necessary therefore to exclude averages containing transitional samples. Patlak used the variance within the averaging window to determine whether the average should be included in the histogram

$$\sigma_i^2 = \frac{1}{n-1} \sum_{j=i-n+1}^{i} [I_j - \text{Avg}(I_i)]^2 \quad [8.6]$$

Figure 8.2 (c) shows the running average amplitude histogram of the test data record, computed according to Patlak's method. The result of the variance calculation is shown inset. When the 16 point sample window is completely contained within an open or closed state the variance is close to a minimum determined by the background noise of the recording. However, whenever the 16 point window spans a transition between states the variance is much higher. The exclusion level can be set using an estimate for the background variance determined from a long and unambiguous closed interval.

Patlak's moving average can be as effective at reducing the background noise as the direct averaging after transition detection. The process, however, is a conservative one excluding all sample windows containing any portion of a transition and also points which happen to exceed the variance threshold. The process is biased towards long events since an opening or closure must last longer than 16 sample points in order to sustain a sample window which does not include a transition. In Figure 8.2 (c) only 44% of the total number of sample points have been included in the histogram.

Routines for generating each of the three types of amplitude histogram, from a continuous digitized record of single-channel currents stored in binary form in the file 'patch.dat', can be found in Listing 8.1. Routine (a) produces the amplitude histogram by simply binning every A/D sample in the file. A/D samples are extracted from the data file and processed in blocks of 512. Although this produces a slightly more complicated code it is at least an order of magnitude more efficient than extracting one sample at a time. The average amplitude histogram routine (b) makes use of the event transition information stored in the file 'events.lst' generated by the transition detection routine (Listing 8.2) to determine the position of the beginning and end of each channel event within 'patch.dat'. In this case again the software design has been influenced by the need for efficiency. A/D samples are processed in blocks of 512 with additional code to account for the fact the openings and closures start and end at arbitrary points within each block, and may also span more than one block. Routine (c) implements Patlak's running average technique, with averaging windows of up to 256 samples. Note that care has been taken to make the calculations efficient by adding and dropping samples from the running sums used to calculate the average and variance, rather than recomputing the whole average for each new window.

8.11 ANALYSIS OF CHANNEL KINETICS

While amplitude analysis provides a useful means of categorizing single channels it is a relatively static property of the channel. In general, ion channels respond to their environment by changing open time and rate of opening rather than conductance. Much greater insights concerning ion channel function can therefore be obtained by studying channel open and closed times and how these are influenced by membrane potential, the application of neurotransmitters or other substances. As can be seen from the channels in Figure 8.3, channel open and closed times are extremely variable even under constant conditions. The analysis of channel open and closed time intervals within a digitized single channel recording consists of two main stages.

(a) *Transition detection*: where the location and duration of channel openings are determined by the detection of the channel open/close transitions within the current signal.
(b) *Dwell time analysis*: where the time intervals for each channel opening or closure are compiled into dwell time histograms for statistical analysis.

8.12 TRANSITION DETECTION

It might seem, at first sight, that the detection of the transitions between channel states should be a straightforward task, well suited to automation. Looking at records of single-channel currents it is usually quite obvious to the observer which state the channel is in. However, this underestimates the natural human ability to discriminate pattern from noise and the difficulty of describing this process sufficiently to allow its implementation as a computer algorithm. There is still, after 10 years of effort, some debate as to the merits and limitations of various approaches to this task. Some take the view that the task can only be safely carried out by a human observer, albeit with some computer assistance in measurement (Colquhoun, 1987a; Barry & Quantararo, 1990), whereas others have proposed fully automated systems operating without any human supervision or validation (Sachs, 1983). The issue is an important one given the large number of intervals that must be measured. The visual inspection and measurement of channel openings is a time-consuming and tiring process which, although making use of the human operator's insight, is prone to error as the operator becomes fatigued. On the other hand, a fully automated process can measure tens of thousands of channels in a few hours (e.g. see Blatz & Magleby, 1989). However, a flaw in the automated measurement procedure may introduce subtle but significant errors into the results which may not be immediately obvious.

Part of the reason for the difference of opinion on transition detection procedures is that single-channel currents show a marked variation in quality between different cell and channel types. Five factors interact to determine the difficulty of transition detection:

- background noise
- single-channel current amplitude
- mean channel open/close times
- number of active channels in patch
- number of channel conductance states.

A single-channel current will be readily observable only if its amplitude is larger than the rms background noise level after low pass filtering. Single channels vary markedly in conductance from the very large Ca-activated K channels with conductances of 200 pS (Magleby & Pallotta, 1983a) to the 5HT$_3$-activated cation channel with a conductance of 0.3 pS (Lambert *et al.*, 1989), with many of the channels commonly studied falling within the 10–30 pS range. The background noise depends to large degree on the resistance of the pipette–membrane seal (Corey & Stevens, 1983) with higher resistances producing lower noise, the quality of the seal varying with cell type. Similarly, channels with long-lasting open and closed times can be subjected to a greater degree of filtering without significant distortion.

If more than one channel is active within a patch it is not uncommon to find that two or more channels may be open at once making it impossible to discern the open/close transition of any individual channel. A single channel may also have more than one conducting state, at the very least complicating the algorithm used for transition detection.

The difficulties facing any transition detection procedure can be seen in Figure 8.3 which shows a selection of single-channel currents obtained from an outside-out patch from a bovine chromaffin cell, evoked by the application of acetylcholine. The currents were recorded under reasonably good recording conditions from channels with a predominant current amplitude of around −3.5 pA, as measured by Gaussian fit to the amplitude histogram. As can be seen, they exhibit a range of behaviours from single very brief openings to prolonged bursts of openings.

The event in Figure 8.3 (a) is almost an ideal single-channel current. It is nearly rectangular in shape with a distinct well-defined open period and an amplitude corresponding with the value derived from the amplitude histogram. The open

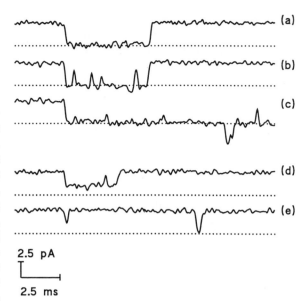

Figure 8.3 Single-channel currents evoked by application of acetylcholine to an outside-out patch from a bovine chromaffin cell. The predominant current amplitude (as measured from the amplitude histogram), −3.5 pA, is indicated by the dotted lines. The traces (a)–(e) are selected to illustrate the typical range of types of openings. (a) A well-resolved relatively long single opening, (b) a burst of openings, (c) a burst during which another channel opens, (d) a channel opening to a sub-conductance state and (e) very short poorly resolved openings. (Records courtesy of Drs J.J. Lambert and J.A. Peters, Dundee.)

period is sufficiently long not to be significantly influenced by the low pass filtering applied to the current. Accurately estimating the open duration of such events is trivial either by eye or by any automated procedure.

However, Figure 8.3 (b) shows another opening of much the same length but this time displaying a series of brief closures within it. In fact, the opening is better regarded as a burst of short openings and closures. This is a common property of ion channels and is observed in many different types of channel ($K_{(Ca)}$, Magleby & Pallotta 1983b; GABA, MacDonald *et al.*, 1989). A longer burst of openings is shown in (c) during which a second channel has opened. The very brief nature of the closures within these bursts clearly pose a detection problem since many of them do not fully reach the zero current baseline and some do not even reach 50% of the unit current level.

A further complication is created by the existence of more than one conductance state. Figure 8.3 (d) shows a clear example of the channel in a less frequently observed open state with an amplitude approximately 60% of the more common level of -3.5 pA level. In some circumstance, such as the very brief openings shown in (e), it may be difficult if not impossible to determine unambiguously which conductance level the opening should be assigned to.

8.13 THRESHOLD CROSSING METHODS

Procedures for transition detection and the measurement of channel dwell times vary in complexity. The most common approach to transition detection is the *threshold crossing* method. It is perhaps the most intuitively obvious method of detecting transitions between channel states and is used in many of the available single-channel analysis packages (IPROC, Axon Instruments pClamp, CED Patch clamp package, PAT Strathclyde Software). Using information from the amplitude histogram, a transition threshold is defined, somewhere between the zero and the unitary current levels of the recording. Points exceeding the threshold level are deemed to indicate that the channel is in an open state, points less than the threshold indicate the closed state. It is common to place the threshold at half the unitary amplitude to avoid any bias towards one or the other of the channel states. Once a threshold level has been defined it is possible to scan sequentially through the digitized record point by point, storing the positions at which the signal crosses the threshold. This list of crossings is then be used to compile an idealized list of channel open and closed intervals.

Figure 8.4 (a) shows the results of half amplitude transition detection when applied to some of the channels discussed earlier. The threshold has been set to -1.75 pA. As can be seen from the first opening, the method works with well-resolved signals with good signal–noise ratios. However, problems arise with very short lasting events which have their amplitudes reduced by low pass filtering. Even though they are quite apparent to the observer, the brief opening and one of the brief closures within the burst have not been detected by the computer algorithm. This has the effect not only of the loss of the undetected events but of falsely increasing the duration of the opposite state. For instance, the two closed intervals on either side of the missed opening are considered as a single long interval. An even worse problem occurs in the case of the final opening where the channel is in a sub-conductance state with an amplitude close to the threshold. A rapid series of false openings and closures have been detected due to the background noise causing transition crossings even though the channel does not change state.

The above problems seriously limit the effectiveness of half amplitude transition detection. In practice it is only safe to use it on well-resolved currents with good signal–noise ratios and no sub-conductance states close to the threshold.

8.14 TWO THRESHOLD METHODS

Some of the deficiencies of the half amplitude method can be remedied by using two detection thresholds, placed close to the zero and open state current amplitudes as shown in Figure 8.4 (b). When the signal amplitude (negative) exceeds the open state threshold (O), the channel is deemed to be in the open state, and in the closed state when less than the closed state (C) threshold. A certain amount of additional logic is needed to deal with sample points lying between the two thresholds. In particular, it is necessary to ensure that the small numbers of intermediate points, occurring solely as part of a transition between states should not be classed as separate events. Such points when they occur are classified as belonging to the preceding state. On the other hand, channel openings to sub-conductance levels lying within the intermediate state should be classified as a separate state. The two conditions can be partially distinguished by the criteria that transition points appear within a pattern of closed–intermediate–open and open–intermediate–closed states while

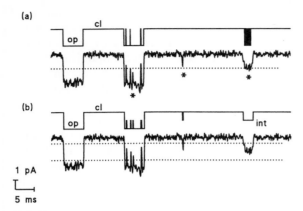

Figure 8.4 Amplitude threshold transition detection. (a) Half amplitude detection method with a single threshold placed half way between open and closed current levels (unitary current amplitude −3.5 pA). The record is categorized into two states – closed and open. The idealised trace is shown above. Detection errors are marked by asterisks. (b) Improved detection of brief events with the two threshold detection method. Thresholds are placed as close as possible above the closed (−0.75 pA) and below the open (−2.75 pA) levels. The record is categorized into three states – closed, intermediate and open.

the true sub-conductance states have the patterns closed–intermediate–closed or open–intermediate–open.

Looking at the idealized trace in Figure 8.4 (b), it can be seen that the two-threshold approach has performed much better than the single half amplitude method, correctly detecting all openings and closures and avoiding the false events associated with the sub-conductance level. It is far from perfect, though. In particular, it fails to detect direct transitions between subconductance states which do not pass through the closed state, as a consequence of the simple logic used to exclude transitional points.

8.15 COMPENSATING FOR ZERO CURRENT BASELINE DRIFTS

So far in our discussion of transition detection it has been assumed that the current amplitude levels used to define the detection thresholds are stable without any long-term drifts. Baseline stability is often related to the quality of the gigaseal with high resistance seals producing highly stable baselines. However, in many cell types good seals are difficult to obtain and some degree of baseline drift cannot always be avoided. In such cases, it becomes necessary to adjust the threshold levels in order to maintain a fixed offset from the zero current level. It is also worth noting that similar considerations apply to the amplitude histograms.

Two different approaches have been proposed for compensating for a drifting baseline. An early suggestion by Sachs (Sachs *et al.*, 1982; Sachs, 1983) was to search for the zero current baseline by repeatedly subtracting a series of trial baselines from each data record being analysed until a significant peak was observed in the number of zero crossings within the signal. The method has the advantage of not requiring any previous assumptions about the location of baseline. However, such procedures can be confused into an erroneous setting of the baseline level by records containing long open states or a rapid burst of openings. Quite a lot of additional logic is required within the method to avoid such problems which makes baseline tracking with this method a slow process.

An alternative approach is to use the average current level during long-lasting channel closed states as estimates of the baseline. The baseline is determined from a digital low pass filter of the sample points within closed intervals with durations longer than a pre-determined minimum (e.g. 5 ms). An initial estimate of the baseline level is required to prime the process, and this can be obtained from the amplitude histogram of the first record within the data file. This method proves to be an efficient and satisfactory means of tracking slow and moderate changes in baseline level. However, it depends on the existence of a sufficient number of long closed intervals to update its estimate of the baseline level. Drifting that occurs during long-lasting bursts of short openings and closures cannot be compensated for. The transition detection and the baseline estimation processes are also linked (a new closed state is detected using a threshold itself determined by previous closed states). Consequently, there is a potential for instability if baseline tracking is lost. This again emphasizes the need for the user to monitor the transition detection process visually.

An implementation of the two threshold detection algorithm with running average baseline tracking can be found in Listing 8.2. The routine scans the digitized signal record from 'patch.dat', looking for crossings of either of its two thresholds. When the channel enters a new state, the type, duration and location within the data file of the preceding state is written to the *event list* file 'events.lst'. This file consists of a series of 20 bytes records defined by a FORTRAN record STRUCTURE statement. At the end of the transition detection run, this file contains a complete list of the open/close states detected, to be used for dwell time histograms or the average current amplitudes.

Even though the two-threshold method is a distinct improvement on the half amplitude method it is still far from a general method applicable to all channel types. Its success here has depended on the fact that there is only one sub-conductance level of significance and that falls centrally within the intermediate range. However, many channels have 3–4 sub-conductance levels (e.g. GABA channels, Bormann *et al.*, 1987). In such circumstances it becomes difficult to find a pair of threshold levels which successfully discriminate channel openings from noise within sub-conductance levels. Other techniques have been proposed such as the AZTEC (Amplitude Zone Time Epoch Coding) method suggested by Vivaudou (1986) as a means for analysing recordings containing multiple channels. The method, orginally developed as a data compression and analysis method for electrocardiogram signals, classifies the signal into a series of slopes and plateaus without using a fixed amplitude threshold level. It avoids many of the problems of the half amplitude threshold method and appears to work well even when the signal is badly corrupted by baseline drift. However, it does not really address the problem of how to deal with multiple subconductance level of different sizes, simply transferring the problem to the analysis of the AZTEC coded signal rather than the raw digitized data. Furthermore, it does not explicitly account for the effects of low pass filtering on the signal.

8.16 TIME COURSE FITTING

The transition detection methods based upon amplitude thresholds perform particularly poorly when applied to a signal with significant numbers of brief openings or closures. Large numbers of these events may be missed completely with the half amplitude methods and, even when the detection of such events has been improved using two thresholds, the duration of the event is not correctly estimated. Since many of these events are often discernible to the operator it is clear that the amplitude threshold methods are not extracting as much information from the signal as possible.

In order to improve the analysis of such signals it is necessary to use more sophisticated signal analysis techniques. The most commonly used of these methods is the method of *time course fitting*, developed by Colquhoun (Colquhoun & Sigworth, 1983; Colquhoun, 1987a). The principle of the method is illustrated in Figure 8.5. An ideal channel opening or closure can be considered to be the sum of two current steps, an opening step from the zero current level to the channel unitary amplitude and a closing step from the open level back to zero as shown in (a) and (b). In a real single-channel current this ideal signal has random background noise superimposed on it and the summed channel+noise signal is subject to low pass filtering both reducing the background noise and rounding the edges of the steps. In the case of a brief opening such as shown in (c) the rising and falling edges of the smoothed steps overlap to the extent that the summed current signal cannot reach full amplitude.

The essence of the time course fitting method is that, if the time courses of the opening and closing steps are known, it is possible to find a summed response which matches the time course of the observed brief event, by adjusting the separation in time between the two steps. Colquhoun first implemented this method on the PDP11 minicomputer using an oscilloscope screen to display the digitized currents and the superimposed step responses. Step separation and amplitude was adjusted manually using a set of potentiometers linked to the computer's A/D converters until the visually best match was obtained on the display screen. He later enhanced the technique to perform the matching process using iterative least squares.

A time course fitting procedure can be readily implemented if a suitable function for defining the step responses can be obtained. The time course curves in Figure 8.5 were obtained using the gaussian digital filter (see Section 4.21) step response function

$$S(t,t_o) = 0.5(1 + \text{erf}(\frac{t-t_o}{\sigma\sqrt{2}}))I_u \quad [8.7]$$

where erf() is the error function, t_o is the time of occurrence of the step and I_u is the unitary current amplitude. The steepness of the transition is represented by the standard deviation σ. A channel opening can be represented by the sum of an opening step $S(t,t_o)$ and a closing step represented by an inverted form $(1 - S(t,t_o + t_d))$

$$I(t) = (S(t,t_o) + (1 - S(t,t_o + t_d)) - 1)I_u \quad [8.8]$$

The duration of a channel event can be obtained by fitting Equation 8.8 to the time course of the single channel current with t_o and t_d as the free parameters. A time course fitting procedure using this method is shown in Listing 8.5. The routine operates in two modes. In mode 1 it fits a single open step response to the leading edge of a long opening (using Equation 8.7) in order to derive estimates of the channel amplitude I_u and the filter standard deviation σ. These values are subsequently used in mode 2 which fits an open/close step response pair (Equation 8.8) in order to estimate the starting time t_o and duration t_d of a brief opening or closure. A Gauss–Newton algorithm E04FDF from the NAG library is used to perform the iterative least-squares fitting. The error function is calculated using an approximation formula from Press et al. (1986).

Figure 8.5 (a) shows a 50-sample record containing the leading edge of a long opening. Applying the routine in mode 1 to this record yielded best-fit values of 0.98 for I_u and 2.38 for σ. Using these values and applying the routine in mode 2 to the brief opening in Figure 8.5 (c) yielded best-fit values of 26.16 for t_o and 3.17 for

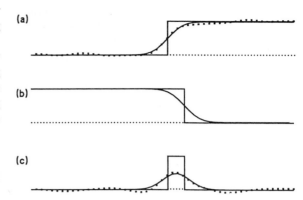

Figure 8.5 Time course fitting used to estimate the duration of a brief opening. (a) A representation of the channel opening step is obtained by fitting a gaussian filter step response (Equation 8.7) to the leading edge of a long opening. (b) The closing step is created by inverting (a). (c) The time course of the brief opening can be synthesized from the sum of (a) and a delayed version of (b). The duration of the opening that produced a brief event can be recovered by fitting the open/close step pair (Equation 8.8) using the previously derived I_u and σ and allowing the starting time t_o and the delay t_d to vary until the best fit is found. The best-fit step response curves and their step transitions estimated using this procedure are shown superimposed on the sampled data points. The duration of the opening in (c) was estimated as 3.17 compared with the actual value of 3.33 (currents are simulations).

t_d. Note that the time units are in fractional units of the sampling intervals.

Further details of the implementation and use of the time course fitting method can be found in Colquhoun & Sigworth (1983) and Colquhoun (1987a). It is clear that the method avoids many of the problems associated with the amplitude threshold methods. Brief time intervals at the limits of the frequency range are correctly estimated, with the range of measurable intervals extended to three fold smaller values (Colquhoun, 1987a). In addition, all dwell times are measured with a greater precision since the intervals are no longer limited to integer multiples of the sampling interval. Unlike the threshold methods, time course fitting is easily extended to deal with currents with multiple sub-conductance levels simply by choosing an appropriate current amplitude level for the fitting procedure.

The main disadvantage with the method is its highly interactive nature, requiring the constant attention of an operator. As it is currently implemented, channel openings are either located by visual inspection of the digitized signal record or at most by scanning with a threshold level set close to the baseline. The operator must then isolate the segment of the data containing the event and, if necessary, select an appropriate amplitude level. Processing events in this way is a tiring and time-consuming process and it is hard to imagine the method being applied to recordings containing the 20 000–30 000 events sometimes processed with threshold methods.

The subjectivity inherent in the operator's role in selecting events for processing also presents its own hazards; as Colquhoun himself has pointed out, it is tempting to over-analyse data. There comes a point for very brief events when it becomes impossible to distinguish between real openings or closures and random noise. One solution to this problem is to include all credible events during the initial measurement phase then at a later stage to discard those with durations less than a rigorously defined lower limit based upon the frequency response and background noise levels (Colquhoun & Sigworth, 1983).

Time course fitting is essentially a method for accurately measuring the duration of channel events. It does not address the problem of detecting and classifying events according to amplitude levels, that being left to the operator. However, at present, it is the most accurate means of estimating channel dwell times and probably the only one capable of dealing with channels with several sub-conductance states. In the case of long well-resolved openings, the range of amplitude levels within a recording can be obtained by a time course fit which also allows the amplitude to vary as well as the channel duration. Unfortunately, for short openings too much information is lost and it becomes impossible to distinguish whether a brief event is small because it is of short duration or opening to a sub-conductance level.

8.17 ANALYSIS OF CHANNEL DWELL TIMES

Irrespective of the particular method of transition detection used, the end result is a list of pairs of open and closed intervals which can then be passed to further analysis stages. The distributions of time intervals for each distinct channel state (open or closed) are represented by dwell time histograms. The range of time intervals for each type of event within the list is divided into a series of contiguous intervals. The number of events falling within each bin is accumulated and plotted in histograms as shown in Figure 8.6.

Dwell time distributions are typically exponential in shape with large numbers of short events and a decreasing rate of occurrence proportional to duration. Distributions often consist of multiple exponential components with different mean event lifetimes, each corresponding to a distinct kinetic state of the channel.

8.18 LINEAR DWELL TIME HISTOGRAMS

Figure 8.6 shows how the same dwell time distribution can be represented in a variety of ways with some of the commonly used types of histogram. For the purposes of illustration, a simulated distribution consisting of three exponential components was generated with mean lifetimes of 1, 10 and 100 ms, 1000 events per component. Such a wide range of times is not unusual particularly for closed time distributions (e.g. Colquhoun & Sakmann, 1983).

The simplest form of histogram is the linearly spaced bins, according to the formula

$$\text{Bin.no} = 1 + \text{int}(t_d/w) \quad [8.9]$$

where w is the width of each bin (ms), and int() is the integer part. Figure 8.6 (a) shows a linear histogram of the test distribution consisting of 100 bins each of width 4 ms. The multi-exponential nature of the distribution can be seen in this form of presentation (the best-fit sum of three exponentials is shown superimposed).

Although this form of representation is perhaps the most readily understood it can have distinct problems. It is difficult to find a single bin width which can satisfactorily represent all components in a widely spaced multi-component distribution. The 4 ms bin histogram in (a) represents the 10 ms and 100 ms components in

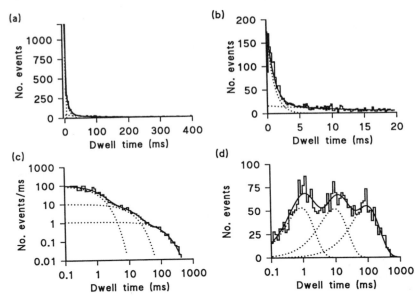

Figure 8.6 Dwell time histograms. Different types of dwell time histogram representing the distribution of times within a simulated data set of three exponential components with mean times of 1, 10 and 100 ms, each containing 1000 events. (a) Linear histogram with 100 × 4 ms width bins. The sum of three exponential pdfs fitted to data shown superimposed, yielding best-fit components (0.87 ± 0.22 ms, 1697 ± 424), (9.31 ± 0.61 ms, 1086 ± 33), (97.73 ± 9.62 ms, 1032 ± 46) (mean dwell time ± s.e., number of events ± s.e.). (b) Linear histogram with 100 × 0.2 ms width bins revealing the fastest component within the distribution. The best-fit two exponential pdf is shown superimposed with components (1.05 ± 0.05 ms, 1108 ± 44), (16.16 ± 3.74, 1355 ± 142). (c) Log-log histogram based upon Magleby's approximate logarithmic binning method with 16 bins per decade and a digital sampling interval of 0.1 ms. Three exponential pdf fit yielded components (0.9 ± 0.07 ms, 1003 ± 50), (9.43 ± 0.84 ms, 980 ± 55), 94.11 ± 6.55 ms, 1037 ± 51). (d) Log-lin histogram based upon Sigworth's method with true logarithmic binning, 64 bins at 16 bins per decade. The best fit of a three exponential pdf yielded (0.98 ± 0.09 ms, 930 ± 68), (9.56 ± 1.2 ms, 971 ± 76), (94.48 ± 9.1 ms, 1038 ± 71).

the test distribution but the 1 ms component is effectively unrepresented in bins beyond the first (0–4 ms). It is clearly not possible to estimate the contribution made by this component accurately. The 10 ms and 100 ms components are correctly estimated both in terms of average time and in number of events, but the 1 ms component has a large error on the average and overestimates the number of events by almost 70%.

To represent the fast component, a much shorter bin width must be used such as the 0.2 ms interval in Figure 8.6 (b). However, this is gained at the price of losing most of the middle and all of the slow component which now exceeds the upper end of the histogram range. A good estimate is obtained of the fast component but the middle component is poorly estimated both in time and number of events. Creating histograms with very large numbers of bins (e.g. 2000 × 0.2 ms) does not help either since most of the bins end up containing no more than one event and consequently are unsuitable for curve fitting.

8.19 LOGARITHMIC DWELL TIME HISTOGRAMS

The solution to the binning problem is to produce a histogram with variable bin widths, starting with narrow bins capable of correctly resolving short intervals and progressively increasing the width of the bins collecting longer intervals so as to maintain an adequate number of events in every bin. Using this method, distributions with components ranging over several orders of magnitude can be represented within a single histogram. Two such methods are in common use, both based upon logarithmically spaced bins but differing slightly in terms of presentation, one first proposed by Magleby (McManus *et al.*, 1988; Blatz & Magleby, 1989) and the other by Sigworth & Sine (1987).

Magleby's method was developed to logarithmically bin dwell time lists obtained using the simple half amplitude threshold crossing technique. In such circumstances, dwell time values are restricted to integer multiples of the digital sampling interval. Event durations are placed into bins using the equation

$$\text{Bin.no} = 1 + \text{int}(bpd \log_{10}(n_d)) \quad [8.10]$$

where n_d is the event duration in units of the digital sampling interval and bpd is the required number of histogram bins per decade (ten-fold change in event duration). When applied to integer data, Equation 8.10 provides only an approximation of logarithmic bin widths. This produces an irregularity which is particularly noticeable for short events. As can be seen from Table 8.1 (a), events with durations in the range 1–9 are placed into nine single sample width bins, the range 10–20 into five double sample bins, 18–27 into three triple-sample bins. Note also that the integer quantization also results in gaps in the bin numbering. Events of duration 1, 2 and 3 samples are mapped to bins 1, 5 and 8. For convenience of presentation these non-contiguous bin numbers are remapped into an integer sequence.

The details of the implementation of the log binning method can be found in McManus *et al.* (1988). As discussed there, the width of each bin and the mapping to contiguous bins can be found by a process of enumeration. After the histogram has been compiled the contents of each bin is then scaled by the bin width and the results plotted using the log–log coordinate.

Figure 8.6 (c) shows the log–log histogram for the three component data set discussed earlier. Note how all components are now equally

Table 8.1 Bin range and width for first 16 bins of a logarithmically binned dwell time histogram. (a) Integer approximation to logarithmic binning (including non-contiguous map number). (b) True logarithmic bins. Times expressed as sampling intervals.

	(a) Integer approximation		(b) True logarithmic bins		
Bin no.	Dwell time	Width	Map no.	Dwell Time	Width
1	1	1	1	0.85–1.15	0.30
2	2	1	5	1.15–1.33	0.18
3	3	1	8	1.33–1.54	0.21
4	4	1	10	1.54–1.78	0.24
5	5	1	12	1.78–2.05	0.27
6	6	1	13	2.05–2.37	0.32
7	7	1	14	2.37–2.74	0.37
8	8	1	15	2.74–3.16	0.42
9	9	1	16	3.16–3.65	0.49
10	10–11	2	17	3.65–4.22	0.57
11	12–13	2	18	4.22–4.87	0.65
12	14–15	2	19	4.87–5.62	0.75
13	16–17	2	20	5.62–6.49	0.87
14	18–20	3	21	6.49–7.50	1.01
15	21–23	3	22	7.50–8.66	1.16
16	24–27	3	23	8.66–9.99	1.33

represented on the histogram, each with a set of bins of appropriate size. The shape of each component is now, however, dramatically different, with the familiar exponential shape from the linear plot converted to a hyperbola. Logarithmic binning also allows accurate and simultaneous curve fitting of all three components. The best-fit curves from a three exponential fit are shown superimposed on Figure 8.6 (c) with both the mean times and number of events in each component correctly estimated to within 7%.

Although the log–log histogram is computationally much superior to the linear histogram, it is quite difficult to interpret visually especially when there are several closely spaced components. With that in mind, Sigworth & Sine (1987) proposed an alternative presentation. They retained logarithmic binning without scaling the bin values by bin width and plotted the bin contents on a linear or square root scale. Figure 8.6 (d) shows the three component data set plotted using this scheme. Each exponential component can now be clearly distinguished as a separate peak.

There is, however, a problem associated with

the use of the log–lin histogram in that it is essential to use true logarithmic bins to achieve the smooth peak shape for exponential components. The approximation used by Magleby in order to apply the method to integer dwell times proves to be unsatisfactory, producing a histogram with peaks with a series of serrated edges whenever an integer change in bin width occurs. A set of true logarithmic bins such as in Table 8.1 (b) are required. Consequently, the log-lin method is restricted to event times which have been obtained with detection methods such as time course fitting which extends the measurement precision in a continous fashion to fractions of the sampling rate. Sigworth, for instance, (Sigworth, 1983) used a transition finding procedure which, although essentially a half amplitude method, interpolates between points to find a more precise (and non integer) estimate of event duration.

In summary, each type of dwell time histogram has both advantages and limitations. The linear histogram is simple to implement and understand but is only suitable for displaying distributions with one or two components differing by less than a factor of 10 in mean lifetime. There are many circumstances where this is sufficient. However, in order to represent and accurately analyse distributions with more than two components, a logarithmic binning method is required. Magleby's log–log histogram has the merit of being applicable to both integer quantized dwell times from simple half amplitude transition detection methods and more sophisticated time course fitting. Sigworth's log–lin method provides the best visual representation of multi-component distributions but is only applicable to non-quantized dwell times from time course fitting or Sigworth's own method.

A subroutine capable of computing any of the three types of dwell time histogram (linear, log–log, log–lin) can be found in Listing 8.4. The routine reads the dwell times stored in the event list file created by the transition detection routine in Listing 8.2 and returns a 4 × n element array containing the lower, middle and upper time points, and number of events contained in each bin.

8.20 MODELLING OF DWELL TIME DISTRIBUTIONS

The dwell time histogram provides a pictorial representation of the dwell time distribution. As usual, to produce a quantitative assessment, it is necessary to fit a mathematical function from which characteristic parameters can be extracted. Like the current amplitude distributions discussed in Section 8.8, this is a distribution of random values and a probability density function is used. Given the exponential shape of the distributions, and for theoretical reasons to be discussed later (Section 8.24), it is natural to use the exponential pdf

$$f(t) = \frac{1}{\tau} \exp\left(\frac{-t}{\tau}\right) \qquad [8.11]$$

$f(t)$ defines the probability of a channel dwell time of duration t occurring. It is characterized by a single parameter τ, which is equal to the mean of the durations within the distribution. A dwell time distribution consisting of multiple exponential components can be represented by a sum of exponential pdfs

$$f(t) = \sum_{i=1}^{n} \frac{a_i}{\tau_i} \exp\left(\frac{-t}{\tau_i}\right), \quad \sum_{i=1}^{n} a_i = 1 \qquad [8.12]$$

where each exponential component is represented by its mean τ_i and its proportion a_i of the total number of events within the distribution. The integral of a pdf over all values must equal unity, therefore the sum of the a_i parameters are constrained to be 1. A distribution with n exponential components is therefore characterized by a set of $2n - 1$ parameters.

Exponential pdfs can be fitted to dwell time distributions by applying the usual iterative least-squares curve-fitting methods to the dwell time histograms. An expression for the number of events y_j contained within histogram bin j is given by

$$y_j = Nw_j \sum_{i=1}^{n} \frac{a_i}{\tau_i} \exp\left(\frac{-t}{\tau_i}\right) \qquad [8.13]$$

where N is the total number of events in the distribution, t_j is the mid-point of the bin and w_j is the width of the bin. The minimum of the sum of squares of the histogram bins

$$\text{Min}\left\{ \sum_{j=1}^{n_b} (Y_j - y_j)^2 \right\} \quad [8.14]$$

is obtained by iteratively adjusting the a_i and τ_i parameters.

As discussed in Chapter 6, iterative curve-fitting procedures often perform poorly when the fitting parameters differ greatly in absolute magnitude. This situation arises here since the τ_i parameters in a three component distribution can easily differ by a factor of 100. This problem can be avoided by using $\lambda = \ln(\tau)$ as the fitted parameter rather than τ itself, so Equation 8.13 becomes

$$y_j = Nw_j \sum_{i=1}^{n} \frac{a_i}{\exp(\lambda_i)} \exp\left(\frac{-t_j}{\exp(\lambda_i)}\right) \quad [8.15]$$

The exponential pdf curves fitted to the data in Figure 8.6 were obtained using this method.

8.21 MAXIMUM LIKELIHOOD METHOD

When fitting probability density functions to distributions an alternative method is available, the *maximum likelihood* method, which for theoretical and practical reasons is preferred to least squares. The term likelihood refers to the probability that a particular pdf is the 'most likely' one to have generated a fixed set of observations, such as a list of dwell times (Colquhoun & Sigworth, 1983). The 'best' function in this sense is the one which has the maximum likelihood.

Assuming an exponential pdf of the form in Equation 8.12, the likelihood of a set of observations $(t_1 \ldots t_m)$ belonging to that distribution is given by the product

$$L(\tau_1 \ldots \tau_n, a_1 \ldots a_n) = f(t_1) \times f(t_2) \times \ldots \times f(t_m) \quad [8.16]$$

more succinctly expressed as

$$L(\tau_1 \ldots \tau_n, a_1 \ldots a_{n-1}) = \prod_{i=1}^{m} f(t_i) \quad [8.17]$$

Like the least-squares function, the likelihood is a function of the τ and a parameters of the exponential pdf. However, the best-fitting parameters are obtained by finding the set which maximizes L. Products of numbers, as in Equation 8.17, are difficult to handle due to the astronomically large values that can be produced, so the log-likelihood function is used instead

$$LL(\tau_1 \ldots \tau_n, a_1 \ldots a_n) = \sum_{i=1}^{m} \log_e(f(t_i)) \quad [8.18]$$

which has a maximum for the same values of a and τ, but is easier to compute. In practice it is also necessary to take into account the fact that events with both very fast and very slow durations can be missing from the list of observations. Therefore an adjusted likelihood is computed

$$LL(\tau_1 \ldots \tau_n, a_1 \ldots a_n) = \sum_{i=1}^{m} \log_e\left(\frac{f(t_i)}{F(t_{hi}) - F(t_{lo})}\right) \quad [8.19]$$

where $F()$ is the exponential distribution function (the integral of the pdf)

$$F(t) = \sum_{i=1}^{n} a_i \exp\left(\frac{-t}{\tau_i}\right) \quad [8.20]$$

The most likely set of parameter values are obtained by repeatedly computing Equation 8.19 and using an iterative function minimization algorithm to find the set of parameters which minimize the equation

$$\text{Min}(-LL(\tau_1 \ldots \tau_n, a_1 \ldots a_n)) \quad [8.21]$$

(expressed as a minimization problem since most readily available software is designed to minimize rather than maximize functions).

One advantage of the maximum likelihood method is that it uses the list of dwell times directly, avoiding the loss of information inherent in the binning process used to produce the dwell time histogram required for the least-squares method. However, the increased precision is paid for by a greatly increased computation time compared to the least-squares method (see Table 8.2). Maximum likelihood estimation can, however, be applied to binned data by substituting into the minimization function the formula

$$LL(\tau_1 \ldots \tau_n, a_1 \ldots a_n) = \sum_{j=1}^{m} Y_j \log_e \left(\frac{w f(t_i)}{F(t_{hi}) - F(t_{lo})} \right)$$

[8.22]

where m is now the number of histogram bins rather than the number of events, Y_j the number of events in bin j with mean time t_j and width w. Sigworth & Sine (1987) have shown that little precision is lost compared to the direct approach if logarithmic binning is used with at least 10 bins per decade.

8.22 PRACTICAL ASPECTS OF LIKELIHOOD MAXIMIZATION

The procedures used to minimize Equations 8.19 or 8.22 are similar to those discussed in Chapter 6 for least-squares minimization. However, there are subtle differences that need to be considered. Least squares minimization problems have an inherent special structure which allow the use of certain techniques which the likelihood function does not. Similarly, least-squares problems are generally part of the class of small residual problems where the minimum is a small number. The equations for the likelihood function may minimize to a large negative number. Techniques which are well suited to least-squares minimization may not be suitable for this more general minimization problem.

The simplex method, discussed in Section 6.7, is probably the most widely used method applied to maximum likelihood estimation (Colquhoun & Sigworth, 1983; Sigworth & Sine, 1987), although quasi-Newton methods have also been used (Wachtel, 1991). The relative accuracy and efficiency of these dwell time fitting methods were compared by fitting a series of simulated distributions using each of the three methods, least squares, direct maximum likelihood and binned maximum likelihood. The least-squares fits were performed using Equation 8.15 with logarithmic binning and the enhanced Levenberg–Marquardt fitting routine SSQMIN (Brown & Dennis, 1972), discussed in Chapter 6. The direct maximum likelihood was found using the simplex routine from Listing 6.2 modified to minimize Equation 8.21 rather than a sum of squares. The binned maximum likelihood was obtained using the same routine applied to Equation 8.22.

The results are shown in Table 8.2. Data set (a) contains the 3000 dwell times generated by the three component exponential pdf used to produce Figure 8.6 (1 ms, 1000; 10 ms, 1000; 100 ms, 1000). This can be regarded as an easy fitting task since the components are well separated in mean duration and contain a large number of events. All of the fitting methods successfully identified both the mean dwell times and the number of events with errors around 6%. However, the direct maximum likelihood method took more than 20 times longer than the binned methods (1524 s) without any improvement in accuracy. The least-squares method (10 s) was also notably faster than the binned maximum likelihood (60 s).

Data set (b) was generated from a distribution with two closely spaced components of equal size (2 ms, 1000; 4 ms, 1000), a situation which produced fitting difficulties for the exponential curves in Section 6.18). Both binned methods produced results with substantial errors, in particular the number of events in the fast component was overestimated by 46%. Interestingly, the direct maximum likelihood method yielded a much better result.

Data set (c) contains a three component distribution with a small intermediate component containing less than 5% of the total number of events (1 ms, 1000; 5 ms, 400; 100 ms, 1000). It is often difficult to estimate such small components, as has been observed by Sigworth & Sine (1987) and Wachtel (1991). The errors for the intermediate component are substantially higher than for the other two.

Overall, no clear distinction could be made in terms of accuracy between the three curve-fitting methods studied. As long as components were separated by at least a factor of five in mean duration and there were at least 1000 events in each component, the parameters of each pdf component could be correctly estimated to within 6%. However, in terms of computing time, both of the binned methods were more than an order

Table 8.2 Comparison of performance of least squares, direct maximum likelihood and binned maximum likelihood methods at fitting two and three exponential pdfs to dwell time data.

	τ_f	N_f	τ_i	N_i	τ_s	N_s	Iterations	Time (s)
(a) True	1	1000	10	1000	100	1000		
Least squares	0.9	1003	9.43	980	94.1	1037	20	10
Direct ML	0.924	987	9.58	981	96.3	1032	397	1524
Binned ML	1.05	948	9.9	996	97.3	1056	452	61
(b) True	2	1000			4	1000		
Least squares	2.16	1405			4.95	606	27	7
Direct ML	1.89	964			3.94	1036	224	451
Binned ML	2.4	1464			4.85	536	234	21
(c) True	1	1000	5	400	100	1000		
Least squares	0.9	999	4.28	131	101.3	995	40	27
Direct ML	0.94	995	6.52	124	102.6	981		
Binned ML	1.05	953	4.9	111	102.7	1036	358	48

of magnitude faster than direct maximum likelihood estimation.

8.23 DETERMINING THE NUMBER OF COMPONENTS

The maximum likelihood and least-squares methods provide the best-fit parameters for any given n-exponential pdf but they do not in themselves yield the best number of components. In practice therefore, it is necessary to evaluate pdfs with different numbers of components and to determine by some additional criteria which appears to be the best choice. This is a model discrimination problem essentially similar to that discussed in Section 6.21 in relation to least-squares curve fitting.

The log-likelihood itself is a measure of goodness of fit, better fits producing larger values. Two trial models with different numbers of exponential components ($m > n$) can be compared by computing their log-likelihood ratio LLR (Horn, 1987)

$$LLR_{m,n} = \text{Max}\{LL(\tau_1 \ldots \tau_m, a_1 \ldots a_m)\} - \text{Max}\{LL(\tau_1 \ldots \tau_n, a_1 \ldots a_n)\}$$

[8.23]

A positive LLR value indicates that the m-exponential model provides a better fit than the n-exponential model. As in all statistical tests, it is necessary to determine whether an observed LLR is significant. As discussed in Section 6.21, models such as the sums of exponentials are nested models (Horn, 1987; Rao, 1973). For such models, significance can be determined from the probabilty p of the observed LLR value occurring purely by chance given the null hypothesis that the two trial models fit equally well. This can be determined from a chi-square distribution with degrees of freedom equal to the difference in the number of parameters between the two models

$$p = \chi^2 (2 \, LLR_{m,n}, m - n)$$

[8.24]

The results of the method in practice can be illustrated by the maximum likelihood fits for a series of one-, two-, three- and four-exponential trial models to the dwell time distribution shown in Figure 8.6. Table 8.3 shows the best fit results, the Max(LL) for each fit, and the LLR between successive fits. Increasing the number of free parameters produces a progressive increase in Max(LL). Large increases are achieved between 1–2 and 2–3 exponentials and the small p values for the LLR show the difference to be significant. However, the improvement achieved by the four- over the three-exponential model is small and p shows it to be non-significant. Consequently, it can be concluded that a three-exponential pdf provides a significantly better fit than one or two exponential, but there is no reason to conclude that there are more than three components within the distribution.

Table 8.3 Maximum likelihood fits and log-likelihood ratio (LLR) between successive fits for 1-, 2- 3- and 4-exponential pdfs to 3000 event simulated data set from Figure 8.6 (1 ms, 1000; 10 ms, 1000; 100 ms, 1000). Significance of LLR (p) calculated using Equation 8.24.

τ_1 (ms)	a_1	τ_2 (ms)	a_2	τ_3 (ms)	a_3	τ_4 (ms)	a_4	Max(LL)	LLR	p
35.1	1.00							−14138		
1.85	0.494	68.3	0.506					−12299	1839	≈ 0.0
0.89	0.321	9.22	0.333	98.2	0.346			−12126	173	≈ 0.0
0.92	0.306	9.11	0.302	89.4	0.318	21.7	0.074	−12125	1	0.54

It should be noted that the statistical tests discussed above are appropriate to the comparison of nested models only. Different criteria are required for non-nested models with different basic mathematical forms, such as exponential models and fractal models (Horn, 1987). In such circumstances Akaike's *asymptotic information criterion* can be used. For two non-nested models with sets of k and l fitted parameters respectively, model k is selected if

$$\text{LLR}_{k,l} > (k - l) \quad [8.27]$$

In effect, this test discriminates in favour of models with fewer parameters. Unfortunately, it is not as simple to determine a significance level for this test as it is was for the nested models. Further details of the issues involved in the discrimination of non-nested models, and an approach to determining the significance using Monte-Carlo simulation methods can be found in Horn (1987) and Korn & Horn (1991).

8.24 KINETIC MODELS OF ION CHANNEL GATING

The single-channel analysis procedures outlined so far allow us to produce a detailed description of channel amplitude and kinetic behaviour. This information can be used to develop **kinetic models** of the ion channel. Such models can be complex, with each channel subconductance level and exponential component within an **open** or **closed** time distribution corresponding to a distinct configurational state of the channel. Channels are rarely so simple as to be describable with only two states (closed and open).

Introductions to channel kinetic modelling can be found in Colquhoun (1987b) or Colquhoun & Hawkes (1983). A large amount of work has been done, since the development of the patch clamp, to establish the mathematical foundations for the development and testing of channel models. Given the complexity of channel behaviour this often requires the use of sophisticated matrix techniques (Colquhoun & Hawkes, 1981, 1983a, 1987; Horn, 1987; Bauer *et al.*, 1987). Consideration has been given to the property of many channels to open as bursts of brief openings and closures (Colquhoun & Hawkes, 1983b). Also, within complex multi-state models with restricted pathways between states, adjacent open and closed times may be correlated in duration (Colquhoun & Hawkes, 1987; Steinberg, 1987; Ball *et al.*, 1988; Blatz & Magleby, 1989; Petracchi *et al.*, 1991)

Most kinetic models are based on the assumption that channel behaviour is describable as a Markov process. In essence, this is a model in which the channel exists in a relatively small number of well-defined states and the probability of leaving that state is constant (in particular, it is not dependent on the time that the channel has been in the state or any previous state). Such a scheme naturally produces dwell time distributions which are sums of exponentials. Leibovitch (1989) proposed an alternative approach where the channel gating is modelled as a fractal process with the probability of leaving a state decreasing with time. The appropriate choice of model and the means for discriminating between them is currently a subject of some debate (McManus *et al.*, 1988; Leibovitch, 1989; Korn & Horn, 1991; Petracchi *et al.*, 1991)

8.25 THE MISSING EVENTS PROBLEM

A particular difficulty associated with the development of channel models is accounting for the imperfections in the recording process, particularly when the single channel currents are subject to low pass filtering. As has been discussed, filtering is essential to distinguish channel events from background noise but causes a selective loss of brief events. Unfortunately, many of the more complex channels, where kinetic modelling is interesting, contain components of short open or closed times. The non-detection of such events is not a serious problem for the estimation of the short duration component itself since it is a simple matter of extrapolation to find how many have been lost. However, the overestimation of the dwell times in the opposite state caused by the concatenation of the two intervals on either side of the missed event is not so easy to handle. Significant distortion of the shape of the dwell time distribution can occur to the extent that they are no longer properly fitted by exponential pdfs. The apparent mean dwell time derived from such fits usually overestimates the true value (Wilson & Brown, 1985).

Much effort has been applied to finding a means for correcting dwell time distributions for the effects of missed events (Roux & Sauve, 1985; Blatz & Magleby, 1986; Ball & Sansom, 1988a, b; Milne et al., 1988; Crouzy & Sigworth, 1990; Hawkes et al., 1991). In the case of simple two or three state channels, where there is no doubt about the underlying model, it is possible to make a simple correction (e.g. Neher, 1983). It is also relatively straightforward, using simulation techniques (Blatz & Magleby, 1986), to determine the effect of missing events on the dwell time histograms for a known channel model. However, the general problem of determining an unknown model from an observed distribution is not so easy, often being unable to yield unique solutions. Hawkes et al. (1991) provide a good review of the strengths and weaknesses of the range of correction methods available.

Another approach, proposed by Magleby & Weiss (1990), is to develop a realistic simulation of the ion channel including background noise and low pass filtering. Using an iterative optimization algorithm, the model parameters are adjusted until the best match is found between the simulated time course and the actual recording. Similarly, different kinetic models can be tried out until the best fitting one is found. As might be expected, the process takes an enormous amount of computing time for even quite simple models (2–3 days on a PDP11/73). A similarly computationally intensive method based upon sophisticated digital signal processing techniques has recently been proposed by Chung et al. (1990).

Correcting exactly for missed events is primarily a problem for those with a deep interest in determining very accurate multi-state channel models. For many purposes it is sufficient to be aware that the problem may exist. Most single channel experimental work is concerned with the effects of external factors such as neurotransmitters, second messengers, or membrane potential on ion channel function. The effects of these upon channel dwell time distributions (e.g. whether they affect closed or open states) can still yield insight into channel function using only simple approximate channel models.

8.26 SINGLE-CHANNEL ANALYSIS SOFTWARE

The widespread interest in the single-channel analysis technique combined with the increasing availability of PCs in the laboratory has resulted in a number of laboratories or companies spending a significant amount of effort in developing software packages for this purpose. An ideal single-channel analysis program would have the following features

- continuous sampling-to-disc for agonist-activated channels
- voltage pulse generation/recording for voltage-activated channels
- gaussian digital filtering
- half amplitude and time course transition detection with baseline tracking
- event list editing facilities

- amplitude histograms
- linear and logarithmic dwell time histograms
- multi-exponential pdf curve fitting
- multi-gaussian amplitude fitting
- flexible import/export of results.

Given the difficulties associated with accurate transition detection discussed earlier, it is important that the software permits the operator to inspect visually and if necessary modify the contents of the event list. For each entry in the list, it should be possible to display the idealized event superimposed on the actual digital record and allow the manual deletion or insertion of events. It is essential to be able to verify that the transition detector is producing meaningful results with the particular class of channels under study. Also, in general, agonist- and voltage-activated single channels require quite different kinds of software.

As in most other areas of electrophysiology most of the available software is for the IBM PC or the PDP 11 minicomputer. Most commercially available electrophysiology software packages have programs or features appropriate for single-channel analysis.

8.27 AXON INSTRUMENTS SOFTWARE

Axon Instrument's pCLAMP suite of programs contains acquisition and analysis software for both agonist- and voltage-activated channels. The core of the single-channel analysis features reside in the FETCHAN program which processes the digitized single-channel records using the half amplitude detection method, and produces an event duration list. Digital records are generated by either FETCHEX using sampling-to-disc for continuous ligand-gated channels or CLAMPEX for voltage-gated channels. The event list file is processed by the pSTAT program to produce dwell time histograms and fit multi-exponential pdf and gaussian fuctions. FETCHAN provides fairly basic transition detection features, in particular it does not support logarithmic binning, nor the maximum likelihood method. However, the widespread use of the pCLAMP program has established it, and particularly its file formats, as a *de facto* standard, and others have produced software which provides enhanced features.

Axon Instruments also distribute a number of analysis programs, developed in research laboratories; in particular, IPROC-2 by Lingle & Sloderbeck of Florida State University and LPROC by Neil *et al.* (1991) of the State University of New York. IPROC-2 provides an enhanced transition detector compared with FETCHAN. It is based upon Sachs' IPROC program (Sachs *et al.*, 1982; Sachs, 1983) and provides the features discussed in his papers, half amplitude transition detection, zero crossings method of baseline tracking, gaussian filters. LPROC, in contrast, is a software tool for the analysis of event list data, created either by FETCHAN or IPROC-2. Unusually, for software of this type, it is command driven, almost a special purpose interpreted language for event analysis.

8.28 STRATHCLYDE SOFTWARE: PAT

The PAT single-channel analysis program is part of the author's Strathclyde Electrophysiology Software suite. PAT was developed to satisfy the needs of a number of laboratories within the UK and is available either from the author or from Dagan Inc (Minneapolis, USA). As for the other programs in the series, the software is written in Microsoft FORTRAN and works with a range of laboratory interfaces; CED 1401, Data Translation DT2801A, DT2821 & DT2812, & National Instruments LAB-PC.

PAT is a single integrated program combining sampling-to-disc, transition detection, amplitude and linear and logarithmic dwell time histogram generation, exponential and gaussian curve fitting. Most of the procedures discussed in this chapter are supported by PAT. At present, PAT uses a two threshold transition detection procedure although time course fitting is under development. PAT is designed essentially for the analysis of agonist-gated channels and therefore itself supports only continuous sampling-to-disc. Voltage-activated channels are handled by importing records which have been acquired using the VCAN program discussed in Section 7.16.

8.29 CAMBRIDGE ELECTRONIC DESIGN SOFTWARE

Cambridge Electronic Design provides single-channel analysis features with their Patch and Voltage Clamp package discussed in Chapter 7. The software has provision for the recording and analysis of both agonist- and voltage-activated channels. Half amplitude transition detection is used and amplitude histograms and linear dwell time histograms are produced. The program uses the direct maximum likelihood method to fit up to three exponential pdfs to dwell time distributions, and least squares to fit gaussian curves to the amplitude histograms.

8.30 OTHER SOFTWARE

Some other single-channel analysis packages are available including Satori marketed by Intracel (Cambridge, UK), using their S200 laboratory interface and RC Electronics (Hollister, USA) software for their own interface. Both of these programs run on the IBM PC family. Software developed by Sigworth, based on his earlier work for the PDP11 but written in the Modula 2 language (Affolter & Sigworth, 1988), is also available for the Atari ST computer, marketed by Instrutech (Mineola, NJ, USA). A version of this software is also available for the Apple Macintosh. Given the large amount of computation sometimes involved in the analysis of single channels, some workers are turning to the use of engineering/scientific workstation computers such as the Sun series (Sun Microsystems) based upon the SPARC microprocessor (Chung et al., 1990). Although the distinction between workstations and personal computers is becoming increasingly blurred, such machines are usually about 2–5 times faster than PCs, and provide better graphics. It is notable in this context that a version of the IPROC program has been ported to run on Sun workstations.

8.31 SUMMARY

The analysis of single-channel currents is, perhaps, the most intellectually demanding of the areas which we have covered so far. The development of precise mathematical models of channel gating, particularly if attempts are made to correct for missed events, has become something of an area for the specialist, employing relatively sophisticated mathematical techniques. Nevertheless, in many cases this degree of detail is unnecessary, and much useful information can be gleaned using only the basic techniques that have been discussed here. It can also be said to be an area where more than the usual caution is required to avoid making errors in the interpretation of results. In particular, even though one of the widely available software packages may be used, there is still a need to have a good grasp of the principles of its operation.

```fortran
c      Listing 8.1 Single-channel current amplitude histograms.
c
c      (a) Histogram of current amplitudes within single-channel current recording.
       subroutine amplitude_histogram(bins,nbins,istart,iend,pa_per_bit)
       real*4 bins(2,nbins)                  ! (Out) Histogram results
       integer*2 nbins                       ! (In) No. of bins in histogram
       integer*2 istart,iend                 ! (In) Range of records to be processed
       real*4 pa_per_bit                     ! (In) Current scale factortsts
       parameter(np=512,miny=0,maxy=4095,izero=2048)
       integer*2 iy(np)                      ! A/D sample value range 0-4095.

c      Open data file containing digitized currents.
       open(unit=1,file='patch.dat',form='binary',access='direct',recl=2*np )
       ibin_width = (maxy - miny +1)/nbins    ! Initialize histogram array.
       do i = 1,nbins                         ! Results stored with
           bins(1,i) = pa_per_bit*float((i-1)*ibin_width + miny -izero)
           bins(2,i) = 0.                     ! bin mean,contents interleaved
       end do

       do irecord = istart,iend
           read( unit=1, rec=irecord ) iy     ! Read A/D sample record
           do i = 1,np
               ix = (iy(i) - miny)/ibin_width + 1   ! Find histogram bins for each
               bins(2,ix) = bins(2,ix) + 1.         ! sample and increment
           end do
       end do
       close(unit=1)
       return
       end

c (b) Histogram of the average amplitudes of each channel state.
       subroutine average_amplitude_histogram(bins,nbins,istart,iend,nskip,
     & pa_per_bit)
       real*4 bins(2,nbins)                  ! (Out) Histogram
       integer*2 nbins                       ! (In) No. of bins in histogram
       integer*2 istart,iend                 ! (In) Range of records to be processed
       integer*2 nskip                       ! (In) Samples ignored at each end of event
       real*4 pa_per_bit                     ! (In) Current scale factor
       parameter(np=512,miny=0,maxy=4095,izero=2048)
       integer*2 iy(np)
       integer*4 sum

       parameter( nbytes_event = 20 )        ! Channel event list data record
       structure / event_record /            ! structure.
           integer*2 level                   ! 0=closed,1=intermediate,2=open state
           integer*2 zero                    ! Zero current level
           integer*4 start                   ! Event start at sample
           integer*4 end                     ! Event end at sample
           real*4 dwell_time                 ! Event duration (samples)
           real*4 average_current            ! Average current during event.
       end structure
       record /event_record/ event

c ... continued
```

```
c Listing 8.1 continued
c
c     Open data file containing digitized currents (512 sample records)
      open(unit=1,file='patch.dat',form='binary',access='direct',recl=2*np )

c     Open data file with lists of detected events (produced by Listing 8.2)
      open(unit=2,file='event.lst',form='binary',access='direct',recl=nbytes_event )

      ibin_width = (maxy - miny +1)/nbins                    ! Initialize
      do i = 1,nbins                                         ! histogram
          bins(1,i) = pa_per_bit*float( (i-1)*ibin_width + miny - izero)
          bins(2,i) = 0.
      end do
      iavailable = 0
      do istate = istart,iend
          read( unit=2, rec=istate ) event                   ! Get event record.
          event.start = event.start + nskip                  ! Avoid including transition
          event.end = event.end - nskip                      ! points.
          if( event.end .ge. event.start ) then
              sum = 0
              ir0 = (event.start/np) + 1                     ! Determine records
              ir1 = (event.end/np) + 1                       ! containing event
              do ir = ir0,ir1
                  if( ir .ne. iavailable ) then              ! If a new record is required
                      read( unit=1, rec=ir ) iy              ! get it from the data file
                      iavailable = ir
                  end if
                  if( ir .eq. ir0 ) then
                      ip0 = event.start - np*(ir0-1)         ! Start at the beginning of
                  else                                       ! channel event
                      ip0 = 1
                  end if
                  if( ir .eq. ir1 ) then                     ! End at the end of the
                      ip1 = event.end - np*(ir1-1)           ! channel event.
                  else
                      ip1 = np
                  end if
                  do i = ip0,ip1
                      sum = sum + iy(i)
                  end do
              end do
              ix = ( (sum/(event.end-event.start+1)) - miny)/ibin_width + 1
              bins(2,ix) = bins(2,ix) + 1.
          end if
      end do
      close(unit=1)
      close(unit=2)
      return
      end

c .. continued ..
```

```fortran
c     Listing 8.1 continued
c     (c) Patlak's running average amplitude histogram using a
c     variance criterion to reject averages containing transition points.
      subroutine running_avg_amplitude_histogram(bins,nbins,
     &         istart,iend,navg,pa_per_bit,var_threshold)
      real*4 bins(2,nbins)              ! (Out) Histogram
      integer*2 nbins                   ! (In) No. of bins in histogram
      integer*2 istart,iend             ! (In) Range of records to be processed
      integer*2 navg                    ! (In) No. of sample in average
      real*4 pa_per_bit                 ! (In) Current scale factor
      integer*4 var_threshold           ! (In) Variance acceptance threshold
      parameter(np=512,miny=0,maxy=4095,izero=2048)
      integer*2 iy(np)
      integer*4 ring(256),sum,sumsq,add,drop,avg,var,bin_width

      open(unit=1,file='patch.dat',form='binary',access='direct',recl=np*2 )
      bin_width = (maxy - miny +1)/nbins           ! Initialize histogram array.
      do i = 1,nbins
          bins(1,i) = pa_per_bit*float((i-1)*bin_width + miny -izero)
          bins(2,i) = 0.
      end do
      do irecord = istart,iend
          read( unit=1, rec=irecord ) iy           ! Read 512 sample record
          if( irecord .eq. istart ) then
              sum = 0                              ! If this is the first record
              sumsq = 0                            ! do the complete running average
              do i = 1,navg                        ! and variance calculation for the
                  add = iy(i)                      ! window
                  sum = sum + add
                  sumsq = sumsq + add*add
                  ring(i) = add
              end do
              inew = navg
              is = navg + 1
          else
              is = 1
          endif

          do i = is,np
              inew = mod( inew, navg ) + 1         ! Keep most recent
              add = iy(i)                          ! window of navg points
              drop = ring(inew)                    ! in ring buffer
              ring(inew) = add
              sum = sum + add - drop               ! Update variance by
              sumsq = sumsq + add*add - drop*drop  ! dropping oldest sample
              avg = sum / navg                     ! and adding most recent
              var = (sumsq - navg*(avg*avg))/navg  ! sample
              if( var .lt. var_threshold ) then    ! Add running average
                  ix = (avg / bin_width ) + 1      ! to histogram if
                  bins(4,ix) = bins(4,ix) + 1      ! variance is less then
              endif                                ! acceptance threshold.
          end do
      end do
      close(unit=1)
      return
      end
```

Analysis of single-channel currents

```
c      Listing 8.2. Two threshold channel transition detection routine.
       subroutine detect_transitions(izero,iupper,ilower,n_records,track_zero)
       integer*2 izero              ! (In) Initial zero current baseline level
       integer*2 iupper             ! (In) Upper transition threshold
       integer*2 ilower             ! (In) Lower transition threshold
       integer*2 n_records          ! (In) No. of records (size=np samples) in data file
       logical track_zero           ! (In) If .true. track drifting baseline.
       parameter( iclosed=0, intermediate=1, iopen=2 )
       parameter( np=512, nring=64, nskip=10, a=0.9 )
       integer*2 ibuffer(np), iring(nring)
       integer*4 isample            ! sample counter (n.b. 4 byte integer)

       parameter( nbytes_event = 20 )
       structure / event_record /   ! Channel event record
           integer*2 state          ! State (iclosed,intermediate,iopen)
           integer*2 zero           ! Zero current baseline
           integer*4 start          ! Starts at sample
           integer*4 end            ! End at sample
           real*4 dwell_time        ! Duration of event (samples)
           real*4 average_current   ! Average current during event
       end structure
       record /event_record/ event

c      Data file containing digitized current signal
       open(unit=1,file='patch.dat',form='binary',access='direct',recl=np*2)

c      Event list file to containg event records
       open(unit=2,file='event.lst',access='direct',form='binary',recl=nbytes_event)
       do i = 1,nring               ! Set baseline history
           iring(i) = izero         ! buffer to initial value
       end do                       ! for baseline supplied in
       ir = 1                       ! subroutine argument list
       event.start = 1
       event.end = 1
       iold_state = iclosed
       icurrent_state = iclosed
       nstate = -1
       isample = 2
       do irecord = 2,n_records
           read( unit=1, rec=irecord ) ibuffer    ! Read current signal
           do i = 1,np
               isample = isample + 1
               ilevel = (ibuffer(i) - izero)      ! Subtract baseline
               if( ilevel .gt. iupper ) then      ! Two threshold transition det
                   inew_state = iopen             ! Points greater than <iupper>
               elseif( ilevel .lt. ilower ) then  ! classed as open, lower than
                   inew_state = iclosed           ! <ilower> as closed. Points
               else                               ! in between as an intermediate
                   inew_state = intermediate      ! state.
               end if

c ... continued ...
```

```
c     Listing 8.2. continued

c           Procedure for tracking a drifting zero baseline.
c           The zero level <izero> is updated with a recursive
c           low pass filter using sample points only when the channel
c           is found to be in the closed state. The first and last
c           <nskip> samples within the closed state are excluded.

            if( track_zero .and. icurrent_state.eq.iclosed ) then
                  ir = mod(ir,nring)+1
                  iring(ir) = ibuffer(i)
                  if( isample-event.end .gt. nskip ) then
                        izero = int(float(iring(mod(ir-nskip+nring,nring)+1))*
     &                        (1.-a) + a*float(izero) )
                  end if
            end if

c           If the channel has entered a new state, write the details of the
c           event preceding the transition to the event list file

            if( inew_state .ne. icurrent_state ) then
c                 Note. Intermediate states which occur within the sequences
c                 closed->intermediate->open  or     open->intermediate-closed
c                 are ignored since they are only part of the transition phase.
c                 The following logic makes this test.
                  if( icurrent_state .eq. intermediate ) then
                        itransition = (inew_state - icurrent_state)
     &                              + (icurrent_state - iold_state)
                  else
                        itransition = 0
                  end if
                  if( itransition .eq. 0 ) then
                        nstate = nstate + 1
                        event.zero = izero
                        event.state = icurrent_state
                        event.end =  isample - 1
                        event.dwell_time = float(event.end - event.start + 1)
                        if( nstate .gt. 0 ) write(unit=2,rec=nstate) event
                        event.start = isample
                        iold_state = event.state
                        ir = mod(ir-nskip+nring,nring)+1
                  end if
                  icurrent_state = inew_state
            end if
         end do
      end do
      close( unit=1 )
      close( unit=2 )
      return
      end
```

```
c Listing 8.3. Estimation of channel open/close time using time course fitting.
c
      subroutine fit_time_course(ibuf,np,ibase,isign,sigma,amplitude,tstart,
     & duration)
      integer ibuf(np)          ! (In) Digitized record of event
      integer np                ! (In) No. of sample points in <ibuf>
      integer ibase             ! (In) Zero current baseline
      integer isign             ! (In) Polarity of current (1,-1)
      real*8 sigma              ! (In/Out) gaussian filter standard dev.
      real*8 amplitude          ! (In/Out) current amplitude
      real*8 tstart             ! (Out) Start of event
      real*8 duration           ! (Out) Duration of event

      parameter(maxpar=3,maxp=512,liw=1,lw=7*maxpar+maxpar*maxpar+
     &    2*maxp*maxpar+3*maxp+maxpar*(maxpar-1)/2)
      real*8 ssq,w(lw),par(maxp)
      integer iw(liw)
      real*8 ydat(maxp),sig,amp   ! Common block to pass data
      common /fitcom/ ydat,sig,amp ! to routine LSFUN1

      ymax = -1D30
      do i = 1,np                 ! Subtract baseline and
                                  ! invert current if
          ydat(i) = (ibuf(i)-ibase)*isign  ! channel is negative-going.
          ymax = max( ydat(i), ymax )      ! Find maximum value
      end do
      do i = 1,np                 ! Scale current to lie
          ydat(i) = ydat(i)/ymax  ! in +/-1 amplitude range
      end do

      amp = isign*(amplitude/ymax)
      sig = sigma*sqrt(2.)
      if( amp .eq. 0. ) then
c         Find current amplitude and gaussian filter standard deviation
c         by fitting an opening step to the leading edge of a long opening.
c         Returns <amplitude> and <sigma> to calling program.
          npar = 3                ! Initial guesses for 3 parameters
          par(1) = np/2           ! Mid-point of rise
          par(2) = 1.             ! sigma*sqrt(2)
          par(3) = 1.             ! amplitude
          ifail = 1
          call e04fdf(np,npar,par,ssq,iw,liw,w,lw,ifail)
          sigma = par(2)/sqrt(2)
          amplitude = (par(3)*ymax)*isign
      else
c         Find starting time and duration of an event using
c         supplied <amplitude> and <sigma> by fitting an open/close step pair
c         Returns <duration> and <tstart> to calling program.
          npar = 2                ! Initial guesses for 2 parameters
          par(1) = np/2           ! starting time = mid-point
          par(2) = np/8           ! Duration = 1/8 of record
          ifail = 1

c continued ...
```

```fortran
c ... Listing 8.3 continued

      call e04fdf(np,npar,par,ssq,iw,liw,w,lw,ifail)
      tstart = par(1)
      duration = par(2)
    end if
    return
    end

    subroutine lsfun1(np,npar,par,residual)
c   Least squares minimization routine for e04fdf()
    integer np                  ! (In) No. of data points
    integer npar                ! (In) No. of parameters
    real*8 par(npar)            ! (In) Function parameter set
    real*8 residual(np)         ! (Out) Residuals
    real*8 s0,s1,x,erfx
    parameter(maxp=512)         ! Data to be fitted supplied via
    real*8 ydat(maxp),sig,amp   ! this common block
    common /fitcom/ ydat,sig,amp

    if( npar .eq. 2 ) then
                                ! Fit open/close step response pair with parameters
      x = 0.                    ! par(1)=start of event, par(2)=duration using
      do i = 1,np               ! <amplitude> and <sig> values in common clock
        x = x + 1.
        call erf( (x - par(1))/sig, erfx )
        s0 = 0.5*( 1.+ erfx )
        call erf( (x - (par(1)+par(2)))/sig, erfx )
        s1 = 1. - 0.5*( 1.+ erfx )
        residual(i) = ydat(i) - (s0 + s1 - 1.)*amp
      end do
    else
c                               ! Fit a single open step response
      x = 0.                    ! with parmeters, par(1) = start
      do i = 1,np               ! par(2)=sigma*2, par(3) = amplitude
        x = x + 1.
        call erf( (x -par(1))/par(2), erfx )
        residual(i) = ydat(i) - 0.5*( 1.+ erfx )*par(3)
      end do
    end if
    return
    end

    subroutine erf(x,erfx)      ! Approximation to error function
    real*8 t,z,y,x,erfx         ! based upon routine ERFCC(X)
    z = dabs(x)                 ! in Press et al. (1986)
    t = 1./( 1. + 0.5*z )
    y = t*dexp( -z*z - 1.26551223 +
   &      t*(1.00002368 + t*(.37409196 + t*(.09678418 +
   &      t*(-.18628806 + t*(.27886807 + t*(-1.13520398 +
   &      t*(1.48851587 + t*(-.82215223 + t*.17087277 )))))))))
    if( x .lt. 0. ) y = 2. - y
    erfx = 1. - y
    return
    end
```

```
c      Listing 8.4. Single-channel dwell time histograms.
c      Routine to compute histogram of open, intermediate, or closed times
c      stored in event.lst file generated by the transition detection routine
c      from Listing 8.2. Linear, logarithmic, or approximate logarithmic
c      bin widths are supported.

       subroutine dwell_time_histogram(bins,nbins,tmax,istart,iend,
     &            iselected_state,dt,itype,bins_per_decade)
       real*4 bins(4,nbins)                ! (Out) Histogram results
       integer*2 nbins                     ! (In) No. of bins
       real*4 tmax                         ! (In) Max. of histogram range (ms)
       integer*2 istart,iend               ! (In) Range of events in histogram
       integer*2 iselected_state           ! (In) 0=closed,1=intermed.,2=open
       real*4 dt                           ! (In) sampling interval (ms)
       integer*2 itype                     ! (In) 1=linear,2=log,3=approx. log.
       real*4 bins_per_decade              ! (In) No. of bins per decade

       parameter( nbytes_event = 20 )      ! Channel event list data record
       structure / event_record /          ! structure
           integer*2 state                 ! 0=closed,1=intermediate,2=open
           integer*2 zero                  ! Zero current level
           integer*4 start                 ! Event start at sample
           integer*4 end                   ! Event end at sample
           real*4 dwell_time               ! Event duration (samples)
           real*4 average_current          ! Average current during event.
       end structure
       record /event_record/ event

       integer*4 i4,i4_lo
       integer*2 imap(256)
       antilog10(x) = exp(2.302585*x)

       if( itype .eq. 1 ) then
           bin_width = float(max(int(tmax/(float(nbins)*dt)),1))
           x = 0.                          ! Linear histogram
           do ibin = 1,nbins               ! fixed bin widths
               bins(1,ibin) = x
               bins(2,ibin) = x + (dt*bin_width)/2.
               bins(3,ibin) = x + bin_width*dt
               x = x + bin_width*dt
               imap(ibin) = ibin
           end do
       else if( itype .eq. 2 ) then
           tmin = log10(dt)                ! Logarithmic histogram.
           bw = 1./bins_per_decade         ! Bin width increasing
           do ibin = 1,nbins               ! logarithmically.
               bins(3,ibin) = float(ibin)*bw + tmin  ! (Sigworth & Sine)
               bins(1,ibin) = bins(3,ibin) - bw
               bins(2,ibin) = bins(3,ibin) - bw/2.
               imap(ibin) = ibin
           end do
       else if( itype .eq. 3 ) then
c ... continued ...
```

```
c Listing 8.4 continued

          iold = 1                                    ! Approximate log. histogram
          i4_lo = 1                                   ! for integer event durations
          i4 = 0                                      ! (McManus et al.)
          ibin = 0
          do while( ibin .le. nbins )                 ! Find bin widths
              i4 = i4 + 1                             ! and mid points
              ix = 1 + int( bins_per_decade*log10(float(i4)) )   ! by direct search
              if( ix .ne. iold ) then                 ! of all possible
                  ibin = ibin + 1                     ! integer values
                  bins(1,ibin) = log10(float(i4_lo)*dt - dt/2.)      ! within range of
                  bins(2,ibin) = log10(float(i4_lo+i4-1)*dt/2.)      ! the histogram.
                  bins(3,ibin) = log10(float(i4)*dt - dt/2.)
                  do i = iold,ix-1
                      imap(i) = ibin
                  end do
                  iold = ix
                  i4_lo = i4
              end if
          end do
      end if
      do i = 1,nbins
          bins(4,i) = 0.
      end do

      open(unit=2,file='event.lst',form='binary',access='direct',recl=nbytes_event)
      n_count = 0
      do istate = istart,iend
        read( unit=2, rec=istate ) event                              ! Get state data
           if( iselected_state .eq. event.state ) then
               n_count = n_count + 1
           if( itype .gt. 1 ) then
                  ix = 1 + int( bins_per_decade*log10(event.dwell_time)) ! Log.
                  ix = imap(ix)                                          ! bins
               else
                  ix = int(event.dwell_time/bin_width + 0.1) + 1          ! Linear
               end if                                                    ! bins
               ix = min(ix,nbins)
               bins(4,ix) = bins(4,ix) + 1.                              ! Increment bin
           end if
      end do
      close(unit=2)

      if( itype .eq. 3 ) then                              ! If approx log. histogram
          do i = 1,nbins                                   ! normalize bin counts for
              if( bins(4,i) .gt. 0. ) then                 ! bin width & convert to logs.
                  bins(4,i) = log10( (bins(4,i)*dt) /
     &                  (antilog10(bins(3,i))-antilog10(bins(1,i))))
              else
                  bins(4,i) = -10.
              end if
          end do
      end if
      return
      end
```

CHAPTER NINE

Analysis of ionic current fluctuations: noise analysis

The stochastic nature of current flow through ion channels has implications concerning the currents recorded using conventional whole cell voltage clamping as well as for the patch recordings discussed in the previous chapter. Whole cell currents are the sum of the currents flowing through all of the individual ion channels within the membrane. Due to the numbers of channels involved, the characteristic rectangular fluctuations of the single channels are not apparent within such signals. Instead the signal appears simply to have random noise superimposed on it. However, a significant amount of information concerning the underlying channel behaviour remains latent within that signal which can be extracted using the appropriate analysis techniques. This chapter is concerned with the methods of recording and analysing this current 'noise' in order to extract information concerning the underlying ion channel behaviour.

Historically, current fluctuation or *noise analysis* predates the development of the patch clamp technique. It provided the first strong, though indirect, evidence of the existence of ion channels (Katz & Miledi, 1972; Anderson & Stevens, 1973), and also a means of estimating the single-channel conductance and the kinetics of channel gating. After the development of the patch clamp it went, at least temporarily, out of fashion since channels could now be directly visualized. However, it is becoming increasingly apparent that noise analysis techniques still have a valuable role to play. Channels are being discovered which have conductances too small to be adequately resolved, even using the patch clamp (Marty *et al.*, 1984; Ascher & Nowak, 1988; Cull-Candy *et al.*, 1988; Lambert *et al.*, 1989). The method has also been applied to the analysis of the current noise during the open periods of single channels (Sigworth, 1985), revealing the existence of additional closely spaced sub-conductance states. In addition the techniques have proved useful in quite different areas, such as the quantal analysis of transmitter release (Ceccarelli *et al.*, 1988).

The theory and practice of noise analysis is extensively discussed by DeFelice (1981),

Eisenberg et al. (1984), Neher & Stevens (1972) and Colquhoun & Hawkes (1977). Analysis of the variance of the fluctuations in whole cell current, measured under voltage clamp, provides information on the single-channel conductance and open-channel probability. Spectral analysis of the frequency components of the variance can be used to calculate rates of channel opening and closure, providing information which can help to discriminate between possible kinetic models for channel gating.

The original work by Katz and others which established the technique used the stationary form of noise analysis. This is a set of techniques based upon the assumption that the ionic current system under study is in a steady state. It is applied mostly to agonist-activated channels where the necessary steady (or at least slowly changing) current level can be maintained for some minutes by the prolonged application of the agonist. Stationary noise analysis is less readily applicable to many voltage-activated currents, such as the Na^+ currents that exist only as transients (although it can be done, e.g. Conti et al., 1976). A similar problem exists when studying agonist-activated channels at high agonist concentrations where desensitization is very rapid (Dilger & Brett, 1990). In these situations, the somewhat more complex *non-stationary* methods can be used which allow the fluctuations about the average time course of a transient current to be analysed.

9.1 RECORDING CURRENT FLUCTUATIONS

Stationary noise analysis has in common with single-channel analysis the need to acquire continuous current records at wide bandwidths perhaps lasting several minutes. It is again convenient to use a tape recorder as an intermediate storage device and sampling-to-disc digitization procedures. A typical steady-state signal consists of a current exhibiting random fluctuations about its mean DC level. A diagram of a system for recording stationary current noise is shown in Figure 9.1.

The amplitude of the fluctuation component of the signal can vary greatly relative to the DC level. Whole cell patch clamp recordings from small cells may fluctuate by as much as 30% of the mean DC level. On the other hand, recordings from large cells such as the skeletal muscle endplates in the original noise studies (Anderson & Stevens, 1973) may exhibit fluctuations of no more than 1%. Given the limited voltage resolution of the commonly used 12 bit A/D converters (see Section 2.2), it is often necessary to separate the two components and provide additional amplification to the small fluctuation component to make it span a sufficient fraction of the input voltage range of the A/D converter. The DC component of the signal can be removed using a high-pass filter. The current output signal from the voltage clamp is therefore split into two channels, a direct coupled DC channel containing the original signal and the high-pass filtered and amplified AC channel which contains only the fluctuations.

Each channel is passed through a low-pass anti-aliasing filter before being digitized by the A/D converter. Unlike most other areas of electrophysiology, the shape of the filter time domain response is of no importance to noise analysis, but it is useful, particularly for the spectral analysis to be discussed in Section 9.6, to have as sharp a filter cut-off as possible. Butterworth filters with their steeper roll-off figures (see Section 3.9) are therefore to be preferred to Bessel filters, even though they introduce ringing into the time response. Typical examples are the Frequency Devices (Haverhill, USA) LPF901 Butterworth filter, Barr & Stroud (Glasgow, UK) and the Kemo VBF8 (Beckenham, UK), both of which can be switched between Bessel and Butterworth characteristics.

The high- and low-pass filters' cut-off frequencies (f_l, f_h) are adjusted to determine the passband for the fluctuations being analysed. These must be set with great care so as to include all the important frequency components within the current fluctuations. Settings are dependent on the kinetics of the particular channels under study and will be discussed later. The sharp cut-off characteristics of the anti-aliasing filter allows the digital sampling rate to be set at the Nyquist rate ($2f_h$).

It is usual to start the recording before the

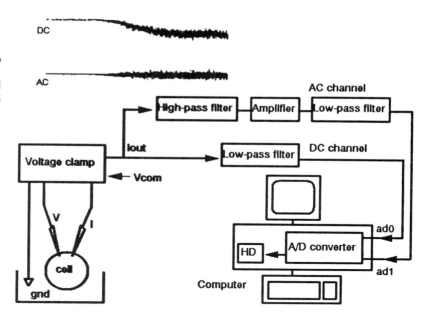

Figure 9.1 Computer system for recording whole cell current noise. The voltage clamp current signal is split into two separate channels, a high gain, high-pass filtered channel isolating the fluctuations and a low-gain DC-coupled channel for mean signal level. Both channels are low-pass filtered with a sharp cut-off Butterworth filter. A typical recording of current noise, GABA-activated Cl$^-$ current from bovine chromaffin cells, is shown inset. (Data courtesy of J.J. Lambert, Dundee University.)

agonist is applied in order to obtain a sample of the background noise (i.e. **fluctuations due to sources other than the ion channels**) within the preparation. As recording continues, the agonist can be applied and the rise of the current to a sustained steady state observed. Figure 9.1 shows a typical recording of stationary current noise from GABA-activated Cl$^-$ channels from bovine chromaffin cells. GABA (10 μM) has been applied ionophoretically at the point shown. The AC channel passband is set at 0.5–100 Hz.

9.2 ANALYSIS OF CURRENT VARIANCE

Analysis of the variance of current fluctuations is essentially a form of amplitude analysis, providing the same kind of information obtained from the amplitude analysis of the single-channel currents. For a population of ion channels, the relationship between mean current and the variance of the current fluctuations about that mean provides information on the magnitude of the current passing through an individual channel and, in some circumstances, the total number of channels in the population. The relationship can be derived from a simple application of the binomial theorem. For a cell containing a population of n ion channels, each capable of passing a current i, the mean steady-state current I for the whole cell is given by

$$I = inp \quad [9.1]$$

where p is the probability of a channel being open at any given time. The variance of the current fluctuations about this mean value is given by

$$\sigma^2 = i^2 n p (1-p) \quad [9.2]$$

This is a parabolic equation where the variance initially rises almost linearly with p, reaches a peak at $p = 0.5$ and falls away to zero again as p approaches 1. These two equations can be combined to provide the relationship between σ^2 and I

$$\sigma^2 = iI - \frac{I^2}{n} \quad [9.3]$$

If p is small ($p < 0.1$) the second term in the above equation can be neglected producing the linear equation

$$\sigma^2 = iI \quad [9.4]$$

i.e. the variance is directly proportional to the mean current. The single-channel current can be calculated as the ratio of current variance over mean current. Equation 9.4 was widely used in the analysis of endplate current noise where, due to desensitization, sustained currents could only be achieved with low agonist concentrations which activated a small fraction of the available channels.

If the cell membrane potential, V_m, and the current reversal potential, V_r, for the ion channel are known, the single-channel conductance γ can be calculated from

$$\gamma = \frac{\sigma^2}{I(V_m - V_r)} \quad [9.5]$$

Note that in circumstances where γ can be calculated in this way, no information can be derived concerning the total number of channels n. However, if σ^2 and I can be determined for values of p greater than 0.1 (two or more p values) then a parabolic equation can be fitted to the data resulting in the estimates for both n and i.

9.3 COMPUTING THE SIGNAL VARIANCE

In order to make use of the variance theory just developed, it is necessary to obtain an estimate of the variance of the digitized noise signals. The signal variance can be calculated by computing the variance of a block of N A/D samples over a period of time T

$$\sigma^2 = \frac{1}{N-1} \sum_{i=1}^{N} (I_i - I_m)^2 \quad [9.6]$$

where I_m is the mean current for the block

$$I_m = \frac{1}{N} \sum_{i=1}^{N} I_i \quad [9.7]$$

Bearing in mind that the current signal fluctuations are spread over a range of frequencies, the digital sampling interval and duration of the variance block must be chosen to ensure that the complete range of frequencies within the signal are represented. The digital sampling interval dt is determined by the desired upper frequency limit f_h, according to the Nyquist criterion

$$dt = \frac{1}{2f_h} \quad [9.8]$$

Once the sampling interval has been fixed in this way, the lower limit of the frequency range f_l is determined by the number of A/D samples in the block

$$N = \frac{1}{f_l \, dt} \quad [9.9]$$

To avoid aliasing problems when the signals are digitized, the cut-off frequencies of the high- and low-pass signal conditioning filters are adjusted to remove frequencies outwith the f_l-f_h range.

Particular care must be taken to ensure that the variance block is long enough since, as will be seen later, the variance associated with ion channel fluctuations is largely concentrated at low frequencies. If the duration of the variance block is too short, large amounts of the variance within the signal can be lost, resulting in an underestimate of the unitary current (Marty et al., 1984). The actual duration required will depend very much on the mean open and closed times of the channel under study. For a simple two state channel, about 91% of the variance is located within frequencies between 1/20 and 10 times the characteristic frequency of channel gating. On this basis, channels with dwell times in the 1 ms range ($f_c = 157$ Hz) require variance blocks of approximately 400 samples digitized at intervals of 0.32 ms, lasting 127 ms while 100 ms channels would require blocks digitized at 32 ms lasting 12.7 s.

Clearly, the correct estimation of the signal variance is dependent upon knowledge of the frequency of channel gating. This can be determined from the power spectrum of the current fluctuations, to be discussed shortly. If this information is not available, then variance results should be treated with caution. A conservative approach should be taken measuring variance over as wide a frequency range as possible. This may mean using variance blocks with 3000–4000 samples, and recordings lasting several minutes.

Figure 9.2 shows the results of applying variance analysis to the digitized current record from Figure 9.1. The continuous record has been

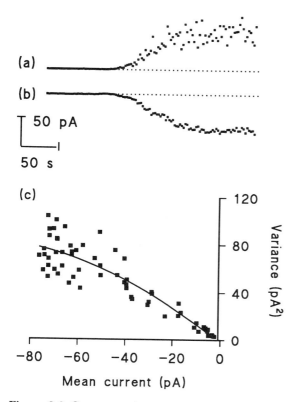

Figure 9.2 Current variance analysis. Current variance (a) and mean current (b) calculated from 116 × 512-sample blocks from a digitized recording of a Cl⁻ current from a bovine chromaffin cell evoked by the application of 10 μM GABA. Filter bandwidth settings 0.2–100 Hz, sampling interval 5 ms. Current variance vs mean current plot (c) from blocks during rising phase of current. The best-fit parabolic curve is shown superimposed which yielded values of −1.61 pA for unitary current and 134 for number of channels in population. (Data courtesy of Professor J.J. Lambert, Dundee University.)

split into 116 variance blocks of 512 samples each. Figure 9.2 (a) shows the mean current calculated for each block using Equation 9.7 and the variance from Equation 9.6 in (b). As the agonist concentration rises around the cell, both the mean current and the current variance increases until a steady state is reached. In practice the recorded variance σ^2 inevitably consists of two components, a background variance σ_b^2 due to instrumentation noise and the ion current variance σ_i^2 due to the ion channel fluctuations such that

$$\sigma^2 = \sigma_b^2 + \sigma_i^2 \qquad [9.10]$$

However, an estimate of σ_b^2 alone can be obtained from the variance of the signal recorded before the application of the agonist. This variance can then be subtracted from σ^2 after the agonist is applied to obtain σ_i^2.

9.4 VARIANCE VS MEAN CURRENT PLOTS

The rising phase of the current record provides a convenient means of studying the relationship between variance and mean current. A plot of current variance (after background subtraction) vs mean current from the blocks during the rising phase is shown in Figure 9.2 (c). The amplitude of the channel unitary current and the number of channels contributing to the current can be estimated by fitting Equation 9.3 using iterative least squares with i and n as free parameters. This results in the hyperbola shown superimposed with values of $i = -1.61$ pA and $n = 134$.

A basic assumption of this approach is, of course, that the channels producing the fluctuations have only one conductance level. As discussed in the previous chapter, this need not be the case and, in such situations, i calculated using Equation 9.3 will be a weighted average of the different states. Further discussion of this problem can be found in Cull-Candy *et al.* (1988).

It is also worth noting that the validity of applying stationary analysis procedures to a changing signal depends very much on the rate of change of the current. In this case the rate of onset of the current is slow, and the current does not change significantly within the period of a single variance calculation. Therefore the current can be treated as a series of steady states, to each of which stationary conditions apply. This need not always be the case, rapidly changing currents will introduce trends within the variance block resulting in a significant overestimate of the variance and erroneous results. Under such conditions it is necessary to resort to the non-stationary analysis methods to be discussed later.

9.5 SPECTRAL ANALYSIS OF CURRENT VARIANCE

Variance analysis provides an estimate of the number and conductance of the channels underlying the observed currents. However, it discards information concerning the rates of the fluctuations and thus can provide no information concerning channel kinetics. The mean open and closed times for the channels underlying the current influence the fluctuation rate. Channels with long open and closed periods, say 100 ms, would be expected to produce noise which fluctuates at slower rates than for channels with 1 ms open and closed times. The distribution of frequencies within the channel noise therefore holds information on channel kinetics.

An estimate of the frequency distribution of the noise variance can be obtained by computing the signal power spectrum (the term 'power' rises from the origin of the technique in electronic engineering). Originally, the power spectrum was calculated by analog means using a bank of narrow bandpass filters to split the signal into its frequency components. However, this process is now more effectively carried out by digital means using digital Fourier analysis techniques. Spectral analysis was revolutionized by the development, by Cooley & Tukey (1965), of the Fast Fourier Transform (FFT). This algorithm provides a rapid means of performing Fourier transforms using the digital computer and has become the standard technique.

The signal power spectrum is calculated in a similar manner to the variance. The record is split into blocks of N samples where N is a power of two (a restriction incurred by the use of the FFT), 512, 1024 and 2048 being common sizes. The power spectrum of an N sample block consists of $N/2 - 1$ frequency components spaced at intervals of $1/(N\, dt)$ Hz from $1/(N\, dt)$ to $1/(2\, dt)$ Hz. An additional component represents the DC level. A power spectrum from a 1024 sample block is shown in Figure 9.3 (a). As is customary, power and frequency are plotted on log–log axes. The spectrum is expressed as the power density; the amount of power (current^2) per unit frequency. In Figure 9.3 the spectral units are pA^2/Hz, although A^2/Hz are also often used.

Single spectra such as this are very noisy, in fact the standard error of the estimate is as large as the signal itself (Bendat & Piersol, 1971). To derive a meaningful estimate it is necessary to compute the average of several individual spectra. It can also be seen that the logarithmic frequency plot results in a bunching of large numbers of the linearly spaced frequency components at the high frequency end of the spectrum. It is convenient to reduce the numbers of high frequency components and further improve the measurement by averaging adjacent components at the higher frequencies.

Figure 9.3 (b) shows the results of combined frequency and block averaging of 16 individual spectra with 512 frequency components reduced to 80. At low frequencies, the spectrum is flat up to a cut-off of around 1 Hz, above which it falls rapidly with increasing frequency until at the highest frequency it is less than 0.2% of the plateau level. The background noise spectrum is also shown. At low frequencies it is more than three orders of magnitude less than the ionic current spectrum. However, since background spectra do not fall as rapidly with frequency (and eventually may actually increase), a point is reached where the background and ion current spectra cross. This determines the practical upper frequency limit of the measurable spectrum.

9.6 ESTIMATING CHANNEL KINETICS FROM THE POWER SPECTRUM

It can be shown that the power spectra of current fluctuations, produced by ion channels with exponential dwell times, can be represented by the sum of one or more Lorentzian functions (Stevens, 1981). A channel with open and closed states with exponential dwell time distributions has the power spectrum

$$S(f) = \frac{S(0)}{1 + (f/f_c)^2} \quad [9.11]$$

A derivation of Equation 9.11 can be found in DeFelice (1981). At low frequencies the function tends to a constant value $S(0)$ (in units of I^2/Hz). At high frequencies the spectrum decays in

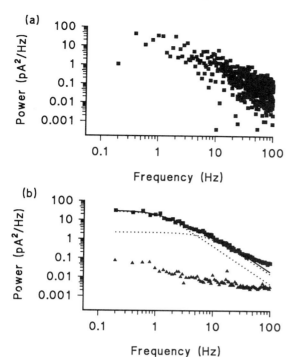

$$\sigma^2 = \frac{f_c S(0)}{2} \quad [9.13]$$

The single channel conductance γ can therefore be calculated from

$$\gamma = \frac{\sigma^2}{I_m(V_m - V_r)} \quad [9.14]$$

where I_m is the mean current during the period of recording of the spectrum. Just as for the variance analysis, if the channels have sub-conductance states, then γ will be a weighted average of the possible levels.

The presence of more than one Lorentzian component within the power spectrum indicates that the underlying channels can exist in more than two states, there being $n-1$ components for n states. This can be seen to be the case for the spectrum in Figure 9.3 (c) where at least two Lorentzians are required to obtain a reasonable fit (shown superimposed). Multi-component Lorentzian fitting has been successfully applied to a number of different systems. It proved useful in the study of the properties of end plate ion channel blocking drugs which convert the normal single Lorentzian spectra of the endplate current into quite distinctive two component spectra (Ruff, 1977). It has also been applied to the analysis of Na^+ currents (Conti et al., 1976). It is, however, a somewhat blunt instrument compared to single-channel analysis. In the case of the GABA channel, for instance, spectra suggest that the channel has at least three states (Barker et al., 1982). However, single-channel analysis reveals the existence of eight states (MacDonald et al., 1989) and four conductance levels (Bormann et al., 1987).

Figure 9.3 Spectral analysis. The frequency distribution of ion channel current variance revealed by the power spectrum. (a) A single 512 point power spectrum, range 0.19–100 Hz, computed from a 1024 sample block during the steady-state phase of the GABA current in Figure 9.2. (b) An averaged spectrum (■) from 16 blocks with averaging of adjacent frequency points at high frequencies. After averaging a similarly calculated background noise spectrum (▲) has been subtracted. The best two Lorentzian fits ($S_1(0) = 28.45$ pA^2 Hz^{-1}, $f_1 = 1.2$ Hz; $S_2(0) = 2.26$ pA^2 Hz^{-1}, $f_2 = 8.09$ Hz) are shown superimposed.

proportion to $1/f^2$. The mean open and closed times for the channel determine the cut-off frequency f_c at which $S(f)$ power has declined to $S(0)/2$ (also known as the half-power frequency).

$$\tau = \frac{1}{2\pi f_c} = \frac{1}{1/\tau_o + 1/\tau_c} \quad [9.12]$$

Note that the characteristic time constant τ is a function of both the mean channel open and closed times. If the channel open probability is low, τ_c is much larger than τ_o and can be neglected and τ becomes a measure of τ_o.

Since the spectrum is the frequency distribution of the signal variance, the variance can be calculated from the integral of the spectrum

9.7 COMPUTING THE POWER SPECTRUM

The procedure for the actual calculation of the power spectrum will now be treated in detail. As discussed in Chapter 3, a signal, normally represented within the time domain as a function of time $y(t)$, can equally well be represented by its Fourier transform

$$Y(f) = \int_{-\infty}^{\infty} y(t) \exp(-2\pi i f t) \, dt \quad [9.15]$$

Note the existence of the imaginary number i^2 ($= -1$). $Y(f)$ is a complex function with real and imaginary components, containing all the amplitude and phase information necessary to specify the original time signal fully. Equation 9.15 properly applies only to continuous analog signals. For a digitized signal expressed as a series of N samples ($y_0 \ldots y_{N-1}$), acquired at intervals dt, the equivalent discrete Fourier transform (DFT) can be computed

$$Y_j = \sum_{k=0}^{N-1} Y_k \exp\left(\frac{-2\pi i j k}{N}\right), \quad j = 0 \ldots N-1 \quad [9.16]$$

which contains N complex frequency components. The power spectrum is the square of the amplitudes of the DFT frequency components averaged over the N samples. Only $N/2$ of the DFT's components are unique, so taking this into account the single-sided power spectrum consists of the $N/2 + 1$ discrete components

$$S_0 = \frac{2\,dt}{N^2} \operatorname{Re}(Y_0)^2$$

$$S_k = \frac{2\,dt}{N^2}[\operatorname{Re}(Y_k)^2 + \operatorname{Im}(Y_k)^2], \quad f_k = \frac{k}{N\,dt}, \quad k=1\ldots N/2-1$$

$$S_{N/2} = \frac{2\,dt}{N^2} \operatorname{Re}(Y_{N/2})^2 \quad [9.17]$$

Each element S_k is a measure of the signal variance contributed by frequencies within a band of width $1/(N\,dt)$ centred on the frequency $k/(N\,dt)$.

The general DFT algorithm is quite slow due to the large number of exponential operations required. However, the FFT algorithm exploits the symmetries in the calculation that occur when N is a power of two to reduce the number of such operations radically, yielding great improvements in performance. FFT algorithms can be readily found within the literature, a well-known source being Munro (1975, 1976). A useful discussion of FFT principles and FFT source code can also be found in Press *et al.* (1986).

Assuming that a digital recording has been made of a current noise signal, high- and low-pass filtered as discussed in Section 9.2, and stored on file as a series of integer A/D samples, a general procedure for calculating the averaged power spectrum can be summarized as follows:

(a) read N A/D samples from file
(b) subtract mean DC level
(c) apply cosine window to data
(d) FFT
(e) calculate power spectrum (Equation 9.16)
(f) add spectrum to average
(g) repeat (a)–(f) until all blocks done
(h) calculate average spectrum
(i) average adjacent high-frequency components
(j) re-scale for effects of cosine window.

A certain amount of pre-processing of the signal in the time domain is necessary before the FFT can be applied. Digital filters, such as the FFT, which act only on a short portion of a signal do not have perfectly sharp passbands. Consequently, each frequency band within the spectrum overlaps its neighbours to some extent. If sharp peaks exist within the spectrum, some of their power leaks into adjacent frequency bands and they appear to be surrounded by several additional peaks. Such *side-lobes* are particularly prominent when the FFT is applied directly to the data block. By applying a smooth taper to the edges of the data block it is possible to alter the filter response to minimize side-lobe effects. The *cosine bell window* function is often used for this purpose, where the first and last 10% of the points within the data block are altered by the formula

$$y_i = \frac{y_i}{2}\left[1 + \cos\left(\frac{10\pi i}{N}\right)\right], \quad i = 1\ldots N/10 \ \& \ i = 9N/10\ldots N \quad [9.18]$$

However, the application of the cosine window, by scaling down the data points at the edges of the data block, effectively reduces the signal variance. It is necessary to account for this and rescale the resulting power spectrum. The choice of data window is not particularly crucial to the analysis of ion channel current noise since the spectra do not usually contain distinct peaks. Other data windows can be used and further details can be found in Press *et al.* (1986).

A power spectrum calculation routine implementing the above procedures can be found in Listing 9.1. It analyses a data file 'noise.dat' which contains the digitized current signal. A 10% cosine window (Equation 9.18) is applied to the data with the sum of the squared window scaling factors accumulated in the process. These are used to obtain the correction factor necessary to account for the loss of variance due to the windowing operation. The FFT is computed by the REALFT routine from Press *et al.* (1986) and takes less than a second to compute a single spectrum. Frequency averaging is implemented in blocks of 16. The 16 lowest frequencies are not averaged, then 16 pairs, 16 groups of four and so on, condensing 512 frequency points into 80.

9.8 NON-STATIONARY NOISE ANALYSIS

There are many circumstances where it is not possible to maintain steady-state current signals for a sufficiently long period to apply stationary analysis techniques. This is particularly so for voltage-activated currents, such as the Na$^+$ current, which are inherently transient phenomena, rising to a peak and then inactivating within a fraction of a second. Similar difficulties can also apply to the study of agonist-activated currents at high agonist concentrations where desensitization of the receptors occurs within fractions of a second (Dilger & Brett, 1990). In either of these conditions, the macroscopic current is changing at a rate of the same order as the ion channel open/close fluctuations. The current is not even approximately constant within the shortest conceivable variance blocks and hence any variance or spectrum calculated using stationary techniques would be meaningless.

However, although voltage-activated currents are transient they are usually repeatable after a short interval to allow re-activation. Hundreds of essentially similar currents can be produced this way. Sigworth (1981a,b) exploited this feature of voltage-activated currents to develop alternative methods of variance analysis, applicable to these non-stationary currents.

In stationary analysis repeated estimates of the current variance and power spectrum are obtained by making the assumption that the

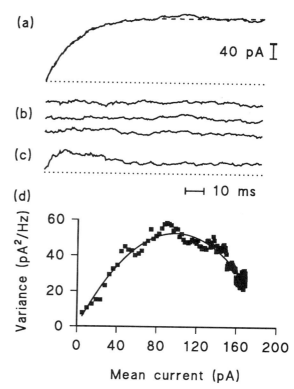

Figure 9.4 Non-stationary variance analysis applied to a simulated transient currents, generated by the summed activity of 200 × 1 pA channels. (a) Ensemble average (dashed line) of 100 current transients computed using Equation 9.18. A typical current record is shown superimposed. (b) Residual current fluctuations from three records after ensemble average has been subtracted. (c) Variance computed from residual currents from 100 records. (d) Variance–mean current plot, (d) vs (a), with best-fit parabola superimposed, yielding $i = 1.07$ pA, $n = 183$.

signal can be split up into blocks of samples each of which is representative of the underlying stochastic process. This is not the only way of calculating the signal variance. For a series of repetitions of a signal, an estimate of the variance can be obtained by observing the fluctuations at corresponding time points between all the records in the series. For a series of transient currents, all evoked by the same stimulus, the average current time course for a series of N-sample digital records, repeated m times is given by

$$\text{Avg}(i) = \frac{1}{m} \sum_{j=1}^{m} y(i)_j, \quad i = 1 \ldots N \quad [9.19]$$

A collection of repeated records is sometimes known as an *ensemble* and the average is described as the ensemble average. The ensemble variance about each sample point i within the record is given by

$$\text{Var}(i) = \frac{1}{m-1} \sum_{j=1}^{m} [y(i)_j - \text{Avg}(i)]^2, \; i = 1...N$$

[9.20]

Having obtained estimates for Avg(i) and Var(i), Equation 9.3 can be used to calculate the single channel conductance and estimate the number of channels, just as for stationary analysis.

Figure 9.4 illustrates the non-stationary analysis procedure applied to a simulated data set. An ion channel simulation has been used to generate transient current records consisting of 200 channels with 1 pA unit amplitude, and mean open and closed times of 100 ms and 40 ms. All channels start off closed at the beginning of the record and the current rises to a steady-state of 143 pA, as shown in Figure 9.4 (a). One hundred such records were generated and the ensemble average, computed using Equation 9.19, is shown as the dashed line. Subtracting the ensemble average from each record creates a set of 100 residual fluctuation records, three of which are shown in (b). The residual variance computed from the complete set of 100 residual records is shown in (c).

A variance–mean current plot Figure 9.4 (d) is obtained by plotting the corresponding points of the residual variance record from (c) and the ensemble average from (a). The best-fit parabola (Equation 9.3) is shown superimposed which produced estimates of 1.07 pA for the unit current and 183 for the number of channels.

It is, of course, crucial to the correctness of the analysis that no systematic trend is occurring throughout the series of m records; such as if the peak amplitude of the current were fading with repetition. It is possible to compensate for the effect of weak trends in the current by splitting the series of records into short sub-series within each of which the trend is negligible, applying Equations 9.18 and 9.19 to these, and then averaging the results. Sigworth (1981a) has used sub-series of $m = 4$ or 6, but as few as $m = 2$ have been used (Conti et al., 1984).

9.9 NON-STATIONARY KINETIC ANALYSIS

The ensemble variance simply provides amplitude information, but the method can also be extended to provide kinetic information, in fact more so than the stationary analysis. Sigworth (1981b) shows how to compute the covariance

$$C(i, j) = \frac{1}{m-1} \sum_{k=1}^{m} [y_k(i) - \text{Avg}(i)] [y_k(j) - \text{Avg}(j)]$$

[9.21]

which represents the correlation between the variance at samples i and j within the record. $C_{i,j}$ is a two-dimensional function which requires a large amount of computation for all possible values of i and j. Sigworth quotes an execution time of 30 min on a PDP11/34 minicomputer (perhaps 5 min or less on a modern PC). The covariance provides a large amount of kinetic detail which allows inferences about gating mechanisms. Details can be found in Sigworth (1981a).

It is also possible to compute the power spectrum of the residual fluctuations after the ensemble average has been subtracted (Sigworth, 1981c). This is much faster than the covariance but provides less information. The procedure is essentially the same as for stationary analysis but a correction term is required to account for the mean current time course.

9.10 NON-STATIONARY NOISE RECORDING PROCEDURES

Recording procedures for non-stationary analysis are quite different from those for stationary noise, having much more in common with the procedures for voltage-activated currents discussed in Chapter 7. In particular, it is not possible to separate the fluctuations from the DC current component using high-pass filtering. Repeated voltage steps are applied to the cell and a record of each current transient digitized and stored on disc, using the techniques in Sections 7.3–7.4.

Currents where the amplitude of the fluctuations is small compared to the amplitude of the transient current may be difficult to digitize accurately given the finite resolution of the A/D converter. This is one of few situations where the extra resolution of a 16 bit, rather than the usual 12 bit, A/D converter may be required, a typical example being the Data Translation DT2836. Similarly, it may be unwise to record such signals on some of the older FM tape recorders since their signal–noise ratio preserves no more than 12 bits of information. The modern DAT recorders are preferable in this respect using 14–16 bit digital recording themselves.

A detailed illustration of the method applied to the Na^+ current and the kind of results obtained can be found in Sigworth's orginal papers (Sigworth, 1980, 1981a,b). The method has also been applied to the delayed rectifier K^+ channels in frog node of Ranvier (Conti *et al.*, 1984) and cardiac muscle (Bennet *et al.*, 1989). The method is not restricted to voltage-activated currents. It has been used to study the fluctuations during the rapid desensitization currents induced by high concentrations of ACh (Dilger & Brett, 1990). It should be noted that, in such experiments where the agonist is being applied ionophoretically or by pressure ejection, it may be difficult to ensure that exactly the same stimulus is being applied every time. Fluctuations in the agonist level would of course add extra variance to the records and Dilger and Brett used a precise high-speed solution exchange mechanism. Recently, however, it has been suggested that it is possible to compensate for fluctuations in agonist level and apply the technique more widely (Robinson *et al.*, 1991).

9.11 SOFTWARE FOR NOISE ANALYSIS

Noise analysis software is comparatively rare compared to that for single-channel or voltage clamp analysis. Given the past work in this field, and the absolute requirement for a computer, programs undoubtedly exist in many laboratories. However, the technique has never gained sufficient general interest for many of these to be made generally available.

The author's SPAN (SPectral ANalysis) program for the variance and spectral analysis of current noise is available under the same conditions as the other packages discussed previously, either from Dagan Inc or directly. The results within this chapter were generated with this program. It provides variance and spectral analysis of both stationary and non-stationary noise, estimation of single-channel conductance from parabolic fits to variance–mean plots, and Lorentzian fitting to power spectra. Like the others it is written in FORTRAN 77.

RC Electronic's single-channel analysis program is described as having noise analysis features such as power spectral analysis and Lorentzian fitting. The tools for constructing noise analysis software are also available. Axon Instruments' AXOBASIC package provides FFT commands as well as other necessary array manipulation features. A similar set of features can be found in National Instrument's Lab-Windows package. The CED interfaces also have power spectrum commands built into them, with the laboratory interface acting as a special purpose FFT coprocessor.

9.12 SUMMARY

Noise analysis methods provide a means of estimating the microscopic properties of ion channels, such as single-channel conductance and channel density, from the fluctuations in the macroscopic current through large populations of the channels. A certain amount of information on the kinetics of channel gating can also be obtained. Although having been superseded by single-channel methods in some areas, both stationary and non-stationary analysis techniques continue to have valid roles in the analysis of ionic currents. They have proved to be particularly useful in the study of channels with very low (sub-picosiemen) conductance, and where too many channels are active within a patch to permit single-channel methods.

The method is not always supported by commonly available electrophysiological software packages such as pCLAMP or the CED software, with the Strathclyde SPAN program being one of the few expressly designed for this work. It may therefore be necessary to develop spectral analysis software using development environments such as AXOBASIC or Lab-Windows which provide the basic elements such as the FFT routines.

```fortran
c      Listing 9.1. Routine to calculate average power spectrum of current noise.
       subroutine power_spectrum(n,istart,iend,good_block,dt,pa_per_bit,
     & power,freq,nfreq)
       integer*2 n                    ! (In) No. of sample points in FFT block
       integer*2 istart,iend          ! (In) Range of blocks to use for spectrum
       logical good_block(iend)       ! (In) Validity of data blocks
       real*4 dt                      ! (In) sampling interval (s)
       real*4 pa_per_bit              ! (In) pA per ADC bit
       real*4 power(n)                ! (Out) Power (pA2/Hz)
       real*4 freq(n)                 ! (Out) Frequency (Hz)
       integer*2 nfreq                ! (Out) No. of frequency points returned
       integer*2 ibuf(4096)           ! Integer ADC data buffer
       real*4 y(4096)                 ! FFT transform buffer
       integer*4 idc
       parameter(pi=3.14156)

       open(unit=1,file='noise.dat',form='binary', access='direct,' recl=2*n )
c
c      Create array of frequency points and clear power summation array
c
       df = 1./(float(n)*dt)          ! Frequency spacing (Hz)
       nfreq = n/2                    ! No. of frequency points
       do i = 1,nfreq
           freq(i) = float(i)*df
           power(i) = 0.
       end do

       navg = 0
       do irecord = istart,iend
           read( unit=1, rec=irecord ) (ibuf(i),i=1,n)
           if( good_block(irecord) .eqv. .true. ) then
               navg = navg + 1
c
c              Subtract DC level, convert integer ADC samples into REAL in units
c              of pA, and apply 10% cosine bell data window

               idc = 0
               do i = 1,n
                   idc = idc + ibuf(i)
               end do
               idc = idc/n
               window_factor = 0.
               do i = 1,n
                   if( (i.lt.n/10) .or. (i.gt.(n-n/10)) ) then
                       w = 0.5*(1.+cos(10.*pi*float(i-1)/float(n)))
                   else
                       w = 1.
                   end if
                   y(i) = w * float( ibuf(i) - idc )*pa_per_bit
                   window_factor = window_factor + w*w
               end do
               window_factor = window_factor/float(n)

c ... Continued ...
```

```
c Listing 9.1 continued
c           Fourier transform array y(1..n) using REALFT routine from
c           Numerical Recipes, (Press et al.) Replaces time domain data in array
c           y(1..n) with complex frequency domain components.
c           y(1) = DC component (Real only) y(2) = Highest frequency (Real only)
c           y(3),y(4), y(5),y(6), ... etc. complex frequency components

            call realft( y, nfreq, 1 )
c
c           Calculate power spectrum as absolute values of FFT
c           frequency components and add it to summation
c
            do i = 1,nfreq-1
                j = 2*(i-1) + 3
                power(i) = power(i) + y(j)*y(j) + y(j+1)*y(j+1)
            end do
            power(nfreq) = y(2)*y(2)
         end if
      end do
c
c     Scale spectrum, accounting for sampling interval and number of
c     points in FFT, correct for effects of cosine window,
c     and divide by number of records to get average.
c
      do i = 1,nfreq+1
          power(i) = 2.*dt*power(i)/(float(navg)*float(n)*window_factor)
      end do

c     Frequency averaging

      navg = 1                    ! Average increasing numbers of
      iavg = 0                    ! adjacent points, as frequency
      psum = 0.                   ! increases.
      fsum = 0.                   ! 16x1, 16x2, 16x4 etc.
      i = 1
      k = 1
      do while( i .lt. nfreq )
          psum = psum + power(i)
          fsum = fsum + freq(i)
          iavg = iavg + 1
          if( iavg .eq. navg ) then
              power(k) = psum/float(navg)
              freq(k) = fsum/float(navg)
              psum = 0.
              fsum = 0.
              iavg = 0
              k = k + 1
              if( mod(k,16).eq.0) navg = navg*2
          end if
          i = i + 1
      end do
      nfreq = k-1

      close(unit=1)
      return
      end
```

Appendix

An introduction to computers

Digital computers utilize a range of technologies from the semiconductors involved in the design and construction of integrated ciruits, through the design of the digital logic circuits of the microprocessor, to the operating system software and applications programs. Fortunately there is no need to understand every detail of their operation fully and generally a small amount of basic knowledge of computing principles proves sufficient to at least operate computers. Various generations of digital computers have differed markedly in size, appearance, cost and performance, but they all work on the same basic set of principles. A digital computer is a machine for manipulating information in the form of integer numbers. Information is fed into the computer, processed in some way desired by the operator and returned. The nature of the processing is determined by the computer program.

A1 THE BINARY NUMBER SYSTEM

A distinctive feature of the modern digital computer is its extensive use of the binary number system for arithmetic operations. Although the decimal number system would have been preferable from the human programmer's point of view, the use of the binary system markedly simpifies the design of computer logic circuitry. The two numbers of the binary system (0, 1) are representable as a switch being ON or OFF, or a voltage level being HIGH or LOW. A number expressed in binary form appears as a series of 0s and 1s. For instance, the decimal number 22 has the binary form 10110, constructed as follows

$$22 = 10110 = 1\times16 + 0\times8 + 1\times4 + 1\times2 + 0\times1$$

Each digit in a binary number is described as a *bit* (binary digit). A *byte* is a binary number with eight bits, capable of representing decimal numbers in the range 0–255. A kilobyte (kbyte) is 1024 bytes and a megabyte (Mbytes) is 1 048 576 bytes.

Binary numbers, although efficient in computer terms, are somewhat inconvenient to express in written form. Consequently, programmers often use two other number systems to represent binary numbers, octal (base eight) and hexadecimal (base 16). Both have the advantage,

An introduction to computers

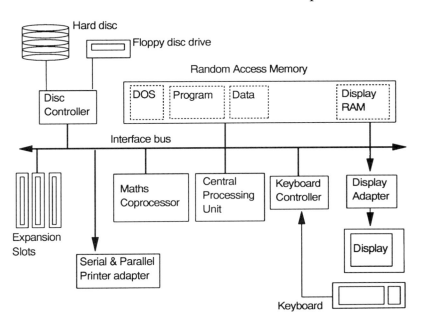

Figure A1 A diagram of the main components of a personal computer. Programs stored in the random access memory are executed by the central processing unit and the maths coprocessor. Data is entered via the keyboard and results displayed on the display screen. Long-term program and data storage is provided by fixed and removable magnetic disc drives. All computer sub-systems are connected and communicate via the interface bus. Additional sub-systems can be added via the bus expansion slots.

compared to decimal, of being powers of two and therefore easily converted to binary. The octal system was extensively used by DEC on the PDP11 minicomputer. However, hexadecimal has proved to be more popular in microcomputer programming. Sixteen distinct digits are required to represent a hexadecimal number, therefore, in addition to the decimal numbers 0–9, the letters A–F are used

    Decimal   0 1 2 3 4 5 6 7 8 9 10 11 12 13 14 15
    Hexadecimal   0 1 2 3 4 5 6 7 8 9   A   B   B   D   E   F

In general the fact that the computer is operating with a binary number system is hidden from the ordinary user, since applications programs (word processors, spreadsheets, etc.) translate and display numbers in decimal form. The same is true for much programming in high level languages (e.g. BASIC, FORTRAN). A capacity to work with binary and hexadecimal is, however, needed for assembler language programming.

A2 COMPUTER SYSTEMS ARCHITECTURE

Although the details of implementation and overall speed and capacity have changed markedly, the essential features of the digital computer remain the same. A block diagram of the main components of a typical personal computer systems is shown in Figure A1. There are two key elements – the *central processing unit* (CPU) and the *random access memory* (RAM).

The CPU is essentially a machine for manipulating numbers. Sets of numbers loaded into the CPU can be added, subtracted, multiplied or divided. In addition, the value of numbers can be compared and also logical AND/OR operations carried out. Each distinct CPU operation is initiated by means of a set of instruction (or operation) codes. Processing operations of any desired complexity can be performed by a series of such CPU instructions. A computer program at its most fundamental is a sequence of machine instructions designed to effect the processing operation desired by the user.

Random access memory is an electronic storage system designed to hold binary numbers which can be rapidly accessed by the CPU. Each location is provided with its own unique address number which the CPU uses when writing and reading data from the memory. Both the CPU instructions and the data to be processed are stored in the RAM. A large part of the CPU instruction set is devoted to data transfer between the CPU and RAM memory. In general, computers with large amounts of RAM can run larger programs and process more data. RAM is quantified in terms of its capacity to store 8 bit binary numbers. A typical laboratory computer might have 1–2 Mbyte of RAM.

CPUs are characterized by their *word size* – the largest binary number which can be processed with a single instruction. A 16 bit CPU, for instance, can handle 16 bit binary numbers (in decimal terms numbers up to 65 535). The CPUs in current computer systems are usually 16 or 32 bit devices.

Instructions are executed by the CPU under the control of a master clock, whose rate determines the execution speed. The frequency of this master clock provides a measure of computing speed. For instance, the original IBM PC of 1981 had a clock frequency of 4.77 MHz while the fastest modern IBM PC compatibles now have 60 MHz clock frequencies. Individual instructions often take different numbers of clock cycles with multiplication or division taking substantially more than addition or subtraction. Average instruction execution speed is also expressed in terms of MIPS (Million Instructions Per Second). For instance, an IBM PC-compatible computer with a 16 MHz 80386 CPU can achieve a 3.6 MIPS performance.

A3 INPUT AND OUTPUT DEVICES

In order to make use of the CPU, a means is required to place data into the RAM and to display the results. A complete computer system must therefore have input and output devices. The standard input device is the keyboard. Each key generates a binary number which can be read by the CPU and stored in RAM. Another common input device is the mouse pointing device which generates a binary number representation of position.

The standard output device is the visual display screen which displays computer-generated text and graphics. The display is generated on a cathode ray tube (CRT) by scanning the electron beam across the screen in a raster fashion similar to that used to generate a TV image. Text and graphics are produced as patterns of dots, illuminated in a variety of colours. Most display screens can produce at least 25 rows of 80 columns of text and graphics with a resolution on 640 horizontal by 480 vertical points, in 16 colours. Display functions are normally handled via a special-purpose CRT controller which forms part of the computer's graphics adapter.

A4 MAGNETIC DISC STORAGE MEDIA

RAM is considered to be the primary data storage medium for the computer since it interacts directly with the CPU. However, the data is only retained in RAM as long as power is applied to the computer system. Therefore a more permanent storage medium is required to retain data during periods when the computer is switched off. A number of secondary storage media exist but the most common one is the magnetic disc. Discs have a magnetizable iron oxide coating similar to that used in magnetic audio tape. The disc rotates underneath a read/write head which can be moved radially across the disc. Information is written on to the disc in the form of binary patterns of magnetic flux laid down in concentric circular tracks which are split radially into groups known as sectors.

Most computers are equipped with two forms of magnetic disc storage, the *floppy disc* and the *hard disc*. Floppy discs consist of thin flexible plastic discs often stored within a rigid case. They are removable and are designed for exchange of data between computer systems that have relatively low storage capacity (360 kbyte – 2.88 Mbyte), slow access times and data transfer times. Typical floppy discs sizes in common use are shown in Table A1.

Computers are also equipped with hard (or Winchester) discs consisting of rigid discs (or stacks of such discs) located within fixed hermetically sealed enclosures. These discs have much higher storage capacities (20 Mbyte – 1000 Mbyte) and faster access and data transfer times. The higher capacities and speeds of the hard disc are achieved by more precise construction of the magnetic disc surface, rotating the disc at higher speeds and moving the read/write as close to the

Table A1 IBM PC floppy disc formats.

Size (inches)	Type	Capacity	Tracks
5.25	Double sided double density	360 k	40
5.25	Double sided high density	1.2 M	80
3.5	Double sided double density	720 k	80
3.5	Double sided high density	1.44 M	80

disc surface as possible. For these reasons the disc must operate with a totally clean environment and therefore the whole assembly of disc, read/write head, etc. is sealed within its case.

The details of controlling, reading and writing to the magnetic disc are handled by a disc controller. A high degree of standardization of the interface between controller and disc has made it relatively simple to fit hard discs to the IBM PC family. Until recently, the most common controller type was the ST506, found in most low–medium cost PCs produced before 1990. ST506 is now being replaced by the higher performance IDE (Integrated Drive Electronics) interface. High performance computer systems also use the SCSI (Small Computer Systems Interface) standard or the ESDI (Enhanced Systems Disk Interface). Apple Macintosh computers have always used the SCSI interface.

A5 COMPUTER INTERFACE BUS

The sub-systems of the computer, RAM, keyboard, display, floppy and hard discs, operate under control of the CPU. They communicate and exchange data by means of the 'bus' – a common data transfer pathway. The bus consists of a set of address, data and control lines on to which numbers can be placed in binary form. Data is exchanged between the CPU and a sub-system by the process of the CPU placing the appropriate sub-system address on the address lines, then placing the data to be transferred on to the data lines (if a CPU sub-system transfer is being made).

Many computers, including the IBM PC and **Macintosh** families, have sets of expansion slots **which** allow circuit cards containing additional sub-systems such as analog–digital converters to be attached to the computer bus. The availability of expansion slots is a crucial feature of a laboratory computer. As for the hard discs, a number of standards exist. The success of the IBM PC family has been in no small measure due to the adoption of its expansion bus as a *de facto* industry standard. In fact, it has been renamed the ISA (Industry Standard Architecture) bus.

The ISA bus has performance limitations and in 1988 IBM introduced the PS/2 (Personal System 2) range of computers with the MCA (Microchannel Architecture) bus. Although PS/2 computers are PC compatible in terms of software, MCA bus expansion slots cannot accept ISA bus cards and vice versa. For technical and commercial reasons the MCA bus did not receive wide acceptance and few computers use it. A consortium of manufacturers of IBM PC-compatible computers established the EISA (Extended Industry Standard Architecture) bus with an equally high performance but retaining compatibility with ISA bus cards. Nevertheless, most IBM PC-compatible computers still use the ISA bus.

The Apple Macintosh family of computers have a somewhat confusing range of expansion buses and slots. Low-cost Macs (e.g. Mac SE, Classic, LC) often only have a single slot, which varies in size and specification between types of machine. However, the Macintosh II series is fitted with the relatively standardized NU-BUS.

A6 OPERATING SYSTEMS

The hardware described so far is not sufficient in itself to produce a fully functioning computer. All of the sub-systems are under the control of the CPU, but the CPU requires a program of instructions before it can operate. In order for the user to interact with the computer, a program must exist which at least monitors the keyboard and displays results on the screen. In addition, programs are needed also to control the hard and floppy disc sub-systems in order to load other applications programs. These functions are carried out by a set of programs known as the *operating system*.

However, a problem still exists in how to load the operating system into the computer in the first place. The operating system may be stored on magnetic disc but it cannot be accessed unless a program already exists to handle the disc sub-system – this is the bootstrap problem, as in how to make the computer pull itself up by its bootstraps. This function is performed by the basic input/output system (BIOS), a program permanently stored in the computer in the form of read only memory (ROM). When the computer is switched on, the CPU is made to start executing instructions from a fixed location in the ROM memory which contains a small program sufficient to load the rest of the operating system.

The nature of the operating system determines how the computer appears to the user and more than one operating system may exist for a given computer system. The most common operating system for the IBM PC family is MS-DOS (Microsoft Disc Operating System) but several others exist including OS/2 and UNIX. Operating systems can be categorized as being single-task systems such as MS-DOS with the capability of running only one program at a time or multi-task systems capable of running several programs simultaneously (OS/2, UNIX). Multi-task operating systems may also be multi-user systems (UNIX) supporting several users sitting at different computer terminals.

The importance of the operating system in determining the character of a computer system can be illustrated by a comparison of the IBM PC and the Macintosh computers. MS-DOS is a traditional command-driven system, in which operations are invoked by the user entering the appropriate command phrase. MS-DOS contains over 50 English-like commands, mostly associated with the manipulation of disc files. For instance, the command to display the directory list of the files on disc drive C on the screen is

    dir c:

The main problem with a command-driven OS is that it requires the user to memorize the command vocabulary and the, often quite strict, syntax. In spite of the popularity of the IBM PC family, there can be no doubt that many computer users find MS-DOS commands difficult to remember and use.

The Apple Macintosh has a radically different approach using a *graphical user interface* (GUI). Data files and programs are represented graphically as icons which can be touched and moved about using a mouse pointing device. Programs can be started simply by pointing to the program icon and clicking a button on the mouse. Commands are invoked by selection from menus which can be made to drop down from the top of the screen.

The GUI is widely recognized as being a simpler and more intuitive way to operate a computer than an equivalent command-driven user interface. The success of this approach has resulted in the production of similar GUIs for MS-DOS, such as Microsoft Windows. Current trends suggest that Windows will eventually become a standard feature of MS-DOS.

MS-DOS also has a key limitation in that it can directly address only 1 Mbyte of RAM, due to its use of the 16 bit real addressing mode of the 80x86 family. Modern PCs which use 80386 or 80486 CPUs have the capability of addressing as much RAM as can be physically fitted into the PC (10–20 Mbyte). The Macintosh has never had such a limitation. The trend towards larger and more sophisticated GUI-based programs, making ever increasing demands on RAM, makes a new PC operating system almost essential. The current version of Windows (V3.1, at the time of writing) partially achieves this, in a not entirely satisfactory way. However, new operating systems such as IBM's OS/2 V2.0, and Microsoft Windows NT are promised to be the solution to this problem.

A7 PROGRAMMING LANGUAGES

In general, most personal computer users do not write programs, making use of the many generally available applications packages. However, packages are not always available for many laboratory applications. If these needs are to be met then some skill at actually programming the computer is required. The remainder of this appendix therefore describes some of the common programming languages and issues associated with their use.

As discussed earlier, the CPU executes instructions in the form of binary numerical codes. For instance, the 8086 CPU has more than 80 basic instructions each of which have more than six modes of operation. The code to add the numbers 2 and 3 in the 8086's AX register is in hexadecimal

B8 00 02 05 00 05 2E A3 00 00

Except in the most trivial circumstances, it is unrealistic to program using the raw machine code. Instead, programs are written initially in an abstract form using a computer *programming language*. The program *source code* created this way is then fed into a translation program which generates a file containing the machine code which can be executed by the CPU.

Programming languages can be classified into *low level* languages, whose statements have a structure closely corresponding to machine code, and more abstract and expressive *high level* languages. Each type of microprocessor has its own individual low-level assembly language in which each numerical machine code instruction is represented by a mnemonic indicating its function. For instance, the assembler form of the 8086 machine code addition program takes the form

```
mov ax,2      ;Move the number 2 to the
              AX register
add ax,5      ;Add the number 5 to the AX
mov sum,ax    ;Store the result in memory
```

The assembler source code is translated into machine code by an *assembler* program. Assembler code is much more readable than machine code and many of the details such as the exact memory address used to store data need not concern the programmer. Assembler code gives great control over the operation of the computer and allows very efficient code to be written. However, even simple tasks can take large numbers of assembler instructions. In addition, the code is not portable since different microprocessors have different instruction sets.

In order to avoid the limitations of assembler, higher level languages were developed which were designed to make programming easier by providing powerful instruction sets which could express in one line what would require many assembler instructions. For instance, the addition sum can be expressed

sum = 2 + 5

High-level languages are also designed to be portable having a standard syntax and instruction set independent of the CPU. However, such languages inevitably make less efficient use of any particular CPU and produce larger and slower machine code program than assemblers. Among the first languages developed was FORTRAN (FORmula TRANslator) designed for scientific calculations and COBOL for business transactions. Both FORTRAN and COBOL are examples of *compiled* languages where the program is written then translated as a whole into machine code.

A8 FORTRAN

FORTRAN was developed by John Backus in the mid-1950s as a language for scientific computation. It is therefore one of the oldest programming languages still in general use. It has, however, evolved over the years (FORTRAN 66, FORTRAN 77, FORTRAN 90) gradually incorporating features from other newer languages. An enormous volume of software has been written in FORTRAN, in particular many libraries of numerical analysis routines. This is one reason for its continued popularity.

In recent years it has been somewhat eclipsed by the C language, although rumours that 'FORTRAN is dead' are somewhat exaggerated. The original version of FORTRAN was rightly criticized for encouraging a complex convoluted 'spaghetti' programming style through the use of the GOTO statement. By the early 1970s, it had become apparent that if large programs were to be written, and more importantly maintained, they had to be more readable with cleaner designs. The *structured programming* style arose in which the program was split into small self-contained blocks, with program flow controlled by new block-structured language statements (see Table A2). The FORTRAN language has

Table A2 Programming languages.

a) FORTRAN 90.

i)
```
do i = 1,n
      sum = sum + x(i)
end do
```

ii)
```
if( n .gt. 0 ) then
      sum = sum/n
else
      sum = 0.
end if
```

iii)
```
do while ( n .le. n )
      sum = sum + x(i)
      n = n + 1
end do
```

iv)
```
select case ( key )
case( 'A' )
      write(*,'(1x,a)') 'This is A'
case( 'B' )
      write(*,'(1x,a)') 'This is B'
end select
```

b) Pascal.

i)
```
for i := 1 to n
      begin
              sum = sum + x[i];
      end;
```

ii)
```
if( n > 0 ) then
      begin
      sum := sum/n
      end;
elseif( n = 0 ) then
      begin
      sum := 0
      end;
```

iii)
```
while i < n do
      begin
      sum := sum + x[i]
      end;
```

iv)
```
case key of
      'A' : writeln() 'This is A';
      'B' : writeln() 'This is B';
```

v)
```
repeat
      begin
      sum := sum + x[i];
      end;
until i >= n
```

c) C.

i)
```
for ( i=0; i<n ; i++ )
      {
      sum = sum + x[i];
      }
```

ii)
```
if ( n > 0 )
      {
      sum = sum / n;
      }
else
      {
      sum = 0;
      }
```

iii)
```
while( i < n )
      {
      sum = sum + x[i];
      i++
      }
```

iv)
```
switch ( c ) {
      case 'A' : printf("This is A");
      case 'B' : printf("This is B");
```

v)
```
do
      {
      sum = sum + x[i];
      i++;
      }
      while( i < n );
```

d) Microsoft Quick-Basic

i)
```
for i% = 1 to n%
      sum = sum + x(i)
      next
```

ii)
```
if n% > 0 then
      sum = sum/n%
else
      sum = 0.
end
```

iii)
```
while i% <= n%
      sum = sum + x(i%)
      i% = i% + 1
wend
```

expanded incorporating many of these structured programming features, while retaining the earlier forms for compatibility. In summary, FORTRAN's primary strength is its standardization, allowing FORTRAN programs to be easily ported between computer systems. It is a conservative language, incorporating new features more slowly than some other languages.

Numerous FORTRAN compilers are available for the IBM PC, the most common being Microsoft FORTRAN V5.1. A notable advance in the design of these compilers has been the introduction of products, such as Microway NDP-FORTRAN and Salford FTN-77 that generate code that makes use of the 32 bit operating mode available on PCs with 80386 and 80486 CPUs, thus freeing themselves of the MS-DOS 640 kbyte memory limitation. A detailed introduction to FORTRAN can be found in many textbooks, e.g. Etter (1988), Ellis (1990).

A9 PASCAL

Pascal was developed, by Nicklaus Wirth, to be a rigorously defined language which enforced a structured programming style, rather than simply permitting it like FORTRAN. One aim was to provide a language for teaching purposes that prevented the student falling into bad habits. Pascal's block-structured statements can be seen in Table A2 (b). It can be shown that there is rarely any need for a FORTRAN-like GOTO statement which allows the flow control to jump about within a program. A small number of appropriate block statements are sufficient. For this reason, Pascal programs were much more readable than the older unstructured FORTRAN program.

Pascal is widely used in elementary computer science courses, but is less commonly used at a professional level. In one sense, its strength as a teaching language weakens it as a professional tool. Some of Pascal's features which enforce good programming discipline in the learner tend to restrict the overall versatility of the language. It has some serious deficiencies, in particular it does not support the handling arrays by subroutines at all well (for instance, see Press et al., 1986). There are also problems of standardization. The most widely used version of Pascal is Borland's Turbo Pascal, a highly non-standard dialect. The language has evolved through numerous versions, each with interesting and innovative features, but each significantly incompatible with previous versions. Wirth has also produced newer languages such as Modula 2 and Oberon, based on his experience with Pascal, which remedied some of Pascal's deficiencies and added new features. However, these languages have not achieved the same degree of popularity as Pascal. Some useful textbooks are; Savitch (1990) and Kernighan & Plauger (1981).

A10 THE C LANGUAGE

The C language was developed by Kernighan & Ritchie (1978) at Bell Laboratories. It retains the block structures of Pascal without the limitations of that language. It was designed from the outset to be easily portable between computer systems. C is probably the most common language used, at present, for professional programming projects. Programs such as Microsoft Word, and many other well-known packages are written in C, including the complete UNIX operating system.

If issues of efficiency, versatility and portability were all that mattered, then C would be the obvious choice for programming in the scientific laboratory. Unfortunately, the very richness and power of the language makes it difficult to learn and use. Complex operations expressed in very terse notation is a characteristic of C. Unlike other high-level languages, a significant level of understanding of the basic CPU operation and memory addressing is required to use C correctly. C is a professional tool and, like all such tools, requires an appropriate level of skill which is not quickly attained. Again, many textbooks are available (Hansen, 1989; Kelley & Pohl, 1987).

A11 BASIC

C is an extreme case, but none of the other languages so far discussed are particularly easy to learn. A significant part of the difficulty lies in the complexity of the overall program development process. A text file is first created containing the program source code. The source code is then translated, by a compiler program into intermediate machine code files called object code files. The final executable machine code program is created by linking together one or more such object files with perhaps a library of object code. In addition to simply understanding the vocabulary and syntax of the program language, the programmer must also understand how to operate the text editor, compiler, and linker programs, and the error messages produced.

The BASIC (Beginner's All-purpose Symbolic Instruction Code) language was developed in an attempt to make the development and running of programs a much simpler and more accessible process. BASIC integrates program development and execution into the single environment of an *interpreter*. Instead of converting the complete source code into machine code in one single step, like a compiler would, an interpreter translates and executes the source code on a line by line basis. The interpreter also provides program editing capabilities. With a BASIC interpreter a program can be typed into the interpreter, then executed immediately, simply by typing RUN. If the program is incorrect (as is likely), execution can be stopped, the contents of variables inspected, the faulty lines corrected and the program restarted.

Interpreters have the advantage of being very easy to learn and use but slow, due to the repeated line by line translation of source into machine code. The original BASIC was also a poorly structured language which lacked block-structured control statements and required a number at the beginning of each line. However, the fact that BASIC could provide a compact self-contained programming environment, implementable on systems with a limited amount of RAM, made it a very popular choice for use on the first generation of microcomputers that appeared in the 1970s. Microsoft BASIC (or compatible products) has usually been distributed as a standard part of the MS-DOS operating system. Consequently, many people's first experiences of programming are with BASIC.

Like the other languages, BASIC has evolved with time, incorporating block structures and shedding the need for line numbers. As can be seen from Table A2 (d), recent versions of BASIC, such as Microsoft Quick-Basic, have many of the control structures found in the other languages. BASIC is perhaps the simplest language to learn and use. It also has excellent text string handling features. Nevertheless, while it is an entirely respectable programming language, it cannot match C and FORTRAN in areas such as file handling, and the use of subroutines.

A12 CHOOSING A PROGRAMMING LANGUAGE

The choice of an appropriate programming language for the development of laboratory software is an issue that is sometimes hotly debated. There is however no single, obviously best, choice with many of the languages discussed earlier being adequate for most purposes. Most languages are quite similar, as can be seen from Table A2. However, certain general observations can be made. It is wise to choose a popular language which has proved itself through use over a number of years such as FORTRAN or C. The widespread use of both of these languages means that they are supported on many different computer systems and continue to evolve. In general, FORTRAN and C compilers are among the first to become available for any new computer system. The same cannot be said for other languages such as Pascal or Modula 2. Although each of these languages have played important roles in the development of computer language design, they have never achieved really widespread use.

A distinction can be made between compiled languages, such as FORTRAN or C, and languages like Quick-BASIC which have their origins as interpreted languages. FORTRAN and C are, in a sense, professional languages, in that they have a steep learning curve. A significant amount of time is needed to achieve a basic proficiency. This is particularly the case for C. From the point of view of the laboratory worker, the time spent learning such a language is poorly invested, unless a reasonably large scale of software development in envisaged, where the benefits of these languages start to appear. However, a number of other issues arise when large-scale projects are contemplated. If a large amount of time is invested it is important that some kind of return is obtained, in the form of publications relating to the software, or perhaps commercial distribution of the software itself. However, software written to be used by others must generally be of a higher standard in terms of design, ease of use, and documentation than that produced for local consumption. Overall, a professional attitude to the work must be being adopted, and it must be worthwhile doing this.

On the other hand, the situation often arises where it is clear that a small program will help solve a particular limited problem, only of interest within that laboratory. The overheads involved in learning FORTRAN or C might not be justifiable. In these circumstances, a simpler, more accessible language such as BASIC is to be preferred. An approximate rule of thumb might be that BASIC is satisfactory for programs which extend to no more 1000–2000 lines of source code. Beyond that, it becomes increasingly important to break the program down into appropriately designed sub-modules, within separate source code files. This can be done much more elegantly in FORTRAN or C.

A13 GRAPHICS AND GUI PROGRAMMING

The development and increasing popularity of GUIs such as Windows and the Macintosh operating system are changing the key issues in program development. The actual choice of programming language is no longer the most important issue. GUIs place very strict limitations on programming style and choice of language. Essentially, a language is chosen which is compatible with the API (Applications Programmers' Interface) library of subroutines which allows the programmer to make use of the GUI.

GUIs are often very complex and difficult to program, one of the reasons for the smaller amount of scientific software available for the Macintosh. At present, it is possible to avoid using a GUI, and program for the simpler MS-DOS. However, eventually, as the GUI and the base operating system become more tightly bound together it will become increasingly difficult to do this. The multi-tasking nature of most GUIs also makes the kind of low-level programming of laboratory interfaces more difficult. Solutions to these problems, however, are slowly emerging in the form of integrated programming environments, such as the National Instruments LabWindows and LabView packages mentioned in Chapter 2. These packages insulate the programmer from the multitude of programming details of the GUI, by providing basic data acquisition, graph plotting, menu selection and other facilities in a simple packaged form.

References

Ackmann J.J. (1979) A computer system for neurosurgical patient monitoring. *Comput. Prog. Biomed.* **10**, 81–88.

Acton F.S. (1970) *Numerical methods that work.* Harper & Row, New York. 541p.

Adams P.R. (1977) Voltage jump analysis of procaine action at frog endplate. *J. Physiol.* **268**, 291–318.

Adrian A.H., Chandler W.K. & Hodgkin A.L. (1970) Voltage clamp experiments on striated muscle fibres. *J. Physiol.* **208**, 607–644.

Affolter H. & Sigworth F.J. (1988) High performance MODULA-2 programs for data acquisition and analysis of single-channel events. *Biophys. J.* **53**, 154a.

Agarwal P. & Priemer R. (1979) Microprocessor based digital signal processing system. *Comput. Biol. Med.* **9**, 87–96.

Akaike H. (1974) A new look at the statistical model identification. *IEEE Trans. Automatic Control* **AC-19**, 716–723.

Aldrich R.W. & Stevens C.F. (1987) Voltage-dependent gating of single sodium channels from mammalian neuroblastoma cells. *J. Neuroscience* **7** (2), 418–431.

Anderson C.R. & Stevens C.F. (1973) Voltage clamp analysis of acetylcholine produced end-plate current fluctuations at the frog neuromuscular junction. *J. Physiol.* **235**, 655–691.

Armstrong C.M. & Bezanilla F. (1974) Charge movement associated with the opening and closing of the activation gates of the sodium channel *J. Gen. Physiol.* **63**, 533–552.

Ascher P. & Nowak L. (1988) Quisqualate and kainate-activated channels in mouse central neurones in culture. *J. Physiol.* **399**, 227–245.

Bagust J. (1985) Sample analysis and display (SAD): A microcomputer data acquisition system for laboratory use. In *Microcomputers in the neurosciences* G. Kerkut (ed.). Oxford University Press. Oxford, 118–141.

Ball F.G., Kerry C.J., Ramsey R.L., Sansom M.S.P. & Usherwood P.N.R. (1988) The use of dwell time cross-correlation functions to study single-ion channel gating kinetics. *Biophys. J.* **54**, 309–320.

Ball F.G. & Sansom M.S.P. (1988a) Aggregate Markov processes incorporating time interval omission. *Adv. Appl. Prob.* **20**, 546–572.

Ball F.G. & Sansom M.S.P. (1988b) Single channel autocorrelation functions – the effect of time interval omission. *Biophys. J.* **53**, 819–832.

Barker J.L., McBurney R.M. & MacDonald J.F. (1982) Fluctuation analysis of neutral amino acid responses in cultured mouse spinal cord neurones. *J. Physiol.* **322**, 365–387.

Barry P.H. & Quantararo N. (1990) PNSCROLL,

A software package for graphical interactive analysis of single channel patch clamp currents and other binary file records under mouse control. *Comput. Biol. Med.* **20**, 193–204.

Bauer R.J., Bowman B.F. & Kenyon J.L. (1987) Theory of the kinetic analysis of patch-clamp data *Biophys. J.* **52**, 961–978.

Beam K.G. (1976) A quantitative description of end-plate currents in the presence of two lidocaine derivatives. *J. Physiol.* **258**, 301–322.

Beasley J.D. (1990) *Practical computing for experimental scientists.* Oxford University Press, Oxford. 233p.

Bendat J.S. & Piersol A.G. (1971) *Random data: analysis and measurement procedures.* John Wiley & Sons, New York. 397p.

Bennet P.B., Kass R. & Begenisich T. (1989) Nonstationary fluctuation analysis of the delayed rectifier K channel in cardiac Purkinje fibers: Actions of norepinephrine on single-channel currents. *Biophys. J.* **55**, 731–738.

Bermejo R. & Zeigler H.P. (1989) A microcomputer-based system for multi-channel neurophysiology recording. *Comp. Biol. Med.* **19**, 35–54.

Bevan S. (1976) Sub-miniature end-plate potentials at untreated frog neuromuscular junctions. *J. Physiol.* **258**, 145–155.

Bevan S., Chad J.S., Hollingwood F.S. & Wise J.C.M. (1986). A system to provide control of stimulation and acquisition of ionic currents in cell membranes. *J. Physiol.* **377**, 17P.

Bezanilla F. (1985) A high capacity data recording device based on a digital audio processor and a video cassette recorder. *Biophys. J.* **47**, 437–442.

Bezanilla F. & Armstrong C.M. (1977a) A low cost signal averaging and data acquisition device. *Am. J. Physiol.* **1**(3), C211–C215.

Bezanilla F. & Armstrong C.M. (1977b) Inactivation of the sodium channel: I. Sodium current experiments. *J. Gen. Physiol.* **70**, 549–566.

Black J.L., Isele D.F., Head R.L., Fleming I.R. & Collins D.W.K. (1976) A versatile averaging system for neurophysiology. *Comput. Biol. Med.* **6**, 9–21.

Blatz A.L. & Magleby K.L. (1986) Correcting single channel data for missed events. *Biophys. J.* **49**, 967–980.

Blatz A.L. & Magleby K.L. (1989) Adjacent interval analysis distinguishes among gating mechanisms for the fast chloride channel from rat skeletal muscle. *J. Physiol.* **410**, 561–585.

Bormann J., Hamill O.P. & Sakmann B. (1987) Mechanism of anion permeation through channels gated by glycine and γ-aminobutyric acid in mouse cultured spinal cord neurones. *J. Physiol.* **385**, 243–286.

Bourne J.R. (1981) *Laboratory Minicomputing.* Academic Press, New York. 297p.

Box M.J. (1966) A comparison of several current optimization methods, and the use of transformations in constrained problems. *Comput. J.* **9**, 67–77.

Brock J.A. & Cunnane T.C. (1988) Electrical activity at the sympathetic neureffector junction in the guinea-pig vas deferens. *J. Physiol.* **399**, 607–632.

Brown K.M. & Dennis J.E. (1971) A new algorithm for non-linear least squares curve fitting. In *Mathematical software*, J.R. Rice (ed.) Academic Press. London, 391–396

Brown K.M. & Dennis J.S. (1972) Derivative free analogs of the Levenberg-Marquardt and Gauss algorithms for non-linear least squares approximation. *Numerische Mathematik* **18**, 289–297.

Caceci M.S. & Cacheris W.P. (1984) Fitting curves to data. *Byte* **9**(5), 340–362.

Carmenes R.S. (1991) LSTSQ: A module for reliable constrained and unconstrained non-linear regression. *Comput. Applications Bios. (CABIOS)* **7**, 373–378.

Ceccarelli B., Fesce R., Grohovaz F. & Haimann C. (1988) The effect of potassium on exocytosis of transmitter at the frog neuromuscular junction. *J. Physiol.* **401**, 163–183.

Char B.W., Geddes K.G., Gonnet G.H., Leong B.L., Monagan M.B. & Watt S.M. (1991) *Maple V language reference manual.* Springer-Verlag, New York. 267p.

Chiu, S.Y. (1977) Inactivation of sodium channels: second order kinetics in myelinated nerve. *J. Physiol.* **273**, 573–596.

Chiu S.Y., Ritchie R.M., Rogart R.B. & Stagg D. (1979) A quantitative description of the membrane currents in rabbit myelinated nerve. *J. Physiol.* **292**, 149–166.

Chung S.H, Moore J.B, Xia. L.G, Premkumar L.S. & Gage P.W. (1990) Characterization of single channel currents using digital signal processing techniques based on hidden Markov models. *Phil. Trans. R. Soc. Lond. (Biol.)* **329**, 265–285

Clark N.A. & Molnar, C.E. (1964) The LINC – A description of the laboratory instrument computer. *Ann. N.Y. Acad. Sci.* **115**, 653–658.

Cohen B.A & Myklebust J. (1978) Evoked potential measurement and analysis on a small laboratory computer. *Comput. Prog. Biomed.* **8**, 256–261

Cole K.S. (1949) Dynamic electrical characteristics of the squid axon membrane. *Arch. Sci. Physiol.* **3**, 253–258.

Cole K.S. & Curtis K.J. (1939) Membrane resting and action potentials from the squid giant axon. *J. Cell. Comp. Physiol.* **19**, 135–144.

Cole K.S. & Antosiewicz H.A. (1955) Automatic computation of nerve excitation. *J. Soc. Ind. Appl. Math.* **3**, 153–172.

Cole K.S. & Moore J.W. (1960) Ionic current measurements in the squid giant axon membrane. *J. Gen. Physiol.* **44**, 123–167.

Colquhoun D. (1971) *Lectures on biostatistics.* Clarendon Press, Oxford. 425p.

Colquhoun D. (1987a) Practical analysis of single channel records. In *Microelectrode techniques. The Plymouth workshop handbook*. N.B. Standen, P.T.A. Gray & M.J. Whitaker (eds). Company of Biologists Ltd, Cambridge, pp. 83–104.

Colquhoun D. (1987b) The interpretation of single channel recordings. In *Microelectrode techniques. The Plymouth workshop handbook*. N.B. Standen, P.T.A. Gray & M.J. Whitaker (eds). Company of Biologists Ltd, Cambridge, pp. 105–135.

Colquhoun D. & Hawkes A.G. (1977) Relaxation and fluctuations of membrane currents that flow through drug operated ion channels. *Proc. R. Soc. Lond. B* **199**, 231–262.

Colquhoun D. & Hawkes A.G. (1981) On the stochastic properties of single ion channels. *Proc. R. Soc. Lond B*, **211**, 205–235.

Colquhoun D. & Hawkes A.G. (1983a) On the stochastic properties of bursts of single ion channel openings and clusters of bursts. *Phil. Trans. R. Soc. Lond B*, **300**, 1–59.

Colquhoun D. & Hawkes A.G. (1983b) The principle of the stochastic interpretation of ion-channel mechanisms. In *Single Channel recording*. Sakmann & E. Neher (eds). Plenum Press, New York, pp. 191–263.

Colquhoun D. & Hawkes A.G. (1987) A note on correlations in single ion channel records. *Proc. R. Soc. Lond B*, **230**, 15–52.

Colquhoun D. & Sakmann B. (1983) Bursts of openings in transmitter-activated ion channels. In *Single channel recording*. B Sakmann B. & E. Neher (eds.) Plenum Press, New York. pp. 345–363.

Colquhoun D. & Sakmann B. (1985) Fast events in single-channel currents activated by acetylcholine and its analogs at the frog muscle end-plate. *J. Physiol.* **369**, 501–557.

Colquhoun D. & Sigworth F.J. (1983) Fitting and statistical analysis of single-channel records. In *Single channel recording*. B. Sakmann & E. Neher (eds). Plenum Press, New York. pp. 191–263.

Computers in Medicine & Biology (1964) In *Ann. N.Y. Acad. Sci.* **115**.

Conti F., Hille B., Neumcke B., Nonner W. & Stampfli R. (1976) Measurement of the conductance of the sodium channel from current fluctuations at the node of Ranvier. *J. Physiol.* **262**, 699–727.

Conti F., Hille B. & Nonner W. (1984) Non-stationary fluctuations of the potassium conductance at the node of Ranvier of the frog. *J. Physiol.* **353**, 199–230.

Cooley J.W. & Tukey J.W. (1965) An algorithm for the machine calculation of complex Fourier series. *Math. Comp.* **19**, 297–301.

Cooper J.W. (1977) *The minicomputer in the laboratory*. Wiley-Interscience, New York. 365p.

Corey D.P. & Stevens C.F. (1983) Science and technology of patch-recording electrodes. In *Single channel recording* B. Sakmann & E. Neher (eds). Plenum Press, New York. pp. 191–263.

Cottrell G.A., Duff D.A., Dunbar S.J. & Green K.A. (1985) Some neurobiological applications of the BBC model B microcomputer and Unilab 8-bit interface. *J. Neurosci. Meth.* **11**, 10–20.

Cottrell G.A., Duff D.A. & Lambert J.J. (1983) An inexpensive microcomputer system for analysis of single channel currents. *J. Neurosci. Meth.* **9**, 259–267.

Coughlin R.F. & Villanucci R.S. (1990) *Introductory operational amplifiers and linear ICs: Theory and experimentation*. Prentice-Hall Inc. New Jersey. 460p.

Crouzy S.C. & Sigworth F.J. (1990) Yet another approach to the dwell-time omission problem of single-channel analysis. *Biophys. J.* **58**, 731–743.

Crunelli V., Forda S., Kelly J.S. & Wise J.C.M. (1983) A programme for the analysis of intracellular data recorded from in vitro preparations of central neurones. *J. Physiol.* **340**, 13P

Cull-Candy S.G., Howe J.R. & Ogden D.C. (1988) Noise and single channels activated by excitatory amino acids in rat cerebellar granule neurones. *J. Physiol.* **400**, 189–222.

D'Agrosa L.S. & Marlinghaus K. (1975) Computerized measurements of cardiac transmembrane potentials. *Comput. Biomed. Res.* **8**, 97–104

Daniel W.W. (1987) *Biostatistics: A foundation for analysis in the health sciences*. John Wiley & Sons, New York, 734p.

Daniel W.W. & Coogler C.E. (1975) Beyond analysis of variance: A comparison of some multiple comparison procedures. *Physical Ther.* **55**, 144–150.

Davies R.G. (1971) *Computer programs in quantitative biology*. Academic Press, London, 492p.

DeFelice L.J. (1981) *Introduction to membrane noise*. Plenum Press, New York. 500p.

Dempster J. (1985) A set of computer programs for electrophysiological analysis of end plate current characteristics. *Br. J. Pharmacol.* **85**, 390P.

Dempster J. (1987) A range of computer programs for electrophysiological signal analysis using the IBM PC AT. *J. Physiol.* **392**, 20P.

Dempster J. (1989) The computer analysis of electrophysiological data. In *Microcomputers in physiology: A practical approach*. P.J. Fraser (ed.) IRL Press, Oxford. pp. 51–93.

Dennis J.E. (1977) Non-linear least squares and equations. In *The state of the art in numerical analysis*. Academic Press, London.

Derenzo S.E. (1990) *INTERFACING. A laboratory approach using the microcomputer for instrumentation, data analysis, and control*. Prentice-Hall, Englewood Cliffs, NJ. 466p

Dilger J.P & Brett R.S. (1990) Direct measurement

of the concentration- and time-dependent open probability of the nicotinic acetylcholine receptor channel. *Biophys. J.* **57**, 723–731.

Dodge F.A. (1963) *A study of ionic permeability changes underlying excitation in myelinated nerve fibers of the frog*. PhD. thesis. Rockefeller University Microfilms (No. 64–7333) Ann Arbor, Mich.

Dodge F.A. & Frankenhaeuser B. (1959) Sodium currents in the myelinated nerve fibres of Xenopus laevis investigated with the voltage clamp technique. *J. Physiol.* **143**, 76–90.

Duncan R. (1988) *The MS-DOS Encyclopedia*. Microsoft Press, Redmond, WA. 1570p.

Eggebrecht L.C. (1990) *Interfacing to the IBM Personal Computer*, Howard W. Sams & Co., Indianapolis. 350p.

Eisenberg R.S., Frank M. & Stevens C.F. (eds) (1981) *Membranes, channels, and noise*. Plenum Press, New York. 288p.

Ellis T.M.R. (1990) *FORTRAN 77 PROGRAMMING. With an introduction to the FORTRAN 90 standard*. Addison-Wesley, Reading, Mass. 672p.

Enomoto K. & Maeno T. (1981) A microcomputer system for on-line analysis of the end-plate potential data. *Brain Res. Bull.* **7**, 97–99.

Etter D.M. (1988) *Structured FORTRAN 77 for engineers and scientists*. Addison-Wesley, Reading, Mass. 520p.

Evans C. (1981) *The making of the micro*. Victor Gollancz, London. 113p.

Everitt B.S. (1987) *Introduction to optimization methods and their applications in statistics*. Chapman & Hall, London. 88p.

Finkel A.S. & Redman S.J. (1984) Theory and operation of a single microelectrode voltage. *J. Neurosci. Meth.*. **11**, 101–127.

Frankenhaeuser B. (1960) Quantitative description of sodium currents in myelinated nerve fibres of *Xenopus laevis*. *J. Physiol.* **151**, 491–501.

Fraser P.J. (ed.) (1988) *Microcomputers in physiology: A practical approach*. IRL Press, Oxford. 276p.

Fusi F., Piazzesi G., Amerini S., Mugelli A. & Livi S. (1984) A low cost microcomputer system for automated analysis of intracellular cardiac action potentials. *J. Pharmacol. Meth.* **11**, 61–66.

Gage P.W. & McBurney R.L. (1975) Effects of membrane potential, temperature and neostigmine on the conductance change caused by a quantum of acetylcholine at the toad neuromuscular junction. *J. Physiol.* **244**, 385–407.

Gage P.W., McBurney R.N. & Schneider G.V. (1975) Effects of some aliphatic alcohols on the conductance change caused by a quantum of acetylcholine at the toad end-plate. *J. Physiol.* **244**, 409–429.

Gibson G.A. & Liu Y. (1987) *Microcomputers for engineers and scientists*. Prentice-Hall, Englewood Cliffs, NJ. 482p.

Gill P.E., Murray W. & Wright M.H. (1981) *Practical optimisation*. Academic Press, London. 401p.

Glantz S.A. & Slinker B.K. (1990) *Primer of applied regression and analysis of variance*. McGraw-Hill Inc., New York. 777p.

Gray P. & Freeman S. (1986) A computerized approach to the collection and analysis of cadiac action potentials. *J. Pharmacol. Meth.* **51**, 347–357.

Green B.F., Wolf A.K. & White B.W. (1959) The detection of statistically defined patterns in a matrix of dots. *Am. J. Psychol.* **72**, 503–520.

Hamill O.P., Marty A., Neher E., Sakmann B. & Sigworth F.J. (1981) Improved patch-clamp techniques for high-resolution current recording from cells and cell-free membrane patches. *Pflugers Arch.* **391**, 85–100.

Hansen A. (1989) *A complete guide to mastering the C language*. Addison-Wesley, Reading, Mass. 550p.

Harmatz J.S. & Greenblatt D.J. (1987) A simplex procedure for fitting non-linear pharmacokinetic models. *Comput. Biol. Med.* **17**, 199–208.

Hawkes A.G., Jalali A. & Colquhoun D. (1991) The distribution of open and shut times in a single channel record when brief events cannot be detected. *Philos. Trans. R. Soc. Lond. A* **332**, 511–538.

Hewtt T.T. (1986) Using an electronic spreadsheet simulator to teach neural modelling of visual phenomena. *Collegiate Microcomput.* **IV**(2), 141–151.

Hill I.D. (1985) Algorithm AS66 – The normal integral. In *Applied statistics algorithms*. P. Griffiths & I.D. Hill. (eds). Ellis Horwood Ltd., Chichester. pp. 126–129.

Hille B. (1977) Local anesthetics: Hydrophilic and hydrophobic pathways for drug-receptor reactions. *J. Gen. Physiol.* **69**, 497–515.

Hille B. & Campbell D.T. (1976) An improved vaseline gap voltage clamp for skeletal muscle fibres. *J. Gen. Physiol.* **67**, 265–293.

Hille B. (1984) *Ionic channels of excitable membranes*. Sinauer Associates, Sunderland, Mass. 426p.

Hodgkin A.L. & Huxley A.H. (1939) Action potentials recorded from inside a nerve fibre. *Nature* **144**, 710–711.

Hodgkin A.L. & Huxley A.H. (1952a) Currents carried by sodium and potassium ions through the membrane of the giant axon of Loligo. *J. Physiol.* **116**, 449–472.

Hodgkin A.L. & Huxley A.H. (1952b) The components of the membrane conductance in the giant axon of Loligo. *J. Physiol.* **116**, 473–496.

Hodgkin A.L. & Huxley A.H. (1952c) The dual effect of membrane potential on sodium conductance in the giant axon of Loligo. *J. Physiol.* **116**, 497–506.

Hodgkin A.L. & Huxley A.H. (1952d) A quantitative description of membrane current and its application to conduction and excitation in nerve. *J. Physiol.* **117**, 500–544.

Hof T.D. (1986) A pulse generating and data recording system based on the microcomputer PDP11/23. *Comp. Meth. Progs. Biomed.* **23**, 309–316.

Horn R. (1987) Statistical methods for model discrimination: Applications to gating kinetics and permeation of the acetylcholine receptor channel. *Biophys. J.* **51**, 255–263.

Horn R. (1991) Estimating the number of channels in patch recordings. *Biophys. J.* **60**, 433–439.

Horowitz P. & Hill W. (1980) *The art of electronics*. Cambridge University Press, Cambridge. 716p.

Intel (1987) *Intel Microprocessor and peripheral handbook. Volume 1–Microprocessor*. Intel Corp. Santa Clara, CA.

Ireland C.R. & Long S.P. (eds) (1984) *Microcomputers in biology: A practical approach*. IRL Press, Oxford. 400p.

Jack J.J.B., Noble D. & Tsein R.W. (1975) *Electrical current flow in excitable cells*. Clarendon Press, Oxford, 502p.

Jack J.J., Redman S.J. & Wong K. (1981) The components of synaptic potentials evoked in cat spinal motoneurones by impulses in single group Ia afferents. *J. Physiol.* **321**, 65–96.

Jennrich R.I. & Raston M.L. (1979) Fitting non-linear models to data. *Ann. Rev. Biophys. Bioeng.* **8**, 195–238.

Katz B. & Miledi R. (1972) The statistical nature of the acetylcholine potential and its molecular components. *J. Physiol.* **224**, 655–699.

Katz G.M & Schwarz T.L. (1974) Temporal control of voltage-clamped membranes: an examination of principles. *J. Membr. Biol.* **17**, 275–291.

Kegel D.R., Wolf B.D., Sheridan R.E. & Lester H.A. (1985) Software for electrophysiological experiments with a personal computer. *J. Neurosci. Meth.* **12**, 317–330.

Kehl T.H. & Dunkel L. (1976) Uses of the LM2 in neurobiology. In *Computer technology in neuroscience*. P. Brown (ed.) Hemisphere Publishing Corp., Washington DC.

Kehl T.H., Moss C. & Dunkel L. (1975) LM2 – A logic machine minicomputer. *IEEE Computer* **8**, 12–21.

Kelley A. & Pohl I. (1987) *C by dissection. The essentials of C*. Addison-Wesley, Reading, Mass. 250p.

Kerkut G. (ed.) (1985) *Microcomputers in the neurosciences*, Oxford University Press. Oxford. 269p

Kernighan B.W. & Plauger P.J. (1981) *Software tools in PASCAL*. Addison-Wesley, Reading, Mass. 366p.

Kernighan B.W. & Ritchie D.M. (1978) *The C programming language*. Prentice-Hall. New Jersey. 228p.

Kits K.S, Mos G.J, Leeuwerik F.J. & Wattel C. (1987) Acquisition and analysis of fast single channel kinetic data on an Apple IIe microcomputer. *J. Neurosci. Meth.* **20**, 57–71.

Klein D.L, Jenkins J.M. & Ten Eick R.M. (1983) Dual microcomputer analysis of cardiac transmembrane action potentials. *IEEE Trans. Biomed. Eng.* **30**, 819–825.

Korn S.J. & Horn R. (1991) Discrimination of kinetic models of ion channel gating. *Methods in Neurosciences* Vol. 4, P.M. Conn (ed.) Academic Press, London. pp. 428–456.

Kreibel M.E. & Gross C.E. (1974) Multimodal distribution of frog miniature endplate potentials in adult, denervated and tadpole leg muscle. *J. Gen. Physiol.* **64**, 85–103.

Ktonas P., Weintraub B., Smith J. & Black F.O. (1975) Computer aided nystagmus analysis. *Comput. Prog. Biomed.* **5**, 153–157

Lamb T.D. (1985) An inexpensive digital tape recorder suitable for neurophysiological signals. *J. Neurosci. Meth.* **15**, 1–13.

Lambert J.J, Peters J.A., Hales T.G. & Dempster J. (1989) Properties of 5–HT3 receptors in clonal cell lines studied by patch clamp. *Br. J. Pharmacol.* **97**(1), 27–40.

Leamer E.E. (1983). Model choice and specification analysis. In *Handbook of Econometrics*, Vol. 1. Z. Griliches & M.D. Intriligator (eds). North Holland Pub. Co., Amsterdam. pp. 285–330.

Lee J.D. & Lee T.D. (1982) *Statistics and computer methods in BASIC*. Van Nostrand Reinhold, New York. 198p.

Leibovitch L.S. (1989) Testing fractal and Markov models of ion channel kinetics. *Biophys. J.* **52**, 979–988.

Levenberg K. (1944) A method for the solution of certain problems in least squares. *Q. J. Appl. Math.* **2**, 164–168.

Littler J. & Maher J. (1989) *Computers in the laboratory*. Longman Scientific & Technical, Harlow, Essex. 385p.

Lynn P.A. & Fuerst W. (1989) *Introductory digital signal processing with computer applications*. John Wiley & Sons, Chichester, 371p.

MacDonald R.L., Rogers C.J. & Twyman R.E. (1989) Kinetic properties of the $GABA_A$ receptor main conductance state of mouse spinal cord neurones in culture. *J. Physiol.* **410**, 479–499.

Magleby K.L. & Pallotta B.S. (1983a) Calcium dependence of open and shut interval distributions from calcium-activated potassium channels in cultured rat muscle. *J. Physiol.* **344**, 585–604.

Magleby K.L. & Pallotta B.S. (1983b) Burst kinetics of single calcium-activated potassium channels in cultured rat muscle. *J. Physiol.* **344**, 605–623.

Magleby K.L. & Stevens C.F. (1972) A quantitative description of end-plate currents. *J. Physiol.* **223**, 173–197.

Magleby K.L. & Weiss D.S. (1990) Estimating kinetic

parameters for single channels for simulation. A general method that resolves the missing event problem and accounts for noise. *Biophys. J.* **58**, 1411–1426.

Marlowe S.P. & Mackensie D.A.R. (1989) Distributed computer system for capture, analysis and display of biological data. *Med. Biol. Eng. Comput.* **27**, 371–378.

Marquardt D. (1963) An algorithm for least-squares estimation of non-linear parameters. *SIAM J. Appl Math* **11**, 431–441.

Martin J.D. (1991) *Signals & processes: A foundation course*. Pitman, London, 421p.

Marty A., Tan Y.P. & Trautmann A. (1984) Three types of calcium-dependent channel in rat lacrimal gland. *J. Physiol.* **357**, 293–325.

Matheny A. (1984) Simulation with electronic spreadsheets. *Byte* **9**(3), 411–414.

Mayzner M.S. & Dolan T.R. (eds) (1978) *Minicomputers in sensory and information-processing research*. Lawrence Erlbaum Associates, New Jersey. 280p.

Mayzner M.S. & Goodwin W.R. (1978). Historical perspectives. In *Minicomputers in sensory and information-processing research*. M.S. Mayzner & T.R. Dolan (eds). Lawrence Erlbaum Associates, New Jersey, pp. 1–28.

McCann F.V, Stibitz G.R. & Keller T.M. (1987) A computer method for the acquisition and analysis of patch-clamp single-channel currents. *J. Neurosci. Meth.* **20**, 45–55.

McManus O.B., Blatz A.L. & Magleby K.L. (1987) Sampling, log binning, fitting and plotting durations of open and shut intervals from single channels and the effects of noise. *Pflugers Arch.* **410**, 530–553.

Mendenhall W. & Sincich T. (1989) *Statistics for the engineering and computer sciences*. Collier Macmillan, London. 1036p.

Mezei L.M. (1990) *Practical spreadsheet statistics and curve fitting for scientists and engineers*. Prentice Hall, Englewood Cliffs, NJ. 311p.

Milne R.K., Yeo G.F., Edeson R.O. & Madsen B.W. (1988) Stochastic modelling of a single ion channel: an alternating renewal approach with application to limited time resolution. *Proc. R. Soc. Lond. B* **233**, 247–292.

Mize R.R. (ed.) (1985) *The microcomputer in cell and neurobiology research*. Elsevier, Amsterdam. 481p.

Morales F.R., Boxer P.A., Jervey J.P. & Chase M.H. (1985) A computerized system for the detection and analysis of spontaneously occurring synaptic potentials. *J. Neurosci. Meth* **13**, 19–35.

Morrison R. (1986) *Grounding and shielding techniques in instrumentation*. John Wiley & Sons, New York. 172p

Munro D.M. (1975) Algorithm AS83 Complex discrete fast Fourier transform. *Appl. Stat.* **24**, 153–160.

Munro D.M. (1976) Algorithm AS97 Real discrete fast Fourier transform. *Appl. Stat.* **25**, 166–172.

Neher (1983) The charge carried by single-channel currents of rat cultured muscle cells in the presence of local anaesthetics. *J. Physiol.* **339**, 663–678.

Neher E. & Sakmann B. (1976) Single channel currents recorded from membrane of denervated frog muscle fibres. *Nature* **260**, 799–802.

Neher E. & Stevens C.F. (1972) Conductance fluctuations of ionic pores in membranes. *Ann. Rev. Biophys. Bioeng.* **6**, 345–381.

Nelder J.A. & Mead R. (1965) A simplex method for function minimization. *Comput J.* **7**, 308–313.

Neil J., Xiang Z. & Auerbach A. (1991) List-oriented analysis of single-channel data. In *Methods in Neurosciences* Vol. 4, P.M. Conn (ed.) Academic Press, London. pp. 474–490.

Nicol R., Smith P. & Raggat P.R. (1986) The use of the simplex method for the optimisation of non-linear functions on the laboratory microcomputer. *Comput. Biol. Med.* **16**, 145–152.

Noble D. (1966) Applications of Hodgkin-Huxley equations to excitable tissue. *Physiol. Rev.* **46**, 1–50.

Noble D. (1975) *The initiation of the heartbeat*. Clarendon Press, Oxford. 156p.

Nonner W. (1969) A new voltage clamp method for Ranvier nodes. *Pflugers Arch. ges. Physiol.* **309**, 176–192.

Nyquist H. (1928) Certain topics in telegraph transmission theory. *Trans AIEE*, 47.

O'Neill R. (1971) Algorithm AS47. Function minimization using a simplex procedure. *Appl. Statist.* **20**, 338–345.

Ouchi G.I. (1987) *Personal computers for scientists*. American Chemical Society. Washington DC. 276p.

Park M.R. (1985) A complete digital neuropsychological recording laboratory. In *The microcomputer in cell and neurobiology research*. R.R. Mize. (ed.). Elsevier, Amsterdam, pp. 411–434.

Patlak J.B. (1988) Sodium channel subconductance levels measured with a new variance-mean analysis. *J. Gen. Physiol.* **92**, 413–430.

Pennefather P. & Quastel D.M.J. (1981) Relation between subsynaptic receptor blockade and response to quantal transmitter release at mouse neuromuscular junction. *J. Gen. Physiol.* **78**, 313–343.

Petracchi D., Barbi M., Pellegrini M., Pellegrino M. & Simoni A. (1991) Use of conditional distributions in the analysis of ion channel recordings. *Eur. Biophys. J.* **20**, 31–39.

Pohlmann K.C. (1989) *Principles of digital audio*. Howard Sams. Indianapolis, 474p.

Press W.H., Flannery B.P., Teukolsky S.A. & Vetterling W.T. (1986) *Numerical recipes. The art of scientific computing*. Cambridge University Press, Cambridge. 818p.

Quayle J.M, Smith R.A.J. & Ward T.A. (1986). A low cost system for on-line pulse generation, patch-clamp current sampling and analysis. *J. Physiol.* **371**, 13P

Quint S.R., Howard J.F. & Antoni L. (1983) On-line analysis of neuromuscular bioelectric potentials. *Comput. Meth. Prog. Biomed.* **16**, 3–12.

Rao C.R. (1973) *Linear statistical inference and its applications*, 2nd ed. John Wiley & Sons.

Rawlings J.O. (1988) *Applied regression analysis – A research tool*. Wadsworth & Brooks, Pacific Grove, CA, 533p.

Re L. & Di Sarra B. (1988) Automated on-line system for the acquisition and computation of skeletal muscle end-plate derived signals. *J. Pharmacol. Meth.* **19**, 253–262.

Re L., Di Sarra B., Concettoni C & Gisuti P. (1989) Computerized estimation of spontaneous release at the neuromuscular junction. *J. Pharmacol. Meth.* **22**, 233–242.

Redman S. & Walmsley B. (1983) Amplitude fluctuations in synaptic potentials evoked in cat spinal motoneurones at identified group Ia synapses. *J. Physiol.* **343**, 117–133.

Robinson H.P, Sahara Y. & Kawai N. (1991) Nonstationary fluctuation analysis and direct resolution of single channel currents at post-synaptic sites. *Biophys. J.* **59**, 295–304.

Robinson K. & Giles W. (1986) A data acquisition, display and plotting program for the IBM PC. *Comput. Meth. Prog. Biomed.* **23**, 319–327.

Roux B. & Sauve R. (1985) A general solution to the time interval omission problem applied to single channel analysis. *Biophys. J.* **48**, 149–158.

Ruff R. (1977) A quantitative analysis of local anaesthetic alteration of miniature end-plate currents and end-plate current fluctuations. *J. Physiol.* **264**, 89–124.

Sachs F. (1983) Automated analysis of single-channel records. In *Single channel recording*, B. Sakmann & E. Nether (eds). Plenum Press, New York, pp. 191–263.

Sachs F., Neil J. & Barkakati N. (1982) The automated analysis of data from single ionic channels. *Pflugers Arch.* **395**, 331–340.

Sakmann B. & Neher E. (eds) (1983) *Single-channel recording*. Plenum Press, New York. 501p.

Sargent M. & Shoemaker R. (1987) *The IBM personal computer from the inside out*. Addison-Wesley, Reading, Mass. 496p.

Savitch W.J. (1990) *PASCAL. An introduction to the art and science of programming*. Addison-Wesley, Reading, Mass. 768p.

Schneider M.F. & Dubois J (1986) Effects of benzocaine on the kinetics of normal and batrachotoxin-modified Na channels in frog node of Ranvier. *Biophys. J.* **50**, 523–530.

Schonfeld R.L. (1964) The role of the computer as a biological instrument. *Ann. N.Y. Acad. Sci.* **115**, 915–942.

Schuyler M. (1985) The evolution of spreadsheets. *Microcomput. Informat. management* **2**, 111–123.

Schwartz G. (1978) Estimating the dimensions of a model. *Ann. Statistics* **6**, 461–464.

Searl T., Prior C. & Marshall I.G. (1991) Acetylcholine recycling and release at rat motor nerve terminals studied using (-) vesamicol and troxypyrrolium. *J. Physiol.* **444**, 99–116.

Sigworth F.J. (1980) The variance of sodium current fluctuations at the node of Ranvier. *J. Physiol.* **307**, 97–129.

Sigworth F.J. (1981a) Covariance of nonstationary sodium current fluctuations at the node of Ranvier. *Biophys. J.* **34**, 111–133.

Sigworth F. (1981b) Nonstationary noise analysis of membrane currents. In *Membranes, channels, and noise*, R.S. Eisenberg, M. Frank & C.F. Stevens. (eds) Plenum Press, NY. U.S.A. 21–48.

Sigworth F.J. (1981c) Interpreting power spectra from nonstationary membrane current fluctuations. *Biophys. J.* **35**, 289–300.

Sigworth F.J. (1983) An example of analysis. In *Single channel recording*, B. Sakmann & E. Neher (eds). Plenum Press, pp. 301–321.

Sigworth F.J. (1985) Open channel noise 1. Noise in acetylcholine receptor currents suggests conformational fluctuations. *Biophys. J.* **47**, 709–720.

Sigworth F.J. & Sine S.M. (1987) Data transformations for improved display and fitting of single-channel dwell time histograms. *Biophys. J.* **48**, 149–158.

Sine S.M. & Steinbach J.H. (1986) Activation of acetylcholine receptors on clonal BC3H-1 cells be low concentrations of an agonist. *J. Physiol.* **373**, 129–162.

Smith T.G., Barker J.L., Smith B.M. & Colburn T.R. (1980) Voltage clamping with microelectrodes. *J. Neurosci. Meth.* **3**, 105–128.

Snedecor G.W. & Cochran W.G. (1967) *Statistical methods* 6th edn. Ames Iowa State University Press. 507p.

Soper J.B. & Lee M.P. (1990) *Statistics with LOTUS 1-2-3*. Chartwell-Bratt, Bromley, 225p.

Soucek B. & Carlson A.D. (1976) *Computers in neurobiology and behaviour*. John Wiley & Sons, New York. 324p.

Spencer C.D. (1990) *Digital design for computer data acquisition*. Cambridge University Press, Cambridge, 356p.

Stacy R.W. & Waxman B.D. (1965) *Computers in biomedical research.*, Vols I-III. Academic Press, New York. 1213p.

Standen N.B., Gray P.T.A. & Whitaker M.J. (eds) (1987) *Microelectrode techniques. The Plymouth workshop handbook*. Company of Biologists Ltd, Cambridge, 256p.

Stark L., Sandberg A.A., Stanten S., Willis P.A. & Dickson J.F. (1964) An on-line digital computer used in biological experiments and modelling. *Ann. N. Y. Acad. Sci.* **115**, 738–762.

Steel R.G. & Torrie J.H. (1980) *Principles and procedures of statistics. A biometrical approach*. McGraw-Hill, New York. 633p.

Steinbach A.B. (1968) A kinetic model for the action

of Xylocaine on receptors for acetylcholine. *J. Physiol.* **52**, 162–180.

Steinberg I.Z. (1987) Frequencies of paired open-closed durations of ion channels. Methods of evaluation from single-channel recordings. *Biophys. J.* **52**, 47–55.

Stevens C.F. (1981) Inferences about molecular mechanisms through fluctuation analysis. In *Membranes, channels, and noise*, R.S. Eisenberg, M. Frank & C.F. Stevens (eds). Plenum Press, New York. pp. 1–20.

Stromquist B.R. Pavlides C. & Zelano J. (1990) On-line acquisition, analysis and presentation of neurophysiological data based on a personal microcomputer system. *J Neurosci. Meth.* **35**, 215–222.

Thistead R.A. (1988) *Elements of statistical computing: Numerical computations*. Chapman & Hall, New York. 427p.

Thomson B.G. & Kuckes A.F. (1979) *The IBM PC in the laboratory*. Cambridge University Press. Cambridge, 246p.

Valko P. & Vajda S. (1989) *Advanced scientific computing in BASIC with applications in chemistry, biology and pharmacology*. Elsevier Science Publishers, Amsterdam. 321p.

Van Mastright R. (1977) A short note on the performance of 2 computer programs for the estimation of the parameters of a multiexponential model. *Comput. Biol. Med.* **7**, 249.

Vivaudou M.B., Singer J.J. & Walsh J.V. jr (1986). An automated technique for analysis of current transients in multilevel single-channel recordings. *Pflugers Arch.* **407**, 355–364.

Wachtel R.E. (1988) Use of the BMDP statistical package to generate maximum likelihood estimates for single channel data. *J. Neurosci. Meth.* **25**, 121–128.

Wachtel R.E. (1991) Fitting of single-channel dwell time distributions. In *Methods in neurosciences*, Vol. 4. Academic Press, London, pp. 410–428.

Wilkison D.M. (1991) Digital filtering of potentials on personal computers. In *Methods in Neurosciences* Vol. 4, P.M. Conn (ed.) Academic Press, London. pp. 397–409.

Willming D.A. & Wheeler B.C. (1990) Real-time multichannel neural spike recognition with DSPs. *IEEE Eng. Med. Biol.* 37–39.

Wilson D.L. & Brown A.M. (1985) Effect of limited interval resolution on single channel measurements with application to Ca channels. *IEEE Trans. Biomed. Eng.* **32**, 786–797.

Witkowski F.X & Corr P.B. (1978) Automated analysis of cardiac intracellular transmembrane action potentials. *Comput. Cardiol.* 315–318.

Wolf S. & Smith R.F.M. (1990) *Student reference manual for electronic instrumentation laboratories*. Prentice-Hall, Englewood Cliffs, NJ. 547p.

Wolfram S. (1991) *Mathematica. A system for doing mathematics by computer*. Addison-Wesley, Reading, Mass. 800p.

Yellen G. (1984) Ionic permeation and blockade in Ca^{2+}-activated K^+ channels of bovine chromaffin cells. *J. Gen. Physiol.* **84**, 157–186.

Suppliers

Amplicon Liveline Ltd, Centenary Industrial Estate, Hollingdean Road, Brighton BN2 4AW, UK. Tel. (0273) 570220.

Ariel Corp., 433 River Road, Highland Park, NJ 08904, USA. Tel. (201) 294 2900.

Axon Instruments Inc., 1101 Chess Drive, Foster City, CA 94404, USA. Tel. (415) 571 9400.

Bio-Logic, 4 rue Docteur Pascal, Z.A. du Rondeau, 38130 Echirolles, France. Tel. 76 40 00 67.

Biosoft, 22 Hills Road, Cambridge CB2 1JP, UK. Tel. (0223) 68622.

BMDP – Statistical Software Inc., 1440 Sepulveda Boulevard, Suite 316, Los Angeles, CA 90025, USA. Tel. (213) 479 7799.

Cambridge Electronic Design, Science Park, Milton Road, Cambridge CB4 4FE, UK. Tel. (0223) 420186.

Dagan Corporation, 2855 Park Avenue, Minneapolis, MN 55407, USA. Tel. (612) 827 5959.

Data Translation Inc., 100 Locke Drive, Marlboro, MA, USA. Tel. (508) 481 3700.

Frequency Devices Inc., 25 Locust Street, Haverhill, MA 01830, USA.

Hera Elecktronik GMBH, Wiesenstrasse 71, D–6734 Lambrecht/Pfalz, Germany. Tel. (49) 63 25 80 36.

Hewlett-Packard Corp., 16399 West Bernardo Drive, San Diego, CA 92127, USA.

Instrutech Corp., 51 Charles Street, Mineola, NY 11501, USA. Tel. (516) 742 4020.

Intracel Ltd, Broad Lane, Cottenham, Cambridge CB4 4SW, UK. Tel. (0954) 50957.

Kemo Ltd, 12 Goodwood Parade, Elmers End, Beckenham, Kent BR3 3QZ, UK. Tel. 081 658 3838.

Medical Systems Corp., One Plaza Road, Greenvale, NY 11548, USA. Tel. (516) 621 9190.

Minitab Inc., 3081 Enterprise Drive, State College, PA 16801–2756, USA. Tel. (814) 238 3280.

National Instruments, 6504 Bridge Point Parkway, Austin, TX 78730–5039, USA. Tel. (512) 794 0100.

R.C. Electronics, 5386 Hollister Avenue, Santa Barbara, CA 93111, USA. Tel. (805) 964 6708.

Sun Microsystems, 950 Marina Village Parkway, PO Box 4016, Alameda, CA 94501, USA. Tel. (415) 769 8700.

World Precision Instruments Inc., 375 Quinnipac Avenue, New Haven, CT 06513, USA. Tel. (203) 469 8281.

Index

AC coupling 44
Active filters 48
adc_dma routine 25, 27, 137
adc_interrupt_count routine 21
adc_service routine 21
Adobe Postscript 60
Aliasing 44–5
Alternate hypothesis 83
AMD9513 21
Amplitude histograms 161–2
   average 163–4
   fitting gaussian curves to 162–3
   running average 164
Analog signals
   conditioning 9, 42–54, 159–60
   digital recording 13–32
   digitization 14
   storage 159–60
Analog-to-digital converter (ADC) 4, 6, 9, 10, 14–15, 136–8, 169, 192, 201
Analysis of variance 88–9
   current fluctuations 193–4
   Kruskal–Wallis one-way 91
   mean current 194
   non-stationary 199
   two-way 90
Anti-alias filtering 44, 45
API (Applications Programmers' Interface) library of subroutines 213
Apple II 5, 6, 19
Apple Macintosh 7, 8, 14, 31, 150
Apple microcomputer 5
Applied Statistics Algorithm AS66 81
ASCII code 66–7
Assembler program 209
Asymptotic information criterion (AIC) 124, 178
Atari ST 149

Attenuation response 47
AXOBASIC 201
Axodata program 150
AXOLAB 71
AZTEC coding 169

Band-pass filtering 44
BASIC 5, 211–12, 213
BASIC-23 5, 72
Basic input/output system (BIOS) 208
BBC Micro 6, 7
Bessel filter 49
Best-fit parameters 105, 109, 116, 121, 125, 177
Binary number system 204–5
Bit 204
Bit map 59, 60
BMDP 96
Boltzmann function 139–40
Butterworth filter 49
Byte 204

C language 72, 211–13
Cathode ray tube (CRT) 206
CED1401 6, 29–30, 71, 148
CED1401–plus 30
CED Voltage and Patch Clamp software 147, 148, 181, 191
Cell-attached configuration 158
Cell membrane potential 194
Central processing unit (CPU) 5, 7, 16, 21, 27, 205–6, 209
Central tendency 80
Channel kinetics estimation from power spectrum 196–7
Channel open and closed time intervals 165
Channel state transition detection 165–7

Chi-square test 95, 123, 125
CLAMPAN 65, 148
CLAMPEX 65, 147–8, 180
CLAMPFIT 65, 148
Command voltage 136
Command voltage protocols
   single-step 138–9
   two-step 142–3
Commodore Pet microcomputer 5
Comparator 50
Compiled languages 209
Computer graphics 57–8, 213
Computer interface bus 207
Computer operating systems 207–8
Computers in Biomedical Research 10
Conductance-voltage (G-V) curves 139
Confidence intervals 81
Continuous sampling-to-disc method 25–7
Control ports 16
Cosine bell window function 198
CricketGraph 96
Current amplitude histograms. *See* Amplitude histograms
Current fluctuations, analysis of variance 193–4
Current reversal potential 194
Current time course analysis 140–1
Current variance
   spectral analysis of 196
   vs mean current plots 195
Current-voltage (I-V) curves 139–40, 158
Curve fitting 92, 104–32
   algorithm performance testing 111
   gaussian 163
   iterative non-linear 108–9
   process of 105
   standard error of best-fit parameters 125
Cut-off frequencies 44, 46, 47, 192

dac_isr routine 137
dac_sweep routine 137
Data acquisition system 9
Data ports 16
Data transfer methods 17
  direct memory access (DMA) 21–5, 137
  hard disc drive 26
  interrupt driven 19–21
  programmed 17–20
  shared memory 25
Data transfer phase 17, 18
DC offset removal 43–4
Decimal numbers 205
Degrees of freedom 80, 84
Descriptive statistics 79–80
Differential amplifier 43
Digital computers 204–13
  development of 2–3
  systems architecture 205–6
Digital filters 69–70
  non-recursive 70
  recursive 70
Digital leak current subtraction 143–6
Digital plotter 59
Digital pulse code modulation (PCM) systems 159
Digital recording of analog signals. *See* Analog signals
Digital signal processors (DSP) 71
Digital-to-analog converter (DAC) 4, 6, 136–8
Direct memory access (DMA) data transfer facilities 21–5, 137
Direct search function minimization 109–10
disable_adc_interrupt routine 21
disable_dac routine 137
Disc controllers 207
Disc optimizer program 26
Discrete Fourier transform (DFT) 198
DMA2800 31
Dot matrix printers 59
Double buffer method 25
Double exponential, error profile 120–1
DT2801A 136
DT2821 137
DT2831 137
DT2836 201
DTR1200 9, 160
DTR1600 160
Duncan's test 90
Dwell time analysis 165, 171
Dwell time distributions 171
  for effects of missed events 179
  modelling 174–5
Dwell time histograms
  linear 171–2
  logarithmic 172–4

Electrical equivalent circuit model 134–5
Electrophysiology, history of 1–2
enable_adc_interrupt routine 21

Endplate currents (EPC) 61
  amplitude 92, 93
Ensemble 200
Ensemble average 200
Ensemble variance 200
Enzyme kinetics 108
Error types 83
Error variance 107
Event detection 49–50
Evoked potentials (EVPs) 68
Exponential function 105, 107, 108, 123
  error profile 119–20
Exponential models, best-fit results from 122
Exponentials, 'ill posed' least-squares problems 117–19
Exporting data and results 66
Extension/expansion/contraction strategies 110–12
External trigger input 49–50

Fast Fourier Transform (FFT) 196, 198
Feedback control systems 134
FETCHAN 180
FETCHEX 180
Fig.P 67, 96, 116, 121, 163
Filter cut-off frequency 44, 46, 47, 192
Filters 69–71
  non-recursive 70
  recursive 70
  types and characteristics 46–50
Finite impulse response (FIR) filters 70
Floppy discs 206
FORTRAN 18, 24, 27, 67, 90, 111, 117, 169, 209–13
FORTRAN-IV 5
FORTRAN-77 20
Forward difference formula 115
Frequency distributions, quantitative comparison of 94–5
Frequency histograms 93–4
Frequency response curve 46
Friedman two-way analysis of variance 91
Function minimization
  direct search 109–10
  gradient methods 112, 116–17
  methods 116
  strategy 108

GABA 193, 197
Gain levels 43
Gauss–Newton method 125
Gaussian curves, fitting to amplitude histograms 162–3
Gaussian filter 70
Gaussian functions 162–3
Gaussian probability density function 81
Global minimum 111–12
Goodness-of-fit 105–6
  criteria for 111
Gradient methods 112, 116–17
Graph-plotting programs 96

Graphical user interface (GUI) 150, 208, 213
GRAPHICS.COM 59
Grid search 111

Half-power 47
Hard disc drives 25–6, 206
Hardware interrupt 19
Hessian matrix 113, 114, 125
Hewlett Packard Graphics Language (HPGL) 59, 60
Hewlett Packard laser printers 9
Hewlett Packard plotters 59
Hexadecimal numbers 205
High level languages 209
High-pass filtering 44
Hodgkin–Huxley equations 3–5, 104, 141–3
HP7470A 59
HP7475 59

IBM360 3
IBM701 3
IBM7090 3
IBM PC 6–8, 13, 19, 22, 150, 206
IBM PC AT 6
Infinite impulse response (IIR) filters 70
Initial estimates 115–16
Initialization phase 17
Input DC offset 43
Input/output (I/O) devices 206
Input/output (I/O) ports 16
Inside-out configuration 158
Inspection of signal records 55–7
Integrated circuits 3, 5, 15
Intel 8237A Programmable DMA Controller 22
Intel 8253 Counter/Timer 18
Intel 8259A Programmable Interrupt Controller 19
Interference signals 45
Interference sources 50–2
Interpreter 212
Interrupt-driven data transfer method. *See* Data transfer methods
Ion channel function 165
Ion channel gating, kinetic models of 178
Ion currents
  analysis of fluctuations 191–203
  mathematical models 104–5
IPROC-2 180
IRQ0 19, 137, 138
IRQ1 19
IRQ3 20
IRQ5 19
IRQ7 19–21
Iterative curve-fitting 142
Iterative least-squares curve-fitting methods, comparative performance of 118
Iterative non-linear least squares 108

Index

Jandel Sigmaplot 67, 96, 116
Johnson noise 45

Kalaedagraph 96
Kemo VBF8 192
Kinetic analysis, non-stationary 200
Kinetic models 178
Kolmogorov–Smirnov test 95
Kruskal–Wallis one-way analysis of variance 91

Labmaster A/D converter 21
Labmaster interface expansion cards 6
Laboratory computers
   current state of the art 8–9
   performance benchmarks 7–8
Laboratory data, analysis of 3
Laboratory interfaces 15–16
   CED1401 29–30
   comparison of features and performance 29
   Data Translation 30
   expansion cards 6
   LAB-PC 16–18, 30–1, 137
   Labmaster DMA 23, 29
   Macintosh 31
   miscellaneous systems 31
   programming 16–17
   selection criteria 27–9
   to host computer data transfer 17
LabView 72, 213
LabWindows 31, 72, 213
Laplace transforms 48
Laser printers 59–60
Leak currents, digital subtraction protocols 143–6
Least-squares equations 107
Least-squares goodness-of-fit criterion 106
Least-squares principle 106
Levenberg–Marquardt (L–M) algorithm 114, 117, 119, 121, 125
Limulus 4
LINC 3–4
LINC-8 4
Linear phase 47
Linear regression 107
Linearizing transformations 107–8
Local minimum 111
Log error ratio (LER) 124
Log-likelihood ratio (LLR) 177–8
Lorentzian functions 196
Low level languages 209
Low-pass filtering 44, 46, 47, 69, 160
LPF901 192
LPROC 180

Machine code 209
Macintosh. *See* Apple-Macintosh
MacLab system 71
MACRO II 5
Macro language 86–7
Magnetic disc storage media 206–7

Mainframe computers 3
Maple program 115
Markov process 178
Mathematica program 115
Mathematical models 92, 104–32
   ion currents 104–5
   nested 123
   non-nested 124
   selection criteria 121
Maximum likelihood method 175–6
   practical aspects of 176–7
Mean 80
Membrane equivalent circuit model 134–5
Microcomputers 5–6
Microprocessors, development of 5
Microsoft Windows 8
Miniature endplate currents (MEPCs) 46, 68–70, 93, 94, 105, 107
Miniature endplate potentials (MEPPs) 43
   acceptance/rejection of 63–4
   analysis of 63–4
Minicomputers 3–4
Minitab 96
MIPS (Million Instructions Per Second) 206
Missing events problem 179
MS-DOS 6, 8, 19, 21, 25, 26, 59, 137, 208
Multiple comparison tests 89–92
Multiplexer 16

NAG E04CCF routine 119
NAG E04FDF routine 170
NAG E04GCF routine 114, 117, 120
NAG E04YCF routine 116
NAG Workstation library 114–15
NB-A2000 31
Nernst equation 134
Nested models 123
Newton's method 113
Noise analysis 191–203
   non-stationary 192, 199–200
   software packages 201
   stationary 192
   theory and practice of 191
Noise recording, non-stationary 200–1
Noise sources 45–6
Non-nested models 124
Non-parametric tests 85–7
   comparison with parametric tests 87–8
Normal distribution 80–1, 95
Normal probability density function 81
n-point moving average algorithm 69–70
Null hypothesis 83
Number of components, determining 177–8
Nyquist criterion 194
Nyquist rate 44, 192
Nyquist theorem 44–5

Open channel probability 162
OS/2 208
Outside-out configuration 158

P/4 leak subtraction method 145
P/N method 145, 146
Page description language 60
Paired t-test 85
Pair-wise tests, limitations of 88
Parameter scaling 115–16
Parameter standard error 116
Parametric tests 85, 91
Pascal 211
PAT single-channel analysis program 180
Patlak's running average amplitude 164–5
PC-DOS 6
pClamp 6, 19, 20, 65–7, 71, 147–8, 180
PDP8 4, 7
PDP11 4–8, 13, 169
PDP12 4
Peizer & Pratt's transform 82
Phase response 47
Postscript printers 59–60
Post-synaptic potentials (PSP) 68
Power spectrum
   channel kinetics estimation from 196–7
   computing 197–9
Probability density function 81, 95
Program source code 209
Programmed data transfer. *See* Data transfer methods
Programmed I/O. *See* Data transfer methods, programmed
Programming languages 208–9
   choice of 212
pSTAT 180

Quasi-Newton methods 114, 119
QuickBasic 72, 212

Random access memory (RAM) 7, 21, 23, 205–6
Random access time 26
Random distributions, analysis of 92–4
Random errors 107
Random noise 45
RC filter 47–8
Read only memory (ROM) 208
REALFT routine 199
Repeated measures 90
Residual differences 106
Residual distributions 124–5
Residual errors 107
Residual variance 124
Reversal potential 134
Rotational latency 26
RS/1 96
RT11 operating system 4

Running average amplitude histogram 164
Runs test 124, 125

SAGE air defence system 3
Sampling-to-disc performance 26
SCAN 57, 63, 65–6, 71
Screen dump 59
Shared memory data transfer method 25
Side-lobes 198
SIGAVG (SIGnal AVeraGer) 71
Signal amplification 42–3
Signal analysis 55–76
  primary 55, 65
  secondary 56
  software for automated measurement 64
  software packages 71
Signal analysis see also Signal measurements
Signal averaging 67–8
  detected spontaneous signal 68–9
Signal characteristics, measurement of 60
Signal collection 27
Signal conditioning systems 159–60
  computer-controlled 53–4
  modular 52–3
Signal detection 27
Signal filtering 44
Signal measurements 56, 60, 61, 65–6
  analysis of trends in 92
  automated 63
  calibration of 62
  characteristics 60
  semi-manual, using screen cursors 61
  see also Signal analysis
Signal/noise (S/N) ratio 46
Signal power spectrum. See Power spectrum
Signal quality enhancement 56
Signal records
  acceptance/rejection procedures 63–4
  digitized display 57–8
  hard copies 59
  inspection of 55–7
  validation of 55, 57–8
Signal storage 159–60
Signal variance, computing 194–5
Simplex method 109–11, 116–17, 176

Simultaneous voltage generation and recording 136–8
Single-channel analysis packages 167
Single-channel current amplitude histograms. See Amplitude histograms
Single-channel current analysis, software packages 179–80
Single-channel currents
  analog signal and conditioning 159–60
  analysis of 157–90
  analysis of amplitudes 161
  digitization 160–1
  recording 159
  voltage-activated 161
Smoothing 44–6
Software development 5, 8–10
  environments 72
SPAN (SPectral ANalysis) program 201
Spectral analysis of current variance 196
Spontaneous signals, detection and recording 27
Spreadsheets 78–9, 86–7
SPSS 96
SSQMIN 117
Standard deviation 80
Standard error of best-fit parameters 125
Standard error of the mean 80
Statistical analysis 65, 66, 77–103
  software packages 95–6
Statistical significance 83
Status ports 16
Steepest descent method 112–13
Stimulus isolation unit (SIU) 52
STRUCTURE 169
Student–Neumann–Kuels test 90
Studentized range 90
Student's t-distribution 81–3
Student's t-test 83–5
Sum of squares 111, 112, 123

Tail probability 81
Termination criteria 110
Termination phase 17
Test statistic 83
Tetrodotoxin 144
Threshold crossing methods 167–8
  use of two thresholds 167–8
Throughput 17
Time constants 142
Time course fitting 169–71

Time-dependent parameter changes, analysis of 92
Time-dependent properties 141
Transfer function 48
Transition detection 165
Treatment-type experiments 77–8
TTL (transistor-transistor logic) pulse 50
Tukey's test 90
Two-way analysis of variance 90
Type 1 error 83
Type 2 error 83

UNIVAC1 3
UNIX 67, 208
US FCC regulations 52

Validation of signal records 55, 57–8
Variance 80
Variance analysis. See Analysis of variance
Variance-mean current plot 200
Variance ratio 123
VCAN (Voltage Clamp ANalysis) program 147–9
VisiCalc 5, 78
Voltage-activated currents 60–1
  analysis of 133–56
Voltage clamp 133–5
  analysis module 148
  software packages 146–50
Voltage clamp signals, digital recording 136
Voltage ramp protocols 140

Waveform characteristics 56
Waveform measurements
  analysis of trends in 92
  automated 63
Weighting factor 108
Whirlwind computer 3
White noise 45
Whole cell configuration 158
Wilcoxon rank-sum test 85–7
Windows 208
Wordstar 5

Zero current baseline drift compensation 168–9
Zero reference level 62–3